Lecture Notes in Computer Science 12750

More information about this subseries at http://www.springer.com/series/7408

Alexander Raschke · Elvinia Riccobene ·
Klaus-Dieter Schewe (Eds.)

Logic, Computation and Rigorous Methods

Essays Dedicated to Egon Börger
on the Occasion of His 75th Birthday

 Springer

Editors
Alexander Raschke 🄾
Institut für Softwaretechnik
und Programmiersprachen
University of Ulm
Ulm, Germany

Klaus-Dieter Schewe 🄾
UIUC Institute
Zhejiang University
Haining, China

Elvinia Riccobene 🄾
Dipartimento di Informatica
Università degli Studi di Milano
Crema, Italy

ISSN 0302-9743   ISSN 1611-3349   (electronic)
Lecture Notes in Computer Science
ISBN 978-3-030-76019-9   ISBN 978-3-030-76020-5   (eBook)
https://doi.org/10.1007/978-3-030-76020-5

LNCS Sublibrary: SL2 – Programming and Software Engineering

This Springer imprint is published by the registered company Springer Nature Switzerland AG
The registered company address is: Gewerbestrasse 11, 6330 Cham, Switzerland

From Logic and Computation Theory to Rigorous Methods for Software Engineering

Tribute to Egon Börger on the Occasion of His 75th Birthday

Alexander Raschke[1], Elvinia Riccobene[2], and Klaus-Dieter Schewe[3]

[1] Institute of Software Engineering and Compiler Construction, Ulm University,
Ulm, Germanyalexander.raschke@uni-ulm.de
[2] Università degli Studi di Milano, Milan, Italy
elvinia.riccobene@unimi.it
[3] Zhejiang University, UIUC Institute, Haining, China
kdschewe@acm.org

Egon Börger started his scientific career in philosophy, and he came to computer science via studying in depth mathematical logic. He spent two decades in logic and three decades in computer science. As he phrased it himself, his youth was dedicated to mathematics and logic, his senior years to software engineering, and with beginning seniority he worked on business processes and other applied topics. What from a superficial view may look like a straight line from theory to practice fulfils in fact the principle of "negated negation" in the sense of the German philosopher Georg Wilheim Friedrich Hegel. By confronting theoretical insights with applied problems from which they were originally derived by abstraction, new enhanced challenges are discovered that require deep scientific and thus theoretical investigation on a higher quality level. Egon Börger is known not only as an excellent scientist but also as a virtuoso for playing with the dialectic antipodes of theory and practice, bringing theoretical results into applications and extracting challenging questions from applications in order to shift scientific knowledge to a higher level.

Egon Börger was born on May 13, 1946, in Westfalia in Germany. More than mathematics he loved music, so his ambition was to become a conductor, but his teachers advised against it. So he started to study philosophy at the Université de Paris I (Sorbonne) and the Institut Supérieur de Philosophie in Louvain. Professors Joseph Dopp and Jean Ladrière, followers of the tradition of Robert Feys, aroused his interest in mathematical logic and foundations of mathematics, which in 1966 brought Egon back to his home town, Münster. This was the last place he wanted to study, and even less he wanted to be a mathematician. However, the lively spirit of mathematical logic commemorated by Heinrich Scholz and his famous student and successor Hans Hermes was just too inspiring. Professors Scholz and Hermes were among the few scientists who realised the impact of Turing's solution of Hilbert's fundamental decision problem, which manifested itself in the tight relationship between abstract computing machines and descriptive logical languages.

From the very beginning Egon was part of the Münsteranian school of logic with a strong focus on the exploration of this relationship between logic and computing science, which coined his interest in this area. He became a student of Dieter Rödding who continued the tradition started by Gisbert Hasenjäger, Hans Hermes, and Wilhelm Ackermann, and his Ph.D. thesis was dedicated to the complexity of decision problems.

One year after completing his Ph.D. he followed an invitation by Edoardo Caianello and joined the Università di Salerno as a lecturer to help develop a new institute for computer science. Between 1972 and 1976 he gave many lectures on various computer science topics, through which he gained a deep understanding of the discipline. He returned to Münster, where he completed his habilitation in 1976 with a thesis on complexity theory. Further years as a lecturer and Associate Professor at the Universities of Münster, Udine and Dortmund followed.

The relationship between logic and computing science determined Egon's first monograph on computation theory, logic, and complexity, which was published in German in —1985 (it was later translated into English), and also— his second monograph on the classical decision problem, which he published together with Erich Grädel and Yuri Gurevich in 1997. The first monograph focused on the concept of formal language as carrier of the precise expression of meaning, facts and problems, and formal operating procedures for the solution of precisely described questions and problems. At that time the text was at the forefront of a modern theory of these concepts, paving the way in which they developed first in mathematical logic and computability theory and later in automata theory, theory of formal languages, and complexity theory. In both monographs random access machines played an important role.

After the unexpected death of Dieter Rödding it was expected by many that Egon would become his successor, but official politics at the Universität Münster prevented this. This was to the detriment of the Universität Münster, where soon its excellent standing in logic and foundations of mathematics was devastated. Universität Münster is still one of the few universities in Germany without a decent computer science department and degree program. Fortunately, Egon could still choose between other offers, and he accepted the position of a chair in computer science at the Università di Pisa, which he held until his retirement in 2011, rejecting various offers from other prestigious universities.

Right at the time of his relocation to Pisa he joined forces with Michael M. Richter and Hans Kleine Büning to establish an annual conference on "Computer Science Logic" (CSL), which soon led to the foundation of the European Association for Computer Science Logic (EACSL) with Egon as its first chairman. He held this position until 1997. Together with the North American conference series "Logic in Computer Science" (LiCS), CSL still counts as one of the most prestigious conferences in theoretical computer science focusing on the connections between logic and computing.

From 1985 on, the field of computer science started to stretch out into many new application areas. Distributed computing over networks became possible, database systems facilitated concurrent computation, artificial intelligence ventured from a niche area to a useful technology enabling inferential problem solving in diagnosis, controlling machines through software became possible, etc. Together with his long-standing collaborator Yuri Gurevich he realised that the rapid developments in computing would require radically new methods in computer science logic. It was Yuri who first formulated a "new thesis" moving computations on Tarski structures to the centre, while Egon realised that the idea of "evolving algebras"—now known as

Abstract State Machines (ASMs)—does not only create a new paradigm for the foundations of computing, subsuming the classical theory, but at the same time can be exploited for rigorous systems engineering in practice thereby fulfilling the criteria of a "software engineering" discipline that deserves this name.

One of the first achievements was the definition of an operational semantics of the Warren Abstract Machine for Prolog programs, and this work by Egon led to the formal foundation of a comprehensive abstract semantics of Prolog used by the ISO Prolog standardisation committee. Since then he has systematically and tirelessly pushed experiments to apply ASMs to real-life software systems, in particular industrial software-based systems. Unlike many other renowned researchers in the field of logic and computation, Egon never liked hiding in a snail shell away from the problems that arise in computing applications. He actively sought the challenges arising in practice. Four of his five sabbaticals during his active time in Pisa were spent with companies (IBM, Siemens, Microsoft, and SAP). He organised several Dagstuhl Seminars, acted as co-chair of several summer schools, gave numerous invited talks, and visited international institutions and academies.

He triggered and led the effort of international groups of researchers who developed the ASM method for high-level system design and analysis. At the beginning of 2000 he wrote another monograph, known as the ASM book, establishing the theoretical foundations of the formal method for building and verifying complex software-based systems in an effectively controllable manner, namely by stepwise refinement of abstract ground models to executable code. To provide a forum for ASMs, he started, in 1994, the annual workshop on Abstract State Machines, which later, in 2008, was turned into the international ABZ conference series to promote fruitful integration of state-based formal methods.

Many extensions of the ASM method are due to Egon, who has always been able to identify and capture new potential of the method, referring to new characteristics of modern complex systems. With the intention of exploiting the ASM method to tackle new challenging computational aspects, Egon has worked in many different computer science areas: from programming languages to hardware architectures, software architectures, control systems, workflow and interaction patterns, business processes, web applications, and concurrent systems.

During his long and still very active research carrier, Egon has made significant contributions to the field of logic and computer science. Since 2005 he has been an Emeritus member of the International Federation for Information Processing. In 2007, in recognition of his pioneering work in logic and its applications in computer science, he received the prestigious Humboldt Research Award, and since 2010, he has been a member of Academia Europea. He is the author of more than 30 books and over 200 scientific publications. Among the published books, three underline applications of the ASM method: (1) the book on Java and the Java Virtual Machine (JVM), which provides a high-level description of Java and the JVM together with a mathematical and an experimental analysis, and shows the correctness of a standard compiler of Java programs to JVM code and the security critical bytecode verifier component of the JVM; (2) the book on Subjective Business Process Management, which presents a novel BPM methodology focusing on process actors and their interactions; and (3) the modelling companion with many detailed application examples of the ASM method, in particular in connection with concurrent systems.

Besides his scientific contributions as a logician and computer scientist, we would like to emphasise two prominent characteristics of Egon as a researcher, namely his deep intuition for addressing real, open and challenging problems, and his passionate approach toward solving research problems by looking deeply into problems, understanding them thoroughly, discussing them broadly with other researchers, and consistently working very hard.

Many international computer science communities owe a lot to Egon, be it for his scientific contributions and wide dissemination of his scholarly work, for his open mind, his still active service activity or his tenacious intellectual honesty.

With this Festschrift we want to express special thanks to Egon, a real master who inspired all of us.

April 2021

Alexander Raschke
Elvinia Riccobene
Klaus-Dieter Schewe

Acknowledgement

We, Alexander Raschke, Elvinia Riccobene, and Klaus-Dieter Schewe, would like to thank all contributors to, and authors and reviewers of, this Festschrift. This is a worthy birthday present for a scientist who has inspired all of us.

We are also grateful to Springer Verlag for giving us the opportunity to express our appreciation for Egon through this Festschrift.

Reviewers of this Volume

Contents

Towards Leveraging Domain Knowledge in State-Based Formal Methods . . .   1
Yamine Aït-Ameur, Régine Laleau, Dominique Méry,
and Neeraj Kumar Singh

Some Observations on Mitotic Sets   14
Klaus Ambos-Spies

Moded and Continuous Abstract State Machines   29
Richard Banach and Huibiao Zhu

Product Optimization in Stepwise Design   63
Don Batory, Jeho Oh, Ruben Heradio, and David Benavides

Semantic Splitting of Conditional Belief Bases   82
Christoph Beierle, Jonas Haldimann, and Gabriele Kern-Isberner

Communities and Ancestors Associated with Egon Börger and ASM   96
Jonathan P. Bowen

Language and Communication Problems in Formalization: A Natural
Language Approach   121
Alessandro Fantechi, Stefania Gnesi, and Laura Semini

ASM Specification and Refinement of a Quantum Algorithm   135
Flavio Ferrarotti and Sénen González

Spot the Difference: A Detailed Comparison Between B and Event-B   147
Michael Leuschel

Some Thoughts on Computational Models: From Massive Human
Computing to Abstract State Machines, and Beyond   173
J. A. Makowsky

Analysis of Mobile Networks' Protocols Based on Abstract State Machine . . .   187
Emanuele Covino and Giovanni Pani

What is the Natural Abstraction Level of an Algorithm?   199
Andreas Prinz

The ASMETA Approach to Safety Assurance of Software Systems   215
Paolo Arcaini, Andrea Bombarda, Silvia Bonfanti, Angelo Gargantini,
Elvinia Riccobene, and Patrizia Scandurra

Flashix: Modular Verification of a Concurrent and Crash-Safe Flash
File System   239
  Stefan Bodenmüller, Gerhard Schellhorn, Martin Bitterlich,
  and Wolfgang Reif

Computation on Structures: Behavioural Theory, Logic, Complexity   266
  Klaus-Dieter Schewe

The Combined Use of the Web Ontology Language (OWL) and Abstract
State Machines (ASM) for the Definition of a Specification Language
for Business Processes.   283
  Matthes Elstermann, André Wolski, Albert Fleischmann, Christian Stary,
  and Stephan Borgert

Models and Modelling in Computer Science.   301
  Bernhard Thalheim

A Framework for Modeling the Semantics of Synchronous
and Asynchronous Procedures with Abstract State Machines   326
  Wolf Zimmermann and Mandy Weißbach

Author Index   353

Towards Leveraging Domain Knowledge in State-Based Formal Methods

Yamine Aït-Ameur[1]([⊠]) [ID], Régine Laleau[2] [ID], Dominique Méry[3] [ID],
and Neeraj Kumar Singh[1] [ID]

[1] INPT-ENSEEIHT/IRIT, University of Toulouse, Toulouse, France
{yamine,nsingh}@enseeiht.fr
[2] LACL, Université Paris-Est Créteil, Créteil, France
laleau@u-pec.fr
[3] LORIA, Université de Lorraine and Telecom Nancy, Nancy, France
dominique.mery@loria.fr

1 Introduction

System engineering development processes rely on modelling activities that lead to different design models corresponding to different analyses of the system under consideration. More rigorous and sound models are required as the system becomes more complex and critical. These models are usually richer than natural language written texts, images or videos descriptions. Engineers use various modelling languages to design these models. According to the system modelling language and to the associated analysis techniques, different system requirements verification and validation activities can be carried out.

Although engineers use complex design descriptions to elaborate their models, they still miss to explicitly model relevant information related to the domain of interest. In most of the cases, we observe:

1. that engineers use the available modelling languages to hard encode, on the fly, specific domain properties with their own point of view, using the semantics of the available modelling languages (implicit semantics). Such a process may lead to incomplete and/or inconsistent descriptions, especially in the case of model composition, decomposition, abstraction and/or refinement;
2. the absence of resources allowing engineers to make explicit the domain knowledge of interest (explicit semantics). The main problem is related to the absence, in the design modelling languages, of formalised constructs supporting such knowledge domain descriptions.

The authors thank time that progresses so as we can celebrate birthdays.

This work was partially supported by grants ANR-13-INSE-0001 (The IMPEX Project http://impex.loria.fr), ANR-14-CE28-000 (The FORMOSE Project http://formose.lacl.fr/), ANR-17-CE25-0005 (The DISCONT Project http://discont.loria.fr) and ANR-19-CE25-0010-01 (The EBRP Project https://www.irit.fr/EBRP/) from the French national research agency (Agence Nationale de la Recherche ANR).

A. Raschke et al. (Eds.): Börger Festschrift, LNCS 12750, pp. 1–13, 2021.
https://doi.org/10.1007/978-3-030-76020-5_1

The triptych $(D, S \vdash R)$, where D stands for Domain, S for Specification and R for Requirements, initially proposed by Michael Jackson and Pamela Zave [27] and taken up by Dines Bjørner [5], summarises the ternary relation among different kinds of knowledge and statements in system engineering.

Following the above mentioned work, we advocate the exploitation of domain knowledge in design models. The resulting methodology may be considered as independent of the formal modelling language which will be able to address the three parts of the triptych or at least should be a consistent collection of domain knowledge entities of D. In our case, these three parts must be formally handled by the state-based approach set up for system design.

This paper is organised as follows. Sections 2 and 3 give a brief overview of the importance of domain knowledge in requirements engineering and of ontologies as a suitable model for this kind of knowledge. Section 4 then presents a case study from the aeronautic domain that was used throughout this paper. The impact of our claim in system engineering, when state-based formal methods are used, in particular Event-B, is discussed in Sect. 5 on the defined case study. A conclusion ends this paper.

2   Domain in Requirements Engineering

Elaborating a formal specification of a system requires a preliminary step that consists of identifying and analysing the requirements that the system must satisfy. There exist many requirements engineering methods, such as KAOS [20], i* [26], the Problem Frames approach [16], based on different paradigms but all these methods agree on a point: it is necessary to understand and describe the domain in which the system will take place well in order to express the right requirements and consequently build the right specification that satisfy the requirements. This statement appeared first in [15] and formalised in [16] and [27] by the well-known triptych $(D, S \vdash R)$ where D stands for Domain, S for Specification and R for Requirements. It means that when a specification S is built in some domain D the system specification entails its requirements R.

In his book [5] and following articles [6,7], Dines Bjørner expresses the same idea, when he describes this Triptych Paradigm of Software Engineering *"Before software can be developed, the software developers and the clients contracting this software must understand the requirements. Before requirements can be developed, the software developers and the clients contracting these requirements must understand the domain"*.

It is also commonly accepted that a domain property is a *descriptive* statement about the system and is generally unchanging regardless of how the system behaves. On the contrary, a requirement is a *prescriptive* statement describing a property that the system shall satisfy.

In his book [20], Axel van Lamsweerde goes further and introduces the concepts of *expectation*, a specific case of prescriptive statements that needs to be satisfied by the environment of the system, and *hypothesis*, a *specific* case of descriptive statement satisfied by the environment of the system and subject to change.

From the above seminal work review, we understand that

- explicit domain knowledge formalisation as *descriptive models* shall be conducted before system design
- and, *prescriptive specific system hypotheses* have to be derived from the domain model. Below, we discuss the case of ontologies as models for domain knowledge.

3 Ontology-Based Modelling of Domain Knowledge

Information on the domain or domain knowledge related to the system to be engineered is used at every step of the system development life cycle. In many engineering areas, standards or shared knowledge are defined and designers refer to it in the development steps. When formal methods are set up, developers use domain knowledge, usually assumptions or environment properties, formalised as hypotheses, definitions, axioms, theorems, invariants, etc.

Although domain modelling is a key activity in requirements engineering, it relies on formal knowledge engineering approaches that offer formal semantics and reasoning capabilities. In this setting, formal ontologies play a major role as they make *explicit* the domain knowledge and offer reasoning and external referencing capabilities at different system development steps.

Briefly, according to [12], an ontology is *an explicit specification of a conceptualisation* while [17] considers a domain ontology as a *formal and consensual dictionary of categories and properties of entities of a domain and the relationships that hold among them.* Both definitions focus on the three core following characteristics. Indeed, ontologies and domain knowledge models require to

- be *explicit* in order to allow designers to use and refer to domain knowledge concepts;
- offer *descriptive* domain knowledge characterisations, in the form of *generic* categories (e.g. classes) that go beyond the ones of the system under design;
- derive, through instantiation, *prescriptive* characteristics for specific concepts used by systems models

When applying formal methods in system engineering [2,5], two categories of domain knowledge can be identified. The first one relates to the mathematics and the formal modelling features required by the application domain related to the system under design. Examples of such domain knowledge are continuous features for hybrid systems or for signal processing. The second one deals with specific knowledge related to the application domain possibly referring to the first type of knowledge. Examples of such domain knowledge are models describing standards for engines, energy consumption, units, etc. In both cases, this knowledge has to be explicitly represented.

Ontology modelling languages equipped with logic-based formal semantics have been defined. They advocate the definition of domain concepts as classes associated to operations, relations on classes, instances and constraints and of an instantiation mechanism. Depending on the existing relationships, operators and constraints, different reasoning capabilities are possible. Whatever is the chosen ontology modelling language, it can be represented as axiomatised theories with data-types, axioms, theorems and proof rules [23].

Once ontologies, are formally described as theories with data-types in the hosting formal method, it is possible to allow models to use it to type state variables and use the available operators. This approach is referred to as *model annotation* [3,9,13,24]. Indeed, use of or reference to ontology concepts brings to the designed formal models ontological definitions, axioms, theorems and proof rules.

Our Claim

In this paper, we claim that first, domain knowledge shall be axiomatised explicitly as reusable domain specific axioms and reasoning rules and second, that state-based formal methods offer the adequate framework for such axiomatisation and use in the system design and safety insurance process [4]. In particular, they may be used to formalise domain ontologies and offer the capability to use the ontology concepts (domain knowledge concepts, hypotheses, theorems, proof rules, etc.) is system models formalised using state-based methods.

4   The Nose Gear (NG) Velocity System Case Study

The *Nose Gear (NG) velocity system* [10] is the toy example we selected to illustrate the need of leveraging domain knowledge in the design of formal systems models. Its goal is to model a critical function estimating the ground velocity of an aircraft and the estimated value should be as close as possible to the effective velocity because it is critical, when considering the minimal velocity required for aircraft take off. Indeed, fluid mechanics laws require a minimal speed on ground for an effective take off of an aircraft.

In this case study, we consider the explicit features involved in the system under consideration: *a nose gear velocity update function*. This function is responsible for estimating the velocity of an aircraft while moving on the ground.

This case study is suitable for exposing the need for integrating explicit semantics in formal modelling as it consists of multiple heterogeneous components working together to produce the result. Additionally, it suggests cases for data conversions while estimating the velocity and publishing the computed velocity.

Next subsection presents a brief system description.

4.1   System Description

In general, the velocity is estimated by calculating the elapsed time for a complete rotation of a nose gear wheel. In the NG system, these rotations are monitored by an electro-mechanical sensor connected to a computer. This sensor generates a *click* (also called '*pulse*') for each complete rotation, which consequently generates a *hardware interrupt*. Interrupt service routine (ISR) of the system is then responsible for serving these interrupts. ISR increments a 16-bit counter, *NGRotations*, to capture a complete rotation of the NG wheel, and stores the current time of the update in *NGClickTime* variable. Both *NGRotations* and *NGClickTime* are modelled as 16-bit coded integer variables.

The NG system is equipped with a millisecond counter called *Millisecs*. This counter is incremented once every millisecond and provides a read-only access of its current value to all components in the system.

The system has another component, namely a real time operating system (RTOS), responsible for *invoking* the update function. RTOS makes sure that this function is invoked *at least once* every 500 ms. However, the exact timing of each invocation of this function relative to hardware interrupt is *not* predictable.

Finally, the velocity update function is responsible for estimating the current velocity of the aircraft on the ground. This estimation is based on currently available values of accessible counters. Estimated velocity is stored in a variable called *estimatedGroundVelocity*. Update function has read-only access to *Millisecs* counter along with *NGRotations* counter and the global variable *NGClickTime*. Also, the diameter of the NG wheel is available to the update function as a compile time constant called, *WheelDiameter*. The update function *can* store all the necessary private data required for calculating an estimation of the ground velocity. These values are protected from invocation to invocation.

There is one explicit, and the most important, requirement for this system,

> **EXFUN-1**. *While the aircraft is on the ground, the estimated velocity shall be within 3 km/h of the true velocity of the aircraft at some moment within the past 3 seconds.*

EXFUN-1 is a very important issue on the control of the speed of the aircraft up to some possible deviation due to external actions like the possible deformation of the wheels or any interaction with environment. Along with **EXFUN-1**, we have systematically extracted several other implicit/derived requirements from this system description. The full analysis of the target system can be found in [22] for details on the Event-B models. Below, we study the specific case of computation of the aircraft travelled distance on the runway.

4.2   The Case of the Travelled Distance Computation

In order to highlight the influential importance of the so called *implicit* information that may be needed when using a state-based method and its proof environment, we choose to study the case of the computation of the travelled distance which is combined with time to compute the ground speed for take-off.

NGRotations stores the number of recorded rotations of the wheel while the aircraft is moving on the ground. It is implemented as a 16-bits variable meaning that its real-world value shall be always smaller than $2^{15} - 1 = 32767$.

As a consequence of this design decision, it is possible to relate the *NGRotations* variable and the physical entity, namely the length of the runway required for landing or take off. In our case, the diameter of the considered aircraft wheel is supposed to be constant and equal to 22 in.

Starting from these hypotheses, the *MAXLENGTHRUNWAY* variable, maximum length of a runway encoded by *NGRotations*, can be computed using the value $2^{15} - 1$ and the wheel diameter. *It is worth noticing that, if the aircraft is still on the runway and the travelled distance is over MAXLENGTHRUNWAY, hen the stored value is no more correct as the 16-bit register overflows.*

In addition, we also need to consider the conversion rules of imperial unit system into metric unit system provided by a specific document (standard, table, book, etc.). So, if we consider that 1 in = 2.54 cm, the *MAXLENGTHRUNWAY* value is defined, in meter unit, by the following expression:

$$MAXLENGTHRUNWAY = \pi \times 22 \times 2.54 \times 0.01 \times (2^{15} - 1) = 57494 \quad (1)$$

When encoded in a 16-bits register, *the longest possible length of a runway is 57494.* If we refer to domain knowledge related to airports designs, runways, in particular runways length, are designed according to technical constraints related to the chosen airport position, altitude, land topology, and weather conditions (wind, rain, etc.).

Again, if we refer to additional domain knowledge related to airports descriptions repository (a finite set of civil airports or *instances*) we identify that the value of the longest existing runway is a runway (in China) equal to 5500 m. Observe that this distance is expressed in *meters* while wheel diameters are given in *inches* by wheel designers.

This means that a 16-bits register encodes safely any aircraft travelled distance (no overflow nor register initialisation is possible) since $5500 \leq 57494$. An alternative choice would have been an 8-bit register but in this case the encoded distance with 8 bits is 447 and safety condition is not fulfilled in this case.

From this example, we clearly understand that it is not possible to assert the travelled distance of an aircraft on a runway without external additional knowledge (i.e. *domain knowledge*).

5   Impact on System Engineering

When the $D, S \vdash R$ triptych mentioned in Sect. 1, is formalised within a formal method, the fulfilment of system requirements requires to derive, by a proof using the supported proof system, the requirements R from the specification S and the domain knowledge D. When state-based formal methods are used, the proof uses hypotheses provided by the domain knowledge D and by the specification S and relies on state changes expressed as before-after predicates (BAP). Inductive proofs are set up for invariant preservation or state reachability.

5.1   The Triptych from State-Based Formal Methods Perspective

Most of the state-based methods model state changes using a generalised assignment operation based on the *"becomes such that"* before after-predicate (offering deterministic or non deterministic assignments). This operation is encapsulated in rules in ASM [8], substitutions or events in B and Event-B [1], Hoare triples [14], operations in RSL [11], actions in TLA [19], operations in VDM [18], schemas in Z [25] and so on (alphabetical order). All these methods associate a proof obligation generation mechanism producing proof goals submitted to the underlying method proof system. As mentioned above, these proof obligations use D and S as hypotheses. For example, if we consider the case of Event-B, the proof obligation for event invariant preservation is described as follows.

Let us consider an event recording a state change with

- $A(s, c)$ axiomatised definitions for sets s and constants c
- x a state variable
- $G(x)$ a guard expressed as a Boolean expression
- $BAP(x, x')$ a before-after predicate linking before x and after x' state variables values
- $I(x)$ the invariant expressed as a predicate with x as a free variable.

then, invariant preservation proof obligation for this event is defined as the following expression:

$$A(s, c) \wedge G(x) \wedge BAP(x, x') \wedge I(x) \implies I(x') \tag{2}$$

If we match the above expression with the the triptych, we observe that $A(s, c)$, $G(x) \wedge BAP(x, x')$ and $I(x)$ correspond to part of the domain knowledge D, the specification S and the requirements R respectively. To be more precise, $I(x)$ is related to R as a requirement can be operationalised by several events and an event can participate to the operationalisation of several requirements.

5.2   Formalising Domain Knowledge

Let us consider again the previous proof obligation. During system development, the axiomatisation provided by $A(s, c)$ can be decomposed as

$$A(s, c) \equiv A_{Dom}(s, c) \wedge A_{Meth}(s, c) \wedge A_{Spec}(s, c)$$

where,

- $A_{Meth}(s, c)$ represents the native or core axiomatisation related to the used formal method. It formally characterises the semantics of the formal method. It never changes unless the semantics of the method is changed. For example, in the case of Event-B, all the axiomatic definitions related to set theory belong to this category. Other such axioms may appear when mathematical theories are externally defined and used in a given development. Examples are theories for vectors or differential equations, etc.

– $A_{Dom}(s,c)$ describes the application domain the system under design belongs to. This axiomatisation is *descriptive* and usually stable. It introduces a set of concepts specific to an application domain. Axiomatisation of domain knowledge, standards, certification processes, etc., belongs to this category. They may evolve subject to standard, technology or regulation rules evolutions. Examples of units with imperial or metric unit systems, polar or Cartesian coordinates, car driving rules, railway protocols (e.g. ERTMS), the European Food Law, the world health agency regulations, access control rules, International Civil Aviation Organization (ICAO) regulations, etc., belong to this category of axiomatisation of domain knowledge. In general, it is represented by standardised and formalised ontologies.

– $A_{Spec}(s,c)$ are specific definitions corresponding to the system under design. They introduce *prescriptive* definitions of the system entities and their characteristics.

Note that we do not discuss the issue of aligning semantics if $A_{Dom}(s,c)$ is formalised in a logic other than the one of the formal modelling language used. In this case, we assume that $A_{Dom}(s,c)$, $A_{Meth}(s,c)$ and $A_{Spec}(s,c)$ are all integrated and formalised in the used formal modelling language.

Thus, the previous proof obligation can be rewritten as:

$$A_{Dom}(s,c) \wedge A_{Meth}(s,c) \wedge A_{Spec}(s,c) \wedge G(x) \wedge BAP(x,x') \wedge I(x) \implies I(x') \quad (3)$$

5.3 Illustration on the Nose Gear Case Study

Back to the case study described in Sect. 3, let us consider the action that records the position of an aircraft along the runway at take off step. The requirement states that *every distance travelled by an aircraft on a runway over the world can be safely encoded.*

Context and Definitions. In the Event-B context ngv0, we gather several constants required for the case study. Axioms are tagged with three possible tags DOM when the axiom is related to the Domain, SPEC when it is related to the Specification and METH when the axiom is related to the underlying methodology as a mathematical computation (no such axiom in the example).

CONTEXT ngv0
SETS
    Units
CONSTANTS

PI	// The Constant π
NBITS	// Available bits for registers
Nbits	//Selected number of bits for coding NGRotations
MAXCODEDINT	// Maximum integer coded in Nbits
Circumference	// Length of one wheel revolution
LMAX	// Maximal length for MAXCODEDINT steps

MAXLENGTHRUNWAY // Maximal length of the runway
WheelDiameter // Diameter of a wheel
in // Inch unit of length
cm // Centimeter unit of length
m // Meter unit of length

AXIOMS

DOM_axm0: $PI = 314$
DOM_axm1: $partition(Units, \{in\}, \{cm\}, \{m\})$
DOM_axm2: $MAXLENGTHRUNWAY \in Units \to \mathbb{N}$
DOM_axm3: $MAXLENGTHRUNWAY = \{in \mapsto 216536, m \mapsto 5500,$
$$cm \mapsto 550000\}$$
DOM_axm4: $WheelDiameter \in Units \to \mathbb{N}$
DOM_axm5: $NBITS \subseteq \mathbb{N}$
DOM_axm6: $NBITS = \{8, 16, 32, 64\}$
SPEC_axm0: $MAXCODEDINT > 0$
SPEC_axm1: $Nbits \in NBITS$
SPEC_axm2: $Nbits = 16$
SPEC_axm3: $MAXCODEDINT = 2^{(Nbits-1)} - 1$
SPEC_axm4: $Circumference \in Units \to \mathbb{N}$
SPEC_axm5: $WheelDiameter = \{in \mapsto 22, m \mapsto 254 * 22/10000,$
$$cm \mapsto 254 * 22/100\}$$
SPEC_axm6: $Circumference(m) = PI * WheelDiameter(m)$
SPEC_axm7: $LMAX \in Units \to \mathbb{N} \wedge LMAX(m) =$
$$MAXCODEDINT * Circumference(m)$$
SPEC_axm8: $MAXLENGTHRUNWAY(m) \le LMAX(m)$

END

Checking Selected Domain Definitions. Axioms of the context should be checked as sound with respect to the engineering process. The definition of axioms is generally simple but is very critical, since we have to check that the resulting list of axioms $A(s, c)$ is consistent. One specific axiom SPEC_axm8 is scanned $MAXLENGTHRUNWAY(m) \le LMAX(m)$.

Here, $MAXLENGTHRUNWAY(m)$ is supplied by the knowledge domain of the problem and is available as an axiom in a domain ontology. $LMAX(m)$ is the value of the longest possible length that can be coded using the $Nbits$ 16-bit coded integer.

Classically, one can use a tool for finding a possible axioms inconsistencies by solving constraints over the whole axioms. In our case, we may use the ProB tool [21] to analyse solution existence for the list of axioms. We do not discuss the technique for analysing formal texts and for validating a list of axioms but in the current case, we identify a critical property which is expressed as an *a priori* axiom.

In the introduction of the case study, we have explicitly given the value when $Nbits$ is 16 and when $Nbits$ is 8. Using the effective values, we obtain that the set of axioms is inconsistent when $Nbits = 8$, since in this case $LMAX(m) = 447$ and $MAXLENGTHRUNWAY(m) = 5500$. Hence, the choice of $Nbits$ in the set $NBITS$ of possible registers should be $Nbits \ne 8$.

Computing the Travelled Distance. Let us consider the critical action incrementing the variable ngv (number of wheel revolutions) while recording the position pos on the runway. Recording the position in the pos state variable is in fact a feature of the model (introduced for the proof process) and not of the implementation and $LMAX(m)$ may not safely encoded using $Nbits$ bits since the pos variable is not implemented.

In case pos is actually implemented, an additional condition over $LMAX(m)$ ie $LMAX(m) \leq MAXCODEDINT$ is required.

We describe the following Event-B machine with a single event tick updating both pos and ngv variables at each wheel revolution.

MACHINE ngv1
SEES ngv0
VARIABLES

    ngv   // Number of wheel revolutions
    pos   // Computed position on the runway (travelled distance)

INVARIANTS

    inv1: $pos \in 0 .. MAXLENGTHRUNWAY(m)$
    inv2: $ngv \in 0 .. MAXCODEDINT$
    inv3: $pos = ngv * Circumference(m)$

EVENTS
Initialisation
    begin
        act1: $pos := 0$
        act2: $ngv := 0$
    end
Event tick $\langle$ordinary$\rangle \,\widehat{=}$
    when
        grd1: $pos + Circumference(m) < MAXLENGTHRUNWAY(m)$
    then
        act1: $ngv := ngv + 1$
        act2: $pos := pos + Circumference(m)$
    end
END

Invariant Preservation. The proof obligation of invariant preservation of the tick event is given by the expression of Eq. (3). We obtain

$A_{Dom}(s,c)$	$DOM\_axm0 \wedge \cdots \wedge DOM\_axm6$
$A_{Meth}(s,c)$	are the axioms of defining set theory and first order logic operators
$A_{Spec}(s,c)$	$SPEC\_axm0 \wedge \cdots \wedge SPEC\_axm8$
$G(x)$	$pos + Circumference(m) < MAXLENGTHRUNWAY(m)$
$BAP(x,x')$	$ngv' = ngv + 1 \wedge pos' = ngv' * Circumference(m)$
$I(x)$	$pos \in 0 .. MAXLENGTHRUNWAY(m) \wedge$ $ngv \in 0 .. MAXCODEDINT \wedge$ $pos = ngv * Circumference(m)$

Equation (3) is using the list of axioms $A(s,c) \equiv A_{Dom}(s,c) \wedge A_{Meth}(s,c) \wedge A_{Spec}(s,c)$ and it is clear that the proof obligation can be proved to be correct even if $A(s,c)$ axioms are not consistent.

From the previous discussion, we have justified the soundness of our list of axioms and discharged the associated proof obligations. The invariant property is expressing that both pos and ngv are bounded under the relation $pos = ngv * Circumference(m)$. The variable pos is bounded by a value which is ensuring that ngv is remaining correctly coded when $Nbits = 16$. It is clear that the axioms over the constants are expressing conditions for ensuring the possibility for pos to be increased. The interpretation of the guard $pos + Circumference(m) < MAXLENGTHRUNWAY(m)$ is that the aircraft has enough space for progressing on the runway. However, the relation $pos = ngv * Circumference(m)$ is used as $pos/Circumference(m) = ngv$ which helps to conclude that $ngv \leq MAXCODEDINT$.

The complete Rodin archive checks the proof obligations and contains also some intermediate lemmas.

6 Conclusion

Despite its simplicity, the example presented in this paper demonstrates that explicit structuring and formalising domain knowledge within the logic of the used state-based method contributes to broadening the spectrum of use of these methods, particularly in system design. Furthermore, we think that this domain knowledge can be specified as off-the-shelf domain theories and reused by system designers with minimal formalisation effort.

Thanks. The authors are grateful to the anonymous reviewers. They warmly thank Prof Dines Bjørner for his comments and careful review on the preliminary drafts of this paper.

References

1. Abrial, J.R.: Modeling in Event-B-System and Software Engineering. Cambridge University Press (2010)
2. Aït Ameur, Y., Baron, M., Bellatreche, L., Jean, S., Sardet, E.: Ontologies in engineering: the OntoDB/OntoQL platform. Soft Comput. **21**(2), 369–389 (2017). https://doi.org/10.1007/s00500-015-1633-5
3. Aït Ameur, Y., Méry, D.: Making explicit domain knowledge in formal system development. Sci. Comput. Program. **121**, 100–127 (2016)
4. Ait-Ameur, Y., Nakajima, S., Méry, D. (eds.): Implicit and Explicit Semantics Integration in Proof-Based Developments of Discrete Systems. Springer, Singapore (2021). https://doi.org/10.1007/978-981-15-5054-6
5. Bjørner, D.: Software Engineering 3 - Domains, Requirements, and Software Design. Texts in Theoretical Computer Science. An EATCS Series. Springer, Heidelberg (2006). https://doi.org/10.1007/3-540-33653-2

6. Bjørner, D.: Manifest domains: analysis and description. Formal Asp. Comput. **29**(2), 175–225 (2017)
7. Bjørner, D.: Domain analysis and description principles, techniques, and modelling languages. ACM Trans. Softw. Eng. Methodol. **28**(2), 8:1–8:67 (2019)
8. Börger, E., Stärk, R.F.: Abstract State Machines, A Method for High-Level System Design and Analysis. Springer, Heidelberg (2003). https://doi.org/10.1007/978-3-642-18216-7
9. Chebieb, A., Aït Ameur, Y.: A formal model for plastic human computer interfaces. Front. Comput. Sci. **12**(2), 351–375 (2018). https://doi.org/10.1007/s11704-016-5460-3
10. Critical Systems Labs Inc: Nose Gear (NG) Velocity Example Version 1.1, September 2011. http://www.cl.cam.ac.uk/~mjcg/FMStandardsWorkshop/example.pdf
11. George, C.: The RAISE specification language a tutorial. In: Prehn, S., Toetenel, H. (eds.) VDM 1991. LNCS, vol. 552, pp. 238–319. Springer, Heidelberg (1991). https://doi.org/10.1007/BFb0019998
12. Gruber, T.R.: Towards principles for the design of ontologies used for knowledge sharing. In: Guarino, N., Poli, R. (eds.) Formal Ontology in Conceptual Analysis and Knowledge Representation. Kluwer Academic Publisher's, Boston (1993)
13. Hacid, K., Aït Ameur, Y.: Handling domain knowledge in design and analysis of engineering models. Electron. Commun. Eur. Assoc. Softw. Sci. Technol. **74**, 1–21 (2017)
14. Hoare, C.A.R.: An axiomatic basis for computer programming. Commun. ACM **12**(10), 576–580 (1969)
15. Jackson, M., Zave, P.: Domain descriptions. In: Proceedings of IEEE International Symposium on Requirements Engineering, RE 1993, San Diego, California, USA, 4–6 January 1993, pp. 56–64. IEEE (1993)
16. Jackson, M.A.: Software Requirements and Specifications - A Lexicon of Practice, Principles and Prejudices. Addison-Wesley, New York (1995)
17. Mossakowski, T.: The distributed ontology, model and specification language – DOL. In: James, P., Roggenbach, M. (eds.) WADT 2016. LNCS, vol. 10644, pp. 5–10. Springer, Cham (2017). https://doi.org/10.1007/978-3-319-72044-9_2
18. Jones, C.B.: Systematic Software Development Using VDM. Prentice Hall International Series in Computer Science. Prentice Hall, Upper Saddle River (1986)
19. Lamport, L.: Specifying Systems, The TLA+ Language and Tools for Hardware and Software Engineers. Addison-Wesley (2002)
20. van Lamsweerde, A.: Requirements Engineering - From System Goals to UML Models to Software Specifications. Wiley, New York (2009)
21. Leuschel, M., Butler, M.J.: ProB: an automated analysis toolset for the B method. Int. J. Softw. Tools Technol. Transf. **10**(2), 185–203 (2008). https://doi.org/10.1007/s10009-007-0063-9
22. Méry, D., Sawant, R., Tarasyuk, A.: Integrating domain-based features into Event-B: A nose gear velocity case study. In: Bellatreche, L., Manolopoulos, Y. (eds.) MEDI 2015. LNCS, vol. 9344, pp. 89–102. Springer, Cham (2015). https://doi.org/10.1007/978-3-319-23781-7_8
23. Mossakowski, T., Codescu, M., Neuhaus, F., Kutz, O.: The distributed ontology, modeling and specification language – DOL. In: Koslow, A., Buchsbaum, A. (eds.) The Road to Universal Logic. SUL, pp. 489–520. Springer, Cham (2015). https://doi.org/10.1007/978-3-319-15368-1_21
24. Singh, N.K., Aït Ameur, Y., Méry, D.: Formal ontology driven model refactoring. In: 23rd International Conference on Engineering of Complex Computer Systems, ICECCS 2018, pp. 136–145. IEEE Computer Society (2018)

25. Spivey, J.M.: Z Notation - a Reference Manual. Prentice Hall International Series in Computer Science, 2nd edn. Prentice Hall, Englewood Cliffs (1992)
26. Yu, E.S.K.: Towards modeling and reasoning support for early-phase requirements engineering. In: 3rd IEEE International Symposium on Requirements Engineering (RE 1997), 5–8 January 1997. Annapolis, MD, USA, pp. 226–235. IEEE Computer Society (1997)
27. Zave, P., Jackson, M.: Four dark corners of requirements engineering. ACM Trans. Softw. Eng. Methodol. **6**(1), 1–30 (1997)

Some Observations on Mitotic Sets

Klaus Ambos-Spies$^{(\boxtimes)}$

Universität Heidelberg, Institut für Informatik, 69120 Heidelberg, Germany
`ambos@math.uni-heidelberg.de`

Abstract. A computably enumerable (c.e.) set A is mitotic if it can be split into two c.e. sets A_0 and A_1 which are Turing equivalent to A. Glaßer et al. [3] introduced two variants of this notion. The first variant weakens mitoticity by not requiring that the parts A_0 and A_1 of A are c.e., the second strengthens mitoticity by requiring that the parts A_0 and A_1 are computably separable. Glaßer et al. [3] raised the question whether these variants are nonequivalent to the classical mitoticity notion. Here we answer this question affirmatively. We also show, however, that the weaker mitoticity property is trivial, i.e., that any c.e. set has this property. Moreover, we consider and compare these new variations of mitoticity for the common strong reducibilities in place of Turing reducibility. Finally - not directly related to these results - we show that all weak truth-table (wtt) complete sets are wtt-mitotic, in fact strongly wtt-mitotic.

1 Introduction

For most 'natural' noncomputable computably enumerable (c.e.) problems (sets) the information content of the sets is coded redundantly. For instance, by Myhill's Theorem, any m-complete c.e. set A is a cylinder, i.e., can be split into infinitely many c.e. parts where each part is m-equivalent to the original set A. A weaker form of redundancy – which has been extensively studied in computability theory as well as in computational complexity theory[1] – is mitoticity. (The term is borrowed from the process of mitosis in cell biology – the division of the mother cell into two daughter cells genetically identical to each other.) A c.e. set A is mitotic if it can be split into two c.e. sets A_0 and A_1 which are equivalent to the original set A (hence equivalent to each other). This notion was originally introduced for Turing reducibility – and most of the work on mitotic sets in computability theory deals with this reducibility – but it can be (and has been) adapted to other reducibilities too. (In the following mitotic will refer to

---

[1] First steps to a systematic study of polynomial-time bounded variants of mitotic sets were taken in the author's 1984 paper [1]. This work was done when the author was a postdoc at the University of Dortmund supervised by Egon Börger who greatly encouraged and supported this work. For a survey of the further work on mitoticity in computational complexity theory, see the survey by Glaßer et al. [3]. In the current paper we confine ourselves to the setting of computability theory.

This article is based on some unpublished notes of the author from February 2017.

A. Raschke et al. (Eds.): Börger Festschrift, LNCS 12750, pp. 14–28, 2021.
https://doi.org/10.1007/978-3-030-76020-5_2

mitotic for Turing reducibility and we say r-mitotic if we consider this notion for some other reducibility r.)

The first work on mitotic sets is by Lachlan [4] who showed that this notion is nontrivial, i.e., that there are non-mitotic c.e. sets. A systematic study of the mitotic sets was started by Ladner [5,6]. For instance he showed the following: the mitotic sets are just Trahtenbrot's c.e. autoreducible sets. i.e., the c.e. sets A for which the question whether a number x is an element of A can be answered using $A\backslash\{x\}$ as an oracle; there are (Turing) complete c.e. sets which are non-mitotic; and there are so-called nontrivial completely mitotic degrees, i.e., Turing degrees of noncomputable c.e. sets where all c.e. members are mitotic.

For more results on mitotic sets in computability theory and computational complexity theory we refer to the fairly recent survey [3] by Glaßer et al. There also two new notions of mitotic c.e. sets are introduced, one strengthening the classical notion, one weakening it: in the stronger notion it is required that the c.e. parts (A_0, A_1) of a mitotic splitting of a c.e. set A are computably separable, in the weaker notion the requirement that the parts A_0 and A_1 of the splitting are computably enumerable is dropped. (Some motivation for this new notions comes from computational complexity where corresponding variants of polynomial-time mitoticity have been considered in [1].) Glasser et al. raise the question (see Question 4.2 there), whether these new notions are really new, i.e., not equivalent to the classical notion. Here we will show that this is indeed the case (which, by the equivalence of autoreducibility and mitoticity, also gives a negative answer to Question 3.4 in [3]). In fact, we give stronger separation results for the three variants of mitoticity by considering these notions not only for Turing reducibility but also for the classical strong reducibilities. Some of these separation results were obtained independently by Michler [7] (see below).

Before we state our results more precisely, we first give formal definitions of the mitoticity notions considered in the following. Here we have to give some warning in order to avoid confusion. Glasser et al. [3] call the classical mitotic sets (which, following Ladner, are called mitotic in the standard literature) c.e. mitotic, and call the stronger notion, requiring computable separability of the parts, mitotic. Here we stick to the traditional notation. So the c.e. mitotic sets of [3] are called mitotic here (as usual), and the mitotic sets of [3] are called strongly mitotic here.

Definition 1. *Let r be any reducibility. A c.e. set A is r-mitotic if there are disjoint c.e. sets A_0 and A_1 such that $A = A_0 \cup A_1$ and $A =_r A_0 =_r A_1$.*

Glaßer et al. [3] consider the following variations of the notion of an r-mitotic set.

Definition 2. *(a) A c.e. set A is weakly r-mitotic if there are disjoint (not necessarily c.e.) sets A_0 and A_1 such that $A = A_0 \cup A_1$ and $A =_r A_0 =_r A_1$.*

(b) A c.e. set A is strongly r-mitotic if there is a computable set B such that $A =_r A \cap B =_r A \cap \overline{B}$.

As usual in the literature, we call a c.e. set mitotic if it is T-mitotic, and we call a c.e. set weakly mitotic (strongly mitotic) if it is weakly T-mitotic (strongly T-mitotic). Besides Turing (T) reducibility we consider the following common strong reducibilities: many-one reducibility (m), truth-table reducibility of norm 1 (1-tt), bounded truth-table reducibility (btt), truth-table reducibility (tt) and weak truth-table reducibility (wtt). (Moreover, for technical convenience we also consider the bounded truth-table reducibilities of fixed norm $k \geq 2$ (k-tt) though these reducibilities are not transitive.) We call a reducibility r stronger than a reducibility r' (and r' weaker than r) if, for any sets A and B, $A \leq_r B$ implies $A \leq_{r'} B$; and we call r strictly stronger than r' (and r' strictly weaker than r) if r is stronger than r' but r and r' are not equivalent, i.e., r' is not stronger than r. For the reducibilities considered here, m is stronger than 1-tt, 1-tt is stronger than btt (more generally, for any $k \geq 1$, k-tt is stronger than $(k+1)$-tt and btt), btt is stronger than tt, tt is stronger than wtt, and wtt is stronger than T (in fact, as well known, all these relations are strict – though m and 1-tt coincide on the c.e. sets). Note that, for reducibilities r and r' such that r is stronger than r', any r-mitotic set is r'-mitotic, any strongly r-mitotic set is strongly r'-mitotic, and any weakly r-mitotic set is weakly r'-mitotic. In the following we tacitly use these trivial observations.

The following implications among the mitoticity notions are immediate by definition (for any reducibility r).

$$A \text{ is strongly } r\text{-mitotic} \Rightarrow A \text{ is } r\text{-mitotic} \Rightarrow A \text{ is weakly } r\text{-mitotic} \qquad (1)$$

Glaßer et al. [3] raise the question whether, for Turing reducibility, the converses of these implications hold (see Question 4.2 there). As pointed out above already, in the following we show that this is not the case. In fact, we prove stronger separation results by considering not only Turing reducibility but also the classical strong reducibilities mentioned above. The separation of strong mitoticity and mitoticity was independently obtained by Michler [7], a student of Glaßer.

In Sect. 2, we compare weak mitoticity with mitoticity. In order to separate r-mitoticity from weak r-mitoticity for any reducibility r weaker than 1-tt, we show that, for such r, weak r-mitoticity is trivial, i.e., that *any* c.e. set is weakly 1-tt-mitotic hence weakly r-mitotic. (In fact, any c.e. set A can be split into a c.e. set A_0 and a co-c.e. (or, alternatively, a left-c.e.) set A_1 such that $A =_{1\text{-tt}} A_0 =_{1\text{-tt}} A_1$.) In the setting of the classical strong reducibilities this separation is optimal since, for any reducibility r stronger than m (just as for any positive reducibility) weak r-mitoticity and r-mitoticity coincide since any set which is m-reducible (or positively reducible) to a computably enumerable set is computably enumerable.

The separation of mitoticity and strong mitoticity is given in Sect. 3. As mentioned above already, this separation was independently obtained by Michler [7]. We show that there is a btt-mitotic (in fact, 2-tt-mitotic) set which is not strongly T-mitotic. In this result we cannot replace 2-tt by 1-tt since, as we also show, any 1-tt-mitotic set is strongly T-mitotic (in fact, strongly wtt-mitotic). Still, as our final result of this section shows, there are 1-tt-mitotic sets which are not strongly 1-tt-mitotic. By the coincidence of (strongly) m-mitotic and (strongly) 1-tt-mitotic, this also gives the separation of m-mitoticity and strong m-mitoticity.

In the final Sect. 4, which is not directly related to the above separations, we look at complete sets. As mentioned above already, Ladner [5] has shown that there is a Turing complete set which is not Turing mitotic (hence not strongly Turing mitotic), whereas, by Myhill's theorem, any m-complete set is a cylinder hence strongly m-mitotic (see Glaßer et al. [3]). Here we show that Ladner's result also fails if we replace Turing reducibility by weak truth-table reducibility: any wtt-complete set is strongly wtt-mitotic hence wtt-mitotic.

Our notation is standard. Unexplained notation can be found, for instance, in the introductory part on computability theory of the monograph [2]. We assume the reader to be familiar with the basic techniques of computability theory. Some of the proofs in Sect. 3 use finite-injury priority arguments.

2   Weakly Mitotic Sets

Here we show that any c.e. set A is weaky 1-tt-mitotic hence weakly (Turing) mitotic. The proof of this rather straightforward observation is based on the fact that any infinite c.e. set contains an infinite computable set as a subset.

Theorem 1. *Let A be any c.e. set. Then A is weakly 1-tt-mitotic (hence weakly r-mitotic for any reducibility r weaker than 1-tt). In fact, there is a c.e. set A_0 and a co-c.e. set A_1 such that A_0 and A_1 are disjoint, $A = A_0 \cup A_1$, $A =_m A_0$, and $A =_{\text{1-tt}} A_1$.*

Proof. If A is computable then the claim is trivial. So w.l.o.g. we may assume that $A \neq \omega$ and A is infinite. Fix an infinite computable subset B of A and let $b_0 < b_1 < b_2 < \ldots$ be the elements of B in order of magnitude. Define the splitting $A = A_0 \stackrel{.}{\cup} A_1$ of A by letting A_0 be the c.e. set

$$A_0 = (A \cap \overline{B}) \cup \{b_n : n \in A\}$$

and by letting A_1 be the co-c.e. set

$$A_1 = \{b_n : n \notin A\}.$$

Then, as one can easily check, $A =_m A_0$ and $\overline{A} =_m A_1$ (hence $A =_{\text{1-tt}} A_0 =_{\text{1-tt}} A_1$). □

Since Ladner [5] has shown that there are c.e. sets which are not mitotic (hence not r-mitotic for any reducibility r stronger than T), Theorem 1 gives the desired separation of weak mitoticity and mitoticity.

Corollary 1. *There is a c.e. set which is weakly mitotic but not mitotic. In fact, for any reducibility r weaker than 1-tt-reducibility and stronger than Turing reducibility, there is a c.e. set which is weakly r-mitotic but not r-mitotic.*

Note that any set which is many-one reducible to a c.e. set is c.e. too. Hence weak m-mitoticity and m-mitoticity coincide. So, in Theorem 1 and Corollary 1 we cannot replace 1-tt-reducibility by the stronger m-reducibility. (More generally, for the same reason, we cannot replace 1-tt-reducibility by any positive reducibility.)

Also note that in Theorem 1 we may replace the co-c.e. set A_1 by a left-c.e. set A_1 (see e.g. Downey and Hirschfeldt [2] for the definition). Namely, in the proof of Theorem 1, it suffices to replace A_0 and A_1 by the c.e. set $A_0 = \{b_n : n \in A\}$ and the left-c.e. (and d-c.e.) set

$$A_1 = (A \cap \overline{B}) \cup \{b_n : n \notin A\},$$

respectively. (The proof that A_1 is left-c.e. is as follows. By Theorem 5.1.7 in [2], it suffices to give a computable approximation $\{A_{1,s}\}_{s \geq 0}$ of A_1 such that, for any number x and any stage s such that $x \in A_{1,s} \setminus A_{1,s+1}$, there is a number $y < x$ such that $y \in A_{1,s+1} \setminus A_{1,s}$. Such an approximation is obtained by letting $A_{1,s} = (A_s \cap \overline{B}) \cup \{b_n : n \notin A_s \cup B\}$ for any given computable enumeration $\{A_s\}_{s \geq 0}$ of A. Namely, if $x \in A_{1,s} \setminus A_{1,s+1}$ then $x = b_n$ for some number $n \in \overline{B}$ such that $n \in A_{s+1} \setminus A_s$. So, for $y = n$, $y < x$ and $y \in A_{1,s+1} \setminus A_{1,s}$.) So, for any reducibility r weaker than 1-tt, mitoticity becomes trivial if we relax the condition that the parts of the splitting are c.e. by only requiring the parts to be left-c.e. (or one c.e. and the other left-c.e.).

3   Strongly Mitotic Sets

We now turn to the relations between mitoticity and strong mitoticity for the common reducibilities. First we give the following strong separation theorem which was independently proven by Michler [7].

Theorem 2. *There is a c.e. set A such that A is 2-tt-mitotic but not strongly T-mitotic.*

Below we will show that this is optimal, i.e., that 2-tt cannot be replaced by 1-tt in Theorem 2.

Proof (of Theorem 2). By a finite injury argument, we enumerate c.e. sets A, A_0 and A_1 such that $A = A_0 \,\dot{\cup}\, A_1$, A is 2-tt-mitotic via (A_0, A_1) and A is not strongly T-mitotic. We let $A_{0,s}$ and $A_{1,s}$ be the finite parts of A_0 and A_1, respectively, enumerated by the end of stage s, and we let $A_s = A_{0,s} \cup A_{1,s}$.

It suffices to ensure that the c.e. sets A_0, A_1 and $A = A_0 \cup A_1$ satisfy the following three conditions.

$$A_0 \cap A_1 = \emptyset \tag{2}$$

$$\forall i \leq 1 \left(A_i \leq_{2\text{-tt}} A_{1-i} \ \& \ A_i \leq_{2\text{-tt}} A \ \& \ A \leq_{2\text{-tt}} A_i \right) \tag{3}$$

$$\forall B \left(B \text{ computable} \Rightarrow \left[A \not\leq_T A \cap B \text{ or } A \not\leq_T A \cap \overline{B} \right] \right) \tag{4}$$

As one can easily check, in order to ensure (2) and (3) it suffices to guarantee that, for any number $x \geq 0$, one of the following four clauses holds.

(I) $A_0 \cap \{3x, 3x + 1, 3x + 2\} = \emptyset$
      $A_1 \cap \{3x, 3x + 1, 3x + 2\} = \emptyset$

(II) $A_0 \cap \{3x, 3x + 1, 3x + 2\} = \{\ 3x \qquad\qquad\ \}$
      $A_1 \cap \{3x, 3x + 1, 3x + 2\} = \{\qquad 3x + 1 \qquad\ \}$

(III) $A_0 \cap \{3x, 3x + 1, 3x + 2\} = \{\qquad\qquad\ 3x + 2\ \}$
      $A_1 \cap \{3x, 3x + 1, 3x + 2\} = \{\ 3x \qquad\qquad\ \}$

(IV) $A_0 \cap \{3x, 3x + 1, 3x + 2\} = \{\qquad 3x + 1 \qquad\ \}$
      $A_1 \cap \{3x, 3x + 1, 3x + 2\} = \{\qquad\qquad\ 3x + 2\ \}$

$$(5)$$

On the other hand, these constraints are tame enough in order to allow us to meet the non-mitoticity requirements ensuring condition (4). The key to satisfy this condition is the ability that, given a computable set B, we can expand A in such a way that the expansion affects only one of the parts $A \cap B$ or $A \cap \overline{B}$ whence the other part fails to record this change in A. Note that this can be done in the presence of the above constraints since for any x we may pick $j_0 \neq j_1 \leq 2$ such that $B(3x + j_0) = B(3x + j_1)$ (since we may choose from 3 numbers and B can take only two values) and we may add $3x + j_0$ and $3x + j_1$ to A in such a way that one of the conditions (II) to (IV) is satisfied.

Formally, in order to guarantee (4) we meet the requirements

$$\mathcal{R}_{\langle e, k \rangle} : \text{ If } \psi_e \text{ is total then } A \neq \Phi_{k_0}^{A \cap S_e} \text{ or } A \neq \Phi_{k_1}^{A \cap \overline{S_e}}.$$

(for all numbers e and $k = \langle k_0, k_1 \rangle$) where $\{\psi_e\}_{e \geq 0}$ is an acceptable numbering of the 0-1-valued partial computable functions, $S_e = \{x : \psi_e(x) = 1\}$, and $\{\Phi_e\}_{e \geq 0}$ is an acceptable numbering of the Turing functionals. To show that this suffices to get (4), for a contradiction assume that all requirements are met but (4) fails. By the latter, fix B such that B is computable and such that $A \leq_T A \cap B$ and $A \leq_T A \cap \overline{B}$ hold. Then (by the former) there is a number e such that ψ_e is total and $B = S_e$, whence (by the latter) we may fix numbers k_0 and k_1 such that $A = \Phi_{k_0}^{A \cap S_e}$ and $A = \Phi_{k_1}^{A \cap \overline{S_e}}$. So, for $k = \langle k_0, k_1 \rangle$, requirement $\mathcal{R}_{\langle e, k \rangle}$ is not met contrary to assumption.

In the course of the construction we use computable approximations $\psi_{e,s}$ and $\Phi_{e,s}$ of the partial functions ψ_e and the Turing functionals Φ_e, respectively, where $\psi_{e,s}(x) = \psi_e(x)$ $(\Phi_{e,s}^X(x) = \Phi_e^X(x))$ if the computation of $\psi_e(x)$ $(\Phi_e^X(x))$ converges in $\leq s$ steps and where $\psi_{e,s}(x) \uparrow$ $(\Phi_{e,s}^X(x) \uparrow)$ otherwise. We let $l(e, s)$ be the least x such that $\psi_{e,s}(x) \uparrow$ (note that for total ψ_e, $\lim_{s \to \infty} l(e, s) = \infty$), and let $S_{e,s} = S_e \upharpoonright l(e, s)$ (i.e., $S_{e,s}$ is the longest initial segment of S_e determined by the end of stage s). Moreover, for technical convenience, we assume that

$$\Phi_{e,s}^X(x) \downarrow \Rightarrow e, x, \varphi_e^X(x) < s \qquad (6)$$

holds where $\varphi_e^X(x)$ is the use of $\Phi_e^X(x)$, i.e., the least strict upper bound on the oracle queries occuring in the computation of $\Phi_e^X(x)$. Note that (by the use principle) this ensures

$$\Phi_{e,s}^X(x) \downarrow \ \& \ X \restriction s = Y \restriction s \ \Rightarrow \ \Phi_e^Y(x) = \Phi_{e,s}^Y(x) = \Phi_{e,s}^X(x). \tag{7}$$

Now, the strategy to meet requirement $\mathcal{R}_{\langle e,k \rangle}$ is as follows.

Step 1. At the stage $s+1$ at which the attack starts, appoint $x = s$ as $\mathcal{R}_{\langle e,k \rangle}$-follower.

When x is appointed, none of the numbers $3x, 3x+1, 3x+2$ is in A. These numbers become reserved for the $\mathcal{R}_{\langle e,k \rangle}$-strategy: this strategy will determine which of of these numbers are enumerated into A_0 and A_1, and these are the only numbers which may be enumerated for the sake of this strategy (unless the follower becomes cancelled by the strategy of a higher priority requirement and another attack on $\mathcal{R}_{\langle e,k \rangle}$ with a new follower is made later).

Step 2. Wait for a stage $s' \geq s$ such that

$$3x + 2 < l(e, s'), \tag{8}$$

$$\forall \, j \leq 2 \ (\Phi_{k_0,s'}^{A_{s'} \cap S_{e,s'}}(3x+j) = 0 \ \& \ \varphi_{k_0,s'}^{A_{s'} \cap S_{e,s'}}(3x+j) < l(e,s')), \tag{9}$$

and

$$\forall \, j \leq 2 \ (\Phi_{k_1,s'}^{A_{s'} \cap \overline{S_{e,s'}}}(3x+j) = 0 \ \& \ \varphi_{k_1,s'}^{A_{s'} \cap \overline{S_{e,s'}}}(3x+j) < l(e,s')) \tag{10}$$

hold.

For the least such s' (if any), perform the following action at stage $s' + 1$.

- Pick j_0 and j_1 minimal (in this order) such that $j_0 < j_1 \leq 2$ and $S_e(3x+j_0) = S_e(3x+j_1)$.
- Enumerate $3x + j_0$ and $3x + j_1$ into A_0 and A_1 according to the matching clause (II) to (IV). I.e., if $j_0 = 0$ and $j_1 = 1$ then enumerate $3x$ into A_0 and $3x + 1$ into A_1, if $j_0 = 0$ and $j_1 = 2$ then enumerate $3x$ into A_1 and $3x + 2$ into A_0, and if $j_0 = 1$ and $j_1 = 2$ then enumerate $3x + 1$ into A_0 and $3x + 2$ into A_1. (Since numbers are enumerated into A_0 and A_1 only according to this clause, this ensures that condition (5) – hence conditions (2) and (3) – are satisfied.)
- Cancel the followers of all lower priority requirements $\mathcal{R}_n$, $n > \langle e, k \rangle$. (This ensures that - besides $3x + j_0$ and $3x + j_1$ - no numbers $\leq s'$ are enumerated into A after stage s', unless a higher priority requirement acts later.)

To show that the strategy succeeds to meet $\mathcal{R}_{\langle e,k \rangle}$ provided that no higher priority requirement will act after stage s, distinguish the following two cases.

If there is no stage s' as in Step 2 of the attack then either ψ_e is not total (whence $\mathcal{R}_{\langle e,k \rangle}$ is trivially met) or (by the use principle) there is a number $j \leq 2$ such that $\Phi_{k_0}^{A \cap S_e}(3x + j) \neq 0$ or $\Phi_{k_1}^{A \cap \overline{S_e}}(3x + j) \neq 0$ (where here and in the following inequality includes the case that one side – here the right side – is defined while the other side – here the left side – is undefined). Since in this case none of the numbers $3x, 3x+1, 3x+2$ is enumerated into A, it follows that requirement $\mathcal{R}_{\langle e,k \rangle}$ is met.

If there is such a stage s' then there are numbers $j_0, j_1 \leq 2$ such that $3x+j_0$ and $3x+j_1$ are enumerated into A at stage $s'+1$ where either $\{3x+j_0, 3x+j_1\} \subseteq S_e$ or $\{3x+j_0, 3x+j_1\} \subseteq \overline{S_e}$. Moreover, by the assumption that no higher priority requirement will act after stage s' and by cancellation of the lower priority followers, no other numbers $\leq s'$ will enter A after stage s'. Obviously, this implies

$$A \upharpoonright s' = (A_{s'} \upharpoonright s') \cup \{3x+j_0, 3x+j_1\}.$$

So, if $\{3x+j_0, 3x+j_1\} \subseteq S_e$ then $(A \cap \overline{S_e}) \upharpoonright s' = (A_{s'} \cap \overline{S_e}) \upharpoonright s'$ whence, by (10) and by (7),

$$\Phi_{k_1}^{A \cap \overline{S_e}}(3x+j_0) = \Phi_{k_1,s'}^{A_{s'} \cap \overline{S_e}}(3x+j_0) = \Phi_{k_1,s'}^{A_{s'} \cap \overline{S_{e,s'}}}(3x+j_0) = 0 \neq A(3x+j_0),$$

and, if $\{3x+j_0, 3x+j_1\} \subseteq \overline{S_e}$ then, by the dual argument, $\Phi_{k_0}^{A \cap S_e}(3x+j_0) \neq A(3x+j_0)$. So, in either case $\mathcal{R}_{\langle e,k \rangle}$ is met.

Having explained the ideas underlying the construction, we now give the formal construction (where stage $s = 0$ is vacuous, i.e., $A_{0,0} = A_{1,0} = \emptyset$ and no requirement has a follower at the end of stage 0).

Stage $s+1$. Requirement $\mathcal{R}_{\langle e,k \rangle}$ requires attention if $\langle e, k \rangle \leq s$ and one of the following holds.

(i) There is no $\mathcal{R}_{\langle e,k \rangle}$-follower at the end of stage s.
(ii) $\mathcal{R}_{\langle e,k \rangle}$ has the follower x at the end of stage s and (8)–(10) above hold for s in place of s'.

Fix $\langle e, k \rangle$ minimal such that $\mathcal{R}_{\langle e,k \rangle}$ requires attention. Say that $\mathcal{R}_{\langle e,k \rangle}$ receives attention and becomes active. If (i) holds, appoint s as $\mathcal{R}_{\langle e,k \rangle}$-follower. If (ii) holds pick j_0 and j_1 minimal (in this order) such that $j_0 < j_1 \leq 2$ and $S_e(3x+j_0) = S_e(3x+j_1)$. If $j_0 = 0$ and $j_1 = 1$ then enumerate $3x$ into A_0 and $3x+1$ into A_1, if $j_0 = 0$ and $j_1 = 2$ then enumerate $3x$ into A_1 and $3x+2$ into A_0, and if $j_0 = 1$ and $j_1 = 2$ then enumerate $3x+1$ into A_0 and $3x+2$ into A_1. In any case, for any $n > \langle e, k \rangle$ cancel the follower of requirement $\mathcal{R}_n$ (if any).

Note that the construction is effective whence the sets A_0, A_1 and A are c.e. Moreover, it is obvious that the construction ensures that, for any x, one of the conditions (I) to (IV) in (5) applies. So (2) and (3) are satisfied. It remains to show that all requirements are met. For this sake one first shows, by a straightforward induction on n, that any requirement $\mathcal{R}_n$ requires attention only finitely often. This implies that, for any given number $\langle e, k \rangle$, there is a stage s at which $x = s$ is appointed as $\mathcal{R}_{\langle e,k \rangle}$-follower and x is not cancelled later (since no higher priority requirement acts later). By the above discussion of the strategy to meet $\mathcal{R}_{\langle e,k \rangle}$, it follows that the attack on $\mathcal{R}_{\langle e,k \rangle}$ via the follower x works. So $\mathcal{R}_{\langle e,k \rangle}$ is met. We leave the details to the reader. □

Next we show that Theorem 2 is optimal. For this sake we first observe that m-mitoticity and 1-tt-mitoticity coincide. So it suffices to show that any m-mitotic set is strongly T-mitotic.

Proposition 1. *A c.e. set A is (strongly) m-mitotic if and only if A is (strongly) 1-tt-mitotic.*

Proof. This is immediate by the fact that $\leq_m$ and $\leq_{1\text{-tt}}$ agree on the class $\mathcal{E}$ of c.e. sets.   □

Theorem 3. *Any m-mitotic set is strongly T-mitotic.*

Proof. Assume that A is m-mitotic. Fix c.e. sets A_i $(i = 0, 1)$ such that

$$A = A_0 \cup A_1 \ \& \ A_0 \cap A_1 = \emptyset, \tag{11}$$

and $A =_m A_i$. By the latter, fix computable functions f_i and g_i $(i = 0, 1)$ such that $A_i \leq_m A$ via f_i and $A \leq_m A_i$ via g_i, i.e., such that

$$\forall\, i \leq 1 \,\forall\, x \geq 0 \,(A_i(x) = A(f_i(x))), \tag{12}$$

and

$$\forall\, i \leq 1 \,\forall\, x \geq 0 \,(A(x) = A_i(g_i(x))) \tag{13}$$

hold. It suffices to define a computable set S such that

$$A \cap S \leq_T A \cap \overline{S} \ \& \ A \cap \overline{S} \leq_T A \cap S \tag{14}$$

holds.

The definition of S is based on the following two observations.

For any $x \geq 0$ and any $i \leq 1$, $A(x)$ and $A(f_{1-i}(g_i(x)))$ can be (uniformly) computed from $A(g_i(x))$ and vice versa. $\tag{15}$

For any $x \geq 0$ such that $x \in A$ there is a number $i \leq 1$ such that $g_i(x) \notin \{x, f_{1-i}(g_i(x))\}$. $\tag{16}$

For a proof of (15) fix x and i. Then $A(x)$ and $A(f_{1-i}(g_i(x)))$ can be computed from $A(g_i(x))$ as follows. If $g_i(x) \notin A$ then $g_i(x) \notin A_i$ and $g_i(x) \notin A_{1-i}$. By the former and by (13), $x \notin A$, while, by the latter and by (12), $f_{1-i}(g_i(x)) \notin A$. On the other hand, if $g_i(x) \in A$ then, by enumerating A_0 and A_1, find the unique $j \leq 1$ such that $g_i(x) \in A_j$. Then, by (13), $x \in A$ iff $j = i$ and, by (12), $f_{1-i}(g_i(x)) \in A$ iff $j = 1 - i$. Finally, the procedure for computing $A(g_i(x))$ from $A(x)$ and $A(f_{1-i}(g_i(x)))$ is as follows. By (13), $g_i(x) \in A_i$ iff $x \in A$, and, by (12), $g_i(x) \in A_{1-i}$ iff $f_{1-i}(g_i(x)) \in A$. So $g_i(x) \in A$ iff $A \cap \{x, f_{1-i}(g_i(x))\} \neq \emptyset$.

For a proof of (16) fix x, assume that $x \in A$, and (by (11)) fix the unique $i \leq 1$ such that $x \in A_{1-i}$. Then, by $x \in A$ and by (13), $g_i(x) \in A_i$ whence (by (11)) $g_i(x) \notin A_{1-i}$. So $g_i(x) \neq x$. Moreover, by $g_i(x) \notin A_{1-i}$ and (12), $f_{1-i}(g_i(x)) \notin A$. So $g_i(x) \neq f_{1-i}(g_i(x))$ since $g_i(x) \in A_i \subseteq A$.

Now, for the definition of S, we inductively define computable enumerations $\{S_x\}_{x \geq 0}$ and $\{\overline{S}_x\}_{x \geq 0}$ of S and $\overline{S}$, respectively. Let $S_0 = \overline{S}_0 = \emptyset$. For the definition of S_{x+1} and $\overline{S}_{x+1}$ distinguish the following cases.

Case1: $x \in S_x \cup \overline{S}_x$. Let

$$S_{x+1} = S_x$$
$$\overline{S}_{x+1} = \overline{S}_x \qquad (17)$$

Case 2: $x \notin S_x \cup \overline{S}_x$. Distinguish the following subcases.
Case 2.1: There is a number $i \leq 1$ *such that* $g_i(x) \notin \{x, f_{1-i}(g_i(x))\}$.
Then, for the least such i, let

$$S_{x+1} = \begin{cases} S_x \cup \{x\} & \text{if } f_{1-i}(g_i(x)) \in S_x \cup \overline{S}_x \\ S_x \cup \{x, f_{1-i}(g_i(x))\} & \text{otherwise} \end{cases}$$

$$\qquad (18)$$

$$\overline{S}_{x+1} = \begin{cases} \overline{S}_x & \text{if } g_i(x) \in S_x \cup \overline{S}_x \\ \overline{S}_x \cup \{g_i(x)\} & \text{otherwise} \end{cases}$$

Case 2.2: Otherwise. Let

$$S_{x+1} = S_x \cup \{x\}$$
$$\overline{S}_{x+1} = \overline{S}_x \qquad (19)$$

Obviously, the above ensures that $S_x \subseteq S_{x+1}$, $\overline{S}_x \subseteq \overline{S}_{x+1}$, $x \in S_{x+1} \cup \overline{S}_{x+1}$ and $S_{x+1} \cap \overline{S}_{x+1} = \emptyset$. So, for $S = \bigcup_{x \geq 0} S_x$, the complement $\overline{S}$ of S is given by $\overline{S} = \bigcup_{x \geq 0} \overline{S}_x$. Moreover, by effectivity of the definition of the sets S_x and $\overline{S}_x$, $\{S_x\}_{x \geq 0}$ and $\{\overline{S}_x\}_{x \geq 0}$ are computable enumerations whence S and $\overline{S}$ are c.e. So S is computable.

It remains to show that (14) holds. For this sake, (by computability of S) it suffices to show that, for any x, $A \cap S_x$ can be computed from $A \cap \overline{S}_x$ uniformly in x and vice versa. We give inductive procedures.

For $x = 0$ the task is trivial since $S_0 = \overline{S}_0 = \emptyset$. So fix x and (by inductive hypothesis) assume that there are procedures for computing $A \cap S_x$ from $A \cap \overline{S}_x$ and vice versa.

Then, in order to give an effective procedure for computing $A \cap S_{x+1}$ from $A \cap \overline{S}_{x+1}$, it suffices to give a procedure for computing $A \cap (S_{x+1} \backslash S_x)$ from $A \cap (S_x \cup \overline{S}_{x+1})$ since, by inductive hypothesis, $A \cap S_x$ can be computed from $A \cap \overline{S}_x$ hence from $A \cap \overline{S}_{x+1}$.

So fix $y \in S_{x+1} \backslash S_x$. In order to decide whether $y \in A$ (using $A \cap (S_x \cup \overline{S}_{x+1})$ as an oracle), distinguish the following two cases according to the case applying to the definition of S_{x+1} and $\overline{S}_{x+1}$. (Note that, by $S_{x+1} \neq S_x$, one of the two subcases of Case 2 must apply.) If Case 2.1 applies then, for the least $i \leq 1$ such that $g_i(x)$ is not in $\{x, f_{1-i}(g_i(x))\}$, it holds that $y \in \{x, f_{1-i}(g_i(x))\}$ and $g_i(x) \in S_x \cup \overline{S}_{x+1}$. So $A(y)$ can be computed from $A \cap (S_x \cup \overline{S}_{x+1})$ by (15). Finally, if Case 2.2 applies then $y = x$ and $x \notin A$ by (16).

The algorithm for computing $A \cap \overline{S}_{x+1}$ from $A \cap S_{x+1}$ is similar. First, by inductive hypothesis, we may argue that it suffices to compute $A \cap (\overline{S}_{x+1} \backslash \overline{S}_x)$ from $A \cap (S_{x+1} \cup \overline{S}_x)$. Then given $y \in \overline{S}_{x+1} \backslash \overline{S}_x$, Case 2.1 must apply to the definition S_{x+1} and $\overline{S}_{x+1}$ (since $\overline{S}_{x+1} = \overline{S}_x$ otherwise). So, given $i \leq 1$ minimal such that $g_i(x) \notin \{x, f_{1-i}(g_i(x))\}$, $y = g_i(x)$ and $\{x, f_{1-i}(g_i(x))\} \subseteq S_{x+1} \cup \overline{S}_x$. So $A(y)$ can be computed from $A \cap (S_{x+1} \cup \overline{S}_x)$ by (15).

This completes the proof. $\qquad \square$

Note that the reductions of $A \cap S$ to $A \cap \overline{S}$ and of $A \cap \overline{S}$ to $A \cap S$ given in the proof of Theorem 3 are wtt-reductions. (Namely, by $x \in S_{x+1} \cup \overline{S}_{x+1}$, the uses of the reductions are bounded by $f(x) = \max(S_{x+1} \cup \overline{S}_{x+1}) + 1$.) So, actually, any m-mitotic set is strongly wtt-mitotic. We do not know whether this can be improved by showing that any m-mitotic set is strongly tt-mitotic or even strongly btt-mitotic. We conclude this section, however, by showing that there is an m-mitotic set which is not strongly m-mitotic.

Theorem 4. *There is an* m-*mitotic set A which is not strongly* m-*mitotic (but strongly* 2-tt-*mitotic).*

Proof (idea). The general format of the proof resembles that of the proof of Theorem 2. So we only give the idea of the proof. By a finite injury argument we enumerate a pair of disjoint c.e. sets A_0 and A_1 such that $A = A_0 \cup A_1$ has the required properties.

For any number x we ensure that, for $E_x = \{5x, 5x+1, 5x+2, 5x+3, 5x+4\}$, the splitting $(A_0 \cap E_x, A_1 \cap E_x)$ of $A \cap E_x$ has one of the following three forms.

$$\text{(I)} \quad \begin{aligned} A_0 \cap E_x &= \emptyset \\ A_1 \cap E_x &= \emptyset \end{aligned}$$

$$\text{(II)} \quad \begin{aligned} A_0 \cap E_x &= \{\ 5x, 5x+1 \qquad\qquad\qquad\quad \} \\ A_1 \cap E_x &= \{ \qquad\qquad\quad 5x+2, \qquad 5x+4\ \} \end{aligned} \qquad (20)$$

$$\text{(III)} \quad \begin{aligned} A_0 \cap E_x &= \{ \qquad 5x+1, \qquad 5x+3 \qquad\quad \} \\ A_1 \cap E_x &= \{\ 5x, \qquad\qquad 5x+2 \qquad\qquad\quad \} \end{aligned}$$

Note that this ensures that A is m-mitotic via (A_0, A_1). For instance, $A \leq_m A_0$ via f_0 where $f_0(5x) = f_0(5x+1) = f_0(5x+2) = 5x+1$, $f_0(5x+3) = 5x+3$ and $f_0(5x+4) = 5x$; and $A_0 \leq_m A$ via g_0 where, for fixed $x_0 \notin A$, $g_0(5x) = 5x+4$, $g_0(5x+1) = 5x+1$, $g_0(5x+2) = g_0(5x+4) = x_0$ and $g_0(5x+3) = 5x+3$. Moreover, as one can easily check, (20) ensures that A is strongly 2-tt-mitotic via $(A \cap S, A \cap \overline{S})$ for the computable set S defined by

$$S = \{5x+1 : x \geq 0\} \cup \{5x+3 : x \geq 0\}.$$

In order to make A not strongly m-mitotic, it suffices to meet the requirements

$$\mathcal{R}_{\langle e,k \rangle} : \text{If } \psi_e, \varphi_{k_0} \text{ and } \varphi_{k_1} \text{ are total then there is a number } x \text{ such that}$$
$$A_0(5x) \neq (A \cap S_e)(\varphi_{k_0}(5x)) \text{ or } A_0(5x) \neq (A \cap \overline{S}_e)(\varphi_{k_1}(5x)).$$

for all $e, k \geq 0$ where $k = \langle k_0, k_1 \rangle$, ψ_e and S_e are defined as in the proof of Theorem 2, and $\{\varphi_n\}_{n \geq 0}$ is a standard numbering of the partial computable functions. (Note that the above requirements ensure that, for any computable set S, $A_0 \not\leq_m A \cap S$ or $A_0 \not\leq_m A \cap \overline{S}$. Since, by (20), $A_0 =_m A$, this implies that $A \not\leq_m A \cap S$ or $A \not\leq_m A \cap \overline{S}$.)

The strategy to meet requirement $\mathcal{R}_{\langle e,k\rangle}$ is as follows. We pick a follower x, i.e., reserve the numbers in E_x for this strategy. We then wait for a stage $s > 5x+4$ such that $\varphi_{k_0}(5x)$ and $\varphi_{k_1}(5x)$ are defined at stage s, say $\varphi_{k_0}(5x) = y_0$ and $\varphi_{k_1}(5x) = y_1$, such that $y_0 < s$ and $y_1 < s$, and such that $\psi_e(y_0)$ and $\psi_e(y_1)$ (hence $S_e(y_0)$ and $S_e(y_1)$) are defined at stage s too. (If there is no such stage s then we may argue that one of the functions ψ_e, φ_{k_0} and φ_{k_1} is not total whence requirement $\mathcal{R}_{\langle e,k\rangle}$ is trivially met.) Now, at stage $s + 1$, cancel all lower priority followers (whence - assuming that requirement $\mathcal{R}_{\langle e,k\rangle}$ is not injured later - the only numbers $\leq s$ enumerated into A_0 or A_1 after stage s are the numbers enumerated into these sets by the $\mathcal{R}_{\langle e,k\rangle}$-strategy hence elements of E_x). Moreover, define the splitting $(A_0 \cap E_x, A_1 \cap E_x)$ of $A \cap E_x$ at stage $s + 1$ where the clause used for the definition depends on the distinction of the following cases. (In each case we shortly explain why the corresponding action guarantees that $\mathcal{R}_{\langle e,k\rangle}$ is met where we tacitly assume that $\mathcal{R}_{\langle e,k\rangle}$ is not injured later, i.e., that $A_s(y) = A(y)$ for all $y \leq s$ which are not in E_x.)

Case 1: $y_0 \notin S_e$ or $y_1 \in S_e$. Then apply clause (II) to E_x.
This implies that $A_0(5x) = 1$ whereas $(A \cap S_e)(\varphi_{k_0}(5x)) = 0$ (if $y_0 \notin S_e$) or $(A \cap \overline{S_e})(\varphi_{k_1}(5x)) = 0$ (if $y_1 \in S_e$).
Case 2: Otherwise, i.e., $y_0 \in S_e$ and $y_1 \in \overline{S_e}$. Note that this implies $(A \cap S_e)(\varphi_{k_0}(5x)) = A(y_0)$ and $(A \cap \overline{S_e})(\varphi_{k_1}(5x)) = A(y_1)$. Distinguish the following two subcases.
Case 2.1: $y_0 \notin E_x$ or $y_1 \notin E_x$. Fix i minimal such that $y_i \notin E_x$. If $y_i \in A_s$ then apply clause (I) to E_x; otherwise, apply clause (II) to E_x.

The former implies that $A_0(5x) = 0$ whereas $(A \cap S_e)(\varphi_{k_0}(5x)) = 1$ (if $i = 0$) or $(A \cap \overline{S_e})(\varphi_{k_1}(5x)) = 1$ (if $i = 1$); the latter implies that $A_0(5x) = 1$ whereas $(A \cap S_e)(\varphi_{k_0}(5x)) = 0$ (if $i = 0$) or $(A \cap \overline{S_e})(\varphi_{k_1}(5x)) = 0$ (if $i = 1$).
Case 2.2: Otherwise, i.e., $y_0 \in E_x$ and $y_1 \in E_x$. Then apply clause (III) to E_x.

Note that this gives $A_0(5x) = 0$. Moreover, by case assumption (Case 2), $y_0 \neq y_1$ whence we may fix $i \leq 1$ such that $y_i \neq 5x+4$. It follows by case assumption (Case 2.2) that $y_i \in \{5x, 5x + 1, 5x + 2, 5x + 3\}$ hence (by (III)) $A(y_i) = 1$. So $(A \cap S_e)(\varphi_{k_0}(5x)) = 1$ (if $i = 0$) or $(A \cap \overline{S_e})(\varphi_{k_1}(5x)) = 1$ (if $i = 1$).

This completes the description of the $\mathcal{R}_{\langle e,k\rangle}$-strategy and the explanation why it succeeds. We leave the formal construction of A implementing this strategy to the reader.   □

4 Completeness and Mitoticity

Ladner [5] has shown that there is a Turing complete set which is not mitotic. The following theorem shows that we cannot replace Turing completeness by weak truth-table completeness. In fact any wtt-complete set is strongly wtt-mitotic.

Theorem 5. *Any* wtt-*complete set is strongly* wtt-*mitotic.*

Proof. Let A be wtt-complete and let $\{A_s\}_{s \geq 0}$ be a computable enumeration of A. We show that there is a strictly increasing computable sequence $0 = x_0 < x_1 < x_2 < x_3 < \dots$ such that, for the corresponding partition $\{I_n\}_{n \geq 0}$ of ω into the intervals $I_n = [x_n, x_{n+1})$, there is a constant c such that

$$\forall\, n \geq c\; \forall s \geq 0 \left(A_s \cap (\bigcup_{m < n} I_m) \neq A \cap (\bigcup_{m < n} I_m) \Rightarrow A_s \cap I_n \neq A \cap I_n \right) \quad (21)$$

holds. This implies that, for any infinite computable set Ind, A is wtt-reducible (hence wtt-equivalent) to $A \cap (\bigcup_{n \in Ind} I_n)$. (Namely, in order to compute $A(x)$ for given x from $A \cap (\bigcup_{n \in Ind} I_n)$, fix $n_x \in Ind$ minimal such that $c \leq n_x$ and $x < x_{n_x}$, and, using $A \cap (\bigcup_{n \in Ind} I_n)$ as an oracle, compute s minimal such that $A_s \cap I_{n_x} = A \cap I_{n_x}$. Then $A(x) = A_s(x)$ by (21). Note that this reduction is bounded by the computable function $g(x) = x_{n_x+1}$.) So, in particular, $A =_{\text{wtt}} A \cap B =_{\text{wtt}} A \cap \overline{B}$ for the computable set $B = \bigcup_{n \geq 0} I_{2n}$ whence A is strongly wtt-mitotic.

The definition of the sequence $\{x_n\}_{n \geq 0}$ exploits the following observation. By wtt-completeness of A, we may fix a strictly increasing computable function f such that any c.e. set is f-bounded Turing reducible to A. (Namely, since the universal set $U = \{\langle e, x \rangle : x \in W_e\}$ (where W_e is the eth c.e. set w.r.t. a standard numbering of the c.e. sets) is c.e. and since A is wtt-complete, there is a computable function g such that U is g-bounded Turing reducible to A, and, obviously, there is a computable function h such that W_e is h-bounded Turing reducible to U for all e. Moreover, we may choose g and h to be strictly increasing. Hence the composition $f = h \circ g$ of g and h will give the desired bound f.) Then the strictly increasing computable sequence $\{x_n\}_{n \geq 0}$ is inductively defined by letting

$$x_0 = 0 \quad \text{and} \quad x_{n+1} = f(\langle n, x_n, x_n \rangle).$$

(Here we assume that $\langle x, y, z \rangle = \langle x, \langle y, z \rangle \rangle$ for a computable pairing function $\langle \cdot, \cdot \rangle$ which is strictly increasing in both arguments. So $\langle \cdot, \cdot, \cdot \rangle$ is strictly increasing in its three arguments too, and $\langle e, y, z \rangle \in \omega^{[e]} = \{\langle e, w \rangle : w \geq 0\}$ for all numbers y, z.)

In order to show that the intervals $I_n = [x_n, x_{n+1})$ induced by this sequence satisfy (21), we enumerate an auxiliary c.e. set C. Since A is f-bounded Turing complete, there will be an index e such that $C = \Phi_e^{A,f}$. (Here $\Phi_e^{A,f}$ is the f-bounded variant of the eth Turing functional Φ_e where any oracle query $\geq f(x)$ in $\Phi_e^A(x)$ causes the computation to diverge.) The definition of C on $\omega^{[e]}$ will ensure (21) if we let $c = e$. The idea is as follows. If $n \geq e$ and a number $x \in I_m$ where $m < n$ enters A at a stage s then, at a later stage where C and $\Phi_e^{A,f}$ agree on all numbers $\leq x_{n+1} = 1 + \max I_n$, we enumerate $\langle e, x_n, x \rangle$ into C (note that by choice of e there must be such a stage) thereby forcing A to change below $f(\langle e, x_{n+1}^e, x \rangle) \leq x_{n+1}^e$ after stage s. So, if x is the last number from $\bigcup_{m < n} I_m$ which enters A, then this forces a number from I_n to enter A after the stage at which x is enumerated into A. So the last change of A on $\bigcup_{m \leq n} I_m$ will be on I_n.

For the formal definition of C we use the length function

$$l(e, s) = \mu\, y\, (\forall\, x < y\, (C_s(x) = \Phi_{e,s}^{A_s, f}(x)))$$

where $l(e, s)$ is the length of agreement between C and $\Phi_e^{A, f}$ observed at the end of stage s of the construction. Here C_s is the finite part of C enumerated by the end of stage s, inductively defined as follows. Stage 0 is vacuous, i.e., $C_0 = \emptyset$.

Stage $s + 1$. For any numbers $e, m, n \leq s$ and x such that $e \leq n$, $m < n$, $x \in I_m$, $x \in A_s$, $l(e, s) \geq x_{n+1}$, and $\langle e, x_n, x \rangle \notin C_s$, enumerate $\langle e, x_n, x \rangle$ into C_{s+1}.

This completes the definition of the computable enumeration $\{C_s\}_{s \geq 0}$ of the auxiliary c.e. set C. The formal proof that this definition ensures (21) is indirect. Fix e such that $C = \Phi_e^{A, f}$, let $c = e$, and, for a contradiction, assume that $n \geq e$ witnesses the failure of (21). Then, for the number $x \in \bigcup_{m \leq n} I_m$ which enters A last, $x \in \bigcup_{m < n} I_m$. In other words, there is a number x, an index m, and a stage t such that

$$x \in I_m \quad \text{and} \quad m < n \quad \text{and} \quad x \in A_{t+1} \backslash A_t \quad \text{and} \quad (A \backslash A_{t+1}) \restriction x_{n+1} = \emptyset. \tag{22}$$

Now fix $s > t$ minimal such that $n \leq s$ and $l(e, s) \geq x_{n+1}$. (Such a stage s must exist, since, by $C = \Phi_e^{A, f}$, $\lim_{s \to \infty} l(e, s) = \infty$.) Since $x \notin A_t$, it follows by definition of the enumeration of C that $\langle e, x_n, x \rangle \notin C_{t+1}$. By minimality of s, this implies $\langle e, x_n, x \rangle \notin C_s$ (since $\langle e, x_n, x \rangle$ may enter C at a stage $v + 1$ only if $n \leq v$ and $l(e, v) \geq x_{n+1}$). So all conditions necessary for enumerating $\langle e, x_n, x \rangle$ into C are satisfied at stage $s + 1$, hence

$$C(\langle e, x_n, x \rangle) = C_{s+1}(\langle e, x_n, x \rangle) = 1 \neq 0 = C_s(\langle e, x_n, x \rangle) = \Phi_{e,s}^{A_s, f}(\langle e, x_n, x \rangle) \tag{23}$$

where the last equality holds by $l(e, s) \geq x_{n+1}$ since

$$\langle e, x_n, x \rangle \leq \langle e, x_n, x_n \rangle \leq f(\langle e, x_n, x_n \rangle) \leq x_{n+1} \tag{24}$$

by strict montonicity of $\langle \cdot, \cdot, \cdot \rangle$ and f. Now, by $C = \Phi_e^{A, f}$ and by (23), $A \restriction f(\langle e, x_n, x \rangle) \neq A_s \restriction f(\langle e, x_n, x \rangle)$ hence $A \restriction x_{n+1} \neq A_s \restriction x_{n+1}$ by (24). But, by $s \geq t + 1$, this contradicts the last clause of (22).

This completes the proof of Theorem 5. $\qquad\square$

References

1. Ambos-Spies, K.: P-mitotic sets. In: Börger, E., Hasenjaeger, G., Rödding, D. (eds.) LaM 1983. LNCS, vol. 171, pp. 1–23. Springer, Heidelberg (1984). https://doi.org/10.1007/3-540-13331-3_30
2. Downey, R.G., Hirschfeldt, D.R.: Algorithmic randomness and complexity. Theory and Applications of Computability, p. 855. Springer, New York (2010). https://doi.org/10.1007/978-0-387-68441-3
3. Glaßer, C., Nguyen, D.T., Selman, A.L., Witek, M.: Introduction to autoreducibility and mitoticity. In: Day, A., Fellows, M., Greenberg, N., Khoussainov, B., Melnikov, A., Rosamond, F. (eds.) Computability and Complexity. LNCS, vol. 10010, pp. 56–78. Springer, Cham (2017). https://doi.org/10.1007/978-3-319-50062-1_5

4. Lachlan, A.H.: The priority method. I. Z. Math. Logik Grundlagen Math. **13**, 1–10 (1967)
5. Ladner, R.E.: Mitotic recursively enumerable sets. J. Symbolic Logic **38**, 199–211 (1973)
6. Ladner, R.E.: A completely mitotic nonrecursive r.e. degree. Trans. Amer. Math. Soc. **184**, 479–507 (1973)
7. Michler, L.: Autoreduzierbarkeit und Mitotizität für unbeschränkte Reduktionsbegriffe. Master Thesis (Christian Glaßer, supervisor), Julius-Maximilians-Universität Würzburg, Institut für Informatik (2017)

Moded and Continuous Abstract State Machines

Richard Banach[1(⊠)] and Huibiao Zhu[2]

[1] Department of Computer Science, University of Manchester,
Manchester M13 9PL, UK
`richard.banach@manchester.ac.uk`
[2] Shanghai Key Laboratory of Trustworthy Computing,
MOE International Joint Laboratory of Trustworthy Software,
International Research Center of Trustworthy Software,
East China Normal University, Shanghai, China
`hbzhu@sei.ecnu.edu.cn`

Abstract. In view of the increasing importance of cyber-physical systems, and of their correct design, the Abstract State Machine (ASM) framework is extended to include continuously varying quantities as well as the conventional discretely changing ones. This opens the door to the more faithful modelling of many scenarios where digital systems have to interact with the continuously varying physical world. Transitions in the extended framework are thus either *moded* (catering for discontinuously changing quantities), or *pliant* (catering for smoothly changing quantities). An operational semantics is provided, first for monolithic systems, and this is then extended to give a semantics for systems consisting of several distinct subsystems. This allows each subsystem to undergo its own subsystem-specific mode and pliant transitions. Refinement is elaborated in the extended context for both monolithic and composed systems. The formalism is illustrated using an example of a bouncing tennis ball.

1 Introduction

Conventional model based formal refinement technologies (e.g. [2,3,13,17,54]) are based on purely discrete concepts. These are typically ill suited to modelling applications which are best expressed using continuous mathematics. So there is a mismatch between the continuous modelling needed at the abstract level, and the discrete techniques used close to code in hybrid and cyber-physical (CPS) systems [1,8,15,18,22,29,36–38,47,48,53,55].

A large portion of the work reported in this paper was done while the first author was a visiting researcher at the Software Engineering Institute at East China Normal University. The support of ECNU is gratefully acknowledged.

Huibiao Zhu is supported by National Key Research and Development Program of China (Grant No. 2018YFB2101300), National Natural Science Foundation of China (Grant No. 61872145), and Shanghai Collaborative Innovation Center of Trustworthy Software for Internet of Things (Grant No. ZF1213).

A. Raschke et al. (Eds.): Börger Festschrift, LNCS 12750, pp. 29–62, 2021.
https://doi.org/10.1007/978-3-030-76020-5_3

Hybrid and CPS systems display considerable complexity in their behaviour, which poses challenges for verification techniques. One well respected way of confronting verification complexity is the top-down design and development approach. Supported by suitable formal notions, it allows complex behaviour to be approached in stages, with properties that have been established previously persisting in suitable form as more design detail is added. This makes it eminently suited to confront the challenges posed by complex systems development.

The Abstract State Machine (ASM) approach [11,13] is an established methodology for top-down design and development. It differs from many other formal approaches by having a very liberal type system, based on universal algebra rather than a fixed collection of low level built-in types. Following on from this, the model of state update in ASM is based on the idea of modifying dynamic functions, a generalisation of the idea of state variables (although, most often, this full generality is not needed).

In this paper we present an extension of the ASM formalism that enables us to treat continuously changing quantities fluently, especially at the abstract level. This is essential if we are to model hybrid and cyber-physical systems effectively. We also develop the needed extension of ASM refinement. The ASM extension is based on restricting the continuous behaviours that are permitted to those which can be described, piecewise, by solutions to well posed initial value problems, this being sufficient for most engineering purposes. These fundamental ideas are applied both to monolithic systems, and to systems consisting of several cooperating subsystems.

The rest of this paper is structured as follows. In Sect. 2 we review the essentials of ASM, and then describe the continuous extension. We base this on a discussion of the desired semantic domain first, and then construct the syntax and the desired semantics to map cleanly to it. Section 3 covers a simple example concerning a tennis ball bouncing back and forth over a tennis net. Section 4 discusses the formal operational semantics of the given description. The formal semantics lends itself to defining the semantics of composed (or decomposed) systems, which we also discuss. With the detailed semantics covered, in Sect. 5 we return to the tennis ball example to explore some of its more subtle aspects. Then in Sect. 6 we develop the refinement machinery relevant to Continuous ASM. This is given as a minimal generalisation of the discrete formulation, and is followed by a discussion of compositionality issues. Section 7 returns to the tennis ball and discusses a simple refinement scenario for the example. Section 8 discusses related work, while Sect. 9 concludes.

2   ASM, Discrete and Continuous

In this section we review the essentials of ASM [11,13], and extend the formalism to cope with continuously varying quantities. The advantage of considering such a clean extension of a discrete formalism is that it opens the door to adapting existing tools for the discrete formalism, rather than having to start from scratch.

2.1 Discrete Basic ASM Models

A definitive description of conventional, discrete ASM, is given in [13]. Here, we give an overview sufficient to prepare the ground for the continuous extension.

As noted above, ASM is founded on concepts of universal algebra [25,50]. This starts by defining *signatures*, from which we can generate *algebras* and then look for *models* that satisfy the constraints imposed by the algebras. These static structures constitute a universe within which ASM dynamics runs its course.

Focusing on the concept of basic ASM, the key update notion is carried by *dynamic functions*, functions that get (partly) redefined by updates of the form:

$$f(t_1, \ldots, t_n) := t \tag{1}$$

In (1), the $t_1, \ldots, t_n, t$ are terms evaluated in the current state, i.e. with respect to the current definitions of all the elements of the algebras (static and dynamic); with these values, the dynamic function f at the element of its domain consisting of the values of the tuple $t_1, \ldots, t_n$ is redefined to be the value of t. If f is a nullary dynamic function, then (1) corresponds to updating a variable.

A basic ASM *transition rule* is a construct of the form:

$$\textbf{if } Condition \textbf{ then } Updates \tag{2}$$

where the *Updates* are as in (1), and *Condition* evaluates to a truth value. In practice, the basic form in (2) is enhanced to improve readability by admitting various syntactic sugars: the usual elaborations of the conditional; a **forall** x **with** *cond Rule* form for iterating *Rule* over a collection ranged over by x (with x constrained by *cond*); and a **choose** x **with** *cond Rule* form to allow *Rule* to be nondeterministic. Below, we only need rules of the form:

$$
\begin{aligned}
&\textsc{Op}(\textbf{in } is, \textbf{out } os) \;= \\
&\textbf{if } guard(xs, is) \textbf{ then} \\
&\textbf{choose } xs', os \textbf{ with } rel(xs', xs, is, os) \\
&\textbf{do } xs, os := xs', os
\end{aligned}
\tag{3}
$$

In (3) the rule's name is $\textsc{Op}$, and we have (read-only) inputs **in** is and (write-only) outputs **out** os in the signature of $\textsc{Op}$. For us, a basic ASM is a finite set of such rules.

We say that a rule like (3) is enabled if its *guard* evaluates to true. In most model based development formalisms, given a state of the model, i.e. a valuation that maps each variable to a value in its type, progress is made by selecting *one* of the enabled rules and executing it. The ASM policy though, is that *all* enabled rules are selected, and their updates are performed in parallel. So the sets of enabled rules that arise must define *consistent* sets of updates. If not, execution aborts. A *run* of an ASM system thus starts at an initial state, and continues via a succession of state changes, defined by maximal sets of enabled rules that define consistent update sets.

2.2 Continuous ASM Models

We extend the framework above to the continuous world by first examining the semantic domain. Looking ahead a little, we will be using differential equations (DEs), and therefore, for mathematical consistency, we need to be precise about the semantic domain with respect to which the DEs will be interpreted.

For simplicity, we restrict to the case in which the states are given by valuations of the tuple of variables of the model, i.e. functions from the tuple of variables to the tuple of the variables' types. To extend such models smoothly to include continuously varying phenomena, we partition the variables into two kinds: **mode variables**, whose types are discrete sets, and which only change discontinuously, and **pliant variables**, whose types include topologically dense sets, and which are permitted to evolve both continuously and via discrete changes. In our terminology, discrete ASM just uses mode variables.

We model time as a left-closed interval $\mathcal{T}$ of the reals $\mathbb{R}$, with a finite left endpoint for the initial state, and with a right endpoint which is either finite (and right-open) or infinite, depending on whether the dynamics is finite or infinite. Now, the values of all variables become functions of $\mathcal{T}$.

For a mode variable v, the function is a piecewise constant function, which is constant on each element of a sequence of left-closed right-open intervals. Thus the behaviour of v partitions $\mathcal{T}$ into a sequence of left-closed right-open intervals, $\langle [t_{v,0} \ldots t_{v,1}), [t_{v,1} \ldots t_{v,2}), \ldots \rangle$, on each piece of which the behaviour is constant.

For a pliant variable x, the permitted behaviours are piecewise continuous.[1] These pieces again partition $\mathcal{T}$ into a sequence of left-closed right-open intervals, $\langle [t_{x,0} \ldots t_{x,1}), [t_{x,1} \ldots t_{x,2}), \ldots \rangle$, on each piece of which the behaviour undergoes no discontinuities.

Putting together all the behaviours of all the variables that participate in defining a particular execution of a system, yields a sequence of left-closed right-open intervals, $\langle [t_0 \ldots t_1), [t_1 \ldots t_2), \ldots \rangle$, which is the coarsest partition of $\mathcal{T}$ into such intervals where all discontinuous changes of all the variables of the system during that execution take place at a boundary point t_i.

We note at this juncture that our formulation is by no means the first work on the ASM formalism to consider the notion of time *per se*. In this context we could mention the earlier work in [9,12,16,23,24,26,42,45] for example. While all of these are concerned with time, in all of them the concern is with pure timing, i.e. there is no continuously varying behaviour. So in our formulation, states in all of these works are piecewise constant functions of time.

In a typical interval $[t_i \ldots t_{i+1})$, mode variables will be constant, but pliant variables will change in a continuously varying manner. However, mere continuity still allows for a very wide range of mathematically pathological behaviours.[2] To constrain these, we make the following restrictions and recommendations:

I Zeno: there is a constant δ_{Zeno}, such that for all i needed, $t_{i+1} - t_i \geq \delta_{\mathsf{Zeno}}$.
    N.B. Since the presence or absence of Zeno behaviour is usually a global

---

[1] We mention below that actually, we need *absolute* continuity, not mere continuity alone.

[2] Texts on mathematical analysis are usually replete with relevant examples.

property of a system's reachability relation, this point must be regarded as a recommendation rather than a restriction that is statically enforceable.

II Limits: for every variable x, and for every time $t \in \mathcal{T}$, the left limit $\lim_{\delta \to 0} x(t - \delta)$, written $\overrightarrow{x(t)}$, and the right limit $\lim_{\delta \to 0} x(t + \delta)$, written $\overleftarrow{x(t)}$, (with $\delta > 0$ in each case) both exist, and for every t, $x(t) = \overleftarrow{x(t)}$. (N. B. At the endpoint(s) of $\mathcal{T}$, any missing limit is defined to equal its counterpart.)

III Differentiability: The behaviour of every pliant variable x in the interval $[t_i \ldots t_{i+1})$ is given by the solution of a well posed initial value problem $\mathcal{D} \, xs = \phi(xs, t)$ (where xs is a relevant tuple of pliant variables and $\mathcal{D}$ is the time derivative). "Well posed" means that $\phi(xs, t)$ has Lipschitz constants which are uniformly bounded over $[t_i \ldots t_{i+1})$ bounding its variation with respect to xs, and that $\phi(xs, t)$ is measurable in t.

It is recognised that ASM types can be mathematically complex entities. Therefore it is intended that **I-III** above apply to variables with as general a type as might be needed, provided that the concepts required in **I-III** (such as left/right limits, initial value problem, Lipschitz constants, uniform boundedness, measurability) make sense for them. That said, in the overwhelming majority of cases, the conventional real type $\mathbb{R}$ is sufficient, so we do not consider more complicated possibilities in this paper.

With **I-III** in place, the behaviour of every pliant variable is piecewise *absolutely* continuous [41,52], with the variation being described by a suitable differential equation.

Accompanying the distinction between mode and pliant variables, is a distinction between mode and pliant transitions. Mode transitions are just like conventional ASM transitions in that they record a discrete transition from before-values to after-values of the mode variables, albeit that these are the values of piecewise constant functions of time. A rule for a mode transition OP can be written using familiar ASM notation:

$$\text{OP}(\textbf{in } \overrightarrow{is}, \textbf{out } \overleftarrow{os}) \ =$$

$$\textbf{if } \ guard(\overrightarrow{xs}, \overrightarrow{is}) \ \textbf{ then}$$

$$\textbf{choose } \ \overleftarrow{xs'}, \overleftarrow{os} \ \textbf{ with } \ rel(\overleftarrow{xs'}, \overrightarrow{xs}, \overrightarrow{is}, \overleftarrow{os})$$

$$\textbf{do } \ xs, os \ := \ \overleftarrow{xs'}, \overleftarrow{os} \tag{4}$$

In (4), the overarrows are semantic decorations. These are not part of the syntax, but are included for clarity to indicate which limiting value for a variable (selected from its behaviour as a function of time) is to be taken as being referred to in the rule. This needs to be understood since all runtime executions of OP take place at points of discontinuity in the temporal behaviour of (at least some of the) variables, because rules like OP are intended precisely to define such discontinuities. Note therefore that the choice of left limit for before-values and right limit for after-values (at a given transition point) makes (4) into the kind of instantaneous transition we would expect. Stripping off the overarrows from (4) yields the form one would write to describe a rule in a specific application.

In (4) we single out the inputs is and outputs os, (read-only and write-only respectively), while xs are the state variables (accessed in read/write manner). Note the double decoration of the after-state variables $\overleftarrow{xs}'$. The prime corresponds to the usual syntactic decoration that one would expect to use in distinguishing before-states (unprimed) from after-states (primed), whereas the overarrow indicates the temporal semantic interpretation. Obviously, if the after-values for xs and os are available explicitly, the relevant expression can be assigned in the **do** clause, and the **choose** and **with** clauses can be omitted.

Pliant transitions do the corresponding job for pliant variables. While a mode transition is a single before-/after-value pair, a pliant transition is a family of before-/after-value pairs parameterised by the relevant time interval $[t_i \ldots t_{i+1})$. Moreover, instead of the change from before-values to after-values taking place instantaneously, the before-value can be understood to refer to the initial value at t_i (which, by **II**, equals the right limit at t_i), while the after-value refers to an arbitrary time in the open interval $(t_i \ldots t_{i+1})$, so the before-value and after-value are separated in time. To reflect the constraints that apply to pliant transitions, we write rules for them thus:

$$\text{PLIOP}(\textbf{in } is(t \in (t_{\text{L}(t)} \ldots t_{\text{R}(t)})), \textbf{out } os(t \in (t_{\text{L}(t)} \ldots t_{\text{R}(t)}))) \overset{c}{=}$$
$$\textbf{if } IV(xs(t_{\text{L}(t)})) \wedge guard(xs(t_{\text{L}(t)})) \textbf{ then}$$
$$\textbf{with } rel(xs, is, os, t)$$
$$\textbf{do } xs(t), os(t) := \textbf{solve } DE(xs(t), is(t), os(t), t) \tag{5}$$

In (5), the symbol $\overset{c}{=}$ signals the presence of a rule for a pliant transition, distinguishing it from the instantaneously executed kind. The notations $t_{\text{L}(t)}$ and $t_{\text{R}(t)}$ refer to the beginning and end, respectively, of the time interval during which PLIOP executes. Of course, the values of these cannot be known statically, even disregarding the fact that different invocations of the rule at runtime will require different values for $t_{\text{L}(t)}$ and $t_{\text{R}(t)}$. Therefore all explicitly given references in (5) to variables' time dependencies, and to $t_{\text{L}(t)}$ and $t_{\text{R}(t)}$ values, are semantic decorations, included for readability, and indicated by the shading. They do not form part of the syntactic form of the rule, and so "$t \in (t_{\text{L}(t)} \ldots t_{\text{R}(t)})$" in the declaration of the input "**in** $is(\ldots)$" is redundant (similarly for the output), as is "(t)" in occurrences of "$xs(t)$", etc. In this paper, we have opted to retain the "$var(t)$" way of referring to the time dependent behaviour of pliant variables inside pliant rules, for improved readability.

Given a specific execution of the system, which generates a specific partition of $\mathcal{T}$, for an arbitrary t, we define $\text{L}(t) = \max\{i \,|\, t_i \leq t\}$ and $\text{R}(t) = \min\{i \,|\, t_i > t\}$ (with obvious default for an infinite last interval) which yields the indexes in the partition of $\mathcal{T}$ relevant to the subinterval of $\mathcal{T}$ to which t belongs. This is consistent with the notations $t_{\text{L}(t)}$ and $t_{\text{R}(t)}$ in (5). These devices allow us to refer to the beginning and end of the interval during which the pliant event runs in a generic manner in our meta level discussions. They also permit, despite what has been said above, rules like PLIOP to refer to *relative time* from the beginning of the execution of a transition specified by PLIOP, by using expressions like

$(t - t_{\mathrm{L}(t)})$. This is useful within clauses such as $rel(xs, is, os, t)$, for example. Obviously, fresh syntactic sugar could be introduced to handle this, if desired.

For a specific execution of PLIOP, the inputs is and outputs os are continuously absorbed from and emitted to the environment over the open interval $(t_{\mathrm{L}(t)} \ldots t_{\mathrm{R}(t)})$, as indicated in the signature. (Both must be absolutely right continuous.) Note that the initial values IV and guard $guard$ depend only on the before-value of the state,[3] and not on the input, whereas rel, which expresses any additional constraints that must hold beyond the differential equation DE itself, can depend on all state and input values from the start of the interval $t_{\mathrm{L}(t)}$ up to the current time t. The assignment in (5) says that the after-state and output at t should satisfy the differential equation DE (as well as rel). As for the instantaneous case, if the continuous functions of t to be assigned to xs, os are known explicitly, we can omit the **with** and/or **solve** clauses as appropriate, and just assign xs, os to the relevant expression.

As mentioned earlier, pliant variables can undergo instantaneous discontinuous transitions as well as continuous ones. For such transitions, the structure in (4) is sufficient. We continue to call instantaneous transitions involving both kinds of variable **mode transitions**, introducing the term **pure mode transitions** for the former kind.

A continuous ASM ruleset is **well formed** iff:

- Every enabled mode transition is feasible, i.e. has an after-state, and on its completion enables a pliant transition (but does not enable any mode transition).
- Every enabled pliant transition is feasible, i.e. has a time-indexed family of after-states, and EITHER:
  (i) During the run of the pliant transition a mode transition becomes enabled. It preempts the pliant transition, defining its end. ORELSE
  (ii) During the run of the pliant transition it becomes infeasible: finite termination. ORELSE
  (iii) The pliant transition continues indefinitely: nontermination.

A **run** of a continuous ASM system starts with a mode transition which assigns the initial state of all system variables, and then, pliant transitions alternate with mode transitions. The last transition (if there is one) is a pliant transition (whose duration may be finite or infinite). We thus see that the sequence t_i of times at which discontinuities take place, emerges as the sequence of times at which the first possible preemptions of the pliant transitions by the enabling of mode transitions arises.

3 Example: A Bouncing Tennis Ball

To illustrate our formalism, we consider an idealised tennis rally, in which a pointlike tennis ball of unit mass is being hit back and forth over the tennis net,

---

[3] Normally, we would expect IV to depend on the pliant variables and $guard$ to depend on the mode variables, but there is no need to insist on this formally.

which is of height N. Let the horizontal and vertical components of the ball's velocity be vx and vy, positive for rightwards and upwards motion. Suppose horizontal and vertical positions are measured from the bottom point of the net, positive for rightwards and upwards displacements, and for the tennis ball, these are px and py.

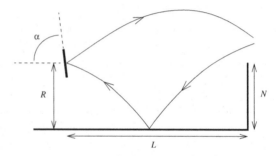

Fig. 1. A single shot in a tennis rally.

We consider a single shot in the rally. As illustrated in Fig. 1, the ball appears from the right, with velocity (vx_{in}, vy_{in}) say (both vx_{in} and vy_{in} being negative), bounces once, and then on its continuing path encounters the player's racquet at height R, having travelled a horizontal distance L. After striking the ball, the racquet gives it a velocity (vx_{out}, vy_{out}). We can model this scenario using continuous ASM rules as follows. The free flight of the ball is governed by a pliant rule:

FLIGHT $\overset{c}{=}$

if $py > 0$ **then with** $py \geq 0$

do $px(t), py(t), vx(t), vy(t)$:= **solve** $[\mathcal{D}px, \mathcal{D}py, \mathcal{D}vx, \mathcal{D}vy] = [vx, vy, 0, -g]$

In FLIGHT we see the usual equations of Newtonian motion for a point mass in first order row-vector form. We use the symbol $\mathcal{D}$ to denote the time derivative in "program-like" situations; g is the acceleration due to gravity. We check, and continually enforce, the constraint that py is non-negative—the ball is not allowed to penetrate the surface of the tennis court. This one rule is enough for all three free-flight episodes of our scenario.

The interactions of the ball with the ground and with the racquet require some mode rules. The simplest is the bounce off the tennis court surface. The following rule will do.

BOUNCE $=$ **if** $py = 0 \land vy < 0$ **then do** vy := $-c\,vy$

Rule BOUNCE assumes that the motion of the pointlike tennis ball in the horizontal direction is unaffected by the bounce, but that the vertical component is reflected, and scaled down by the coefficient of restitution c (where we have $0 < c < 1$).

While the modelling of the bounce can be said to be reasonably realistic, we simplify the interaction with the racquet fairly dramatically, by assuming that the racquet has infinite mass and is infinitely stiff. In this case, the encounter between the ball and racquet can be modelled just like a bounce, i.e. the normal component of the relative velocity is reflected modulo the coefficient of restitution c, and the tangential component remains unaffected.

To model this properly, we need position and velocity variables for the racquet; let these be rpx, rpy, rvx, rvy respectively, and suppose that these variables refer to the precise point of impact on the racquet of the ball. Suppose that at the moment of impact, the racquet is inclined at an angle α to the horizontal, as in Fig. 1. Then a mode rule that will fulfill our requirements is the following.

RACQUET $=$

if $py > 0 \land py = rpy \land px < 0 \land px = rpx \land vx < 0 \land vx.rvx + vy.rvy < 0$ **then**

do

$$vx := -(vx - rvx)(\cos^2(\alpha) + c\sin^2(\alpha)) + (vy - rvy)(1 - c)\cos(\alpha)\sin(\alpha) + rvx,$$
$$vy := (vx - rvx)(1 - c)\cos(\alpha)\sin(\alpha) - (vy - rvy)(\sin^2(\alpha) + c\cos^2(\alpha)) + rvy$$

The guard of RACQUET checks that the ball is above the ground and to the left of the net, and that the ball and racquet are in the same place. The final conjunct of the guard is the inner product of the racquet and ball velocities. Insisting that it is negative ensures that the racquet strikes the ball in such a way that there is a component of the resulting velocity that is opposed to the ball's previous motion—which, if the ball is travelling as we would expect, towards the left, ensures that the ball will travel towards the net after the impact. Beyond that, explaining the assignments to vx and vy in the rule takes us deeper into classical mechanics than is appropriate here, so the details are relegated to an appendix.

Our model is completed with an INIT rule to assign appropriate initial values to all the variable. We do not write it down.

4   Formal Semantics

The account of the Continuous ASM in Sect. 2 was intended to give a picture of our formalism that is conceptually easy to grasp and is clear enough for model building, relying to some extent on the reader's intuition and experience to fill in any gaps (e.g. positing *ab initio* the sequence t_i of times at which discontinuities take place). In this section we give a summary of the formal operational semantics of our formalism. In order to not waste large amounts of space on repeating routine material, we rely heavily on existing work: on [13] (especially Chap. 2.4) for conventional ASM semantics; and on [49] (especially Chapter III §10) for differential equations in the sense of Carathéodory. Given these trusted foundations for discrete and continuous update respectively, the issues we must be most careful about are the handovers between mode and pliant transitions. We discuss these further after presenting the semantics.

One thing that we have not explicitly mentioned hitherto, is that we have been assuming that the system being discussed is defined monolithically, i.e. as a single indivisible syntactic unit. This is in accord with the automata-centric view taken in the majority of work on hybrid systems in the literature (see Sect. 8). However, in rule based formalisms (such as ASM), it is quite common to compose systems out of smaller subsystems—in the ASM case, the simultaneous execution of *all* enabled rules at each step provides a very simple semantics for composing subsystems that just aggregates the subsystems' rulesets.[4]

In this regard, the semantics we sketched in Sect. 2.2 is inadequate. For one thing, we spoke (almost exclusively) of transitions, and did not explore in detail how they might be related to ASM rules, except that intuitively it is clear that rules should *specify* transitions. This also sidesteps the scheduling convention just mentioned. For another thing, we did not consider whether insisting that the system as a whole engaged in the alternation of mode and pliant transitions as we described them, made sense when the system is not monolithic.

The latter point raises an issue not present in the usual discrete world. In the discrete world, when an update is made to some system variables, any variables not mentioned in the syntactic description of the update, conventionally remain at their existing value. This coincides with the natural real time behaviour of variables that have piecewise constant values over time. So there is no observable distinction between leaving such a variable unaltered (to pursue its natural temporal evolution) on the one hand, and updating it to remain at the same constant value on the other hand. The former view is appropriate if the variable belongs to a different subsystem which is unaware of the ongoing update, while the latter view is appropriate if the variable belongs to the system being currently updated, but no change in its value is required.

In the continuous world, in which the values held in system variables may vary in a non-piecewise constant manner over time, the distinction between these two views can become apparent. If a variable that belongs to the subsystem currently being updated (via a pliant transition that is about to start) is not mentioned in the syntactic description of the update, then the policy that its value remains constant throughout the ensuing interval of time during which the new pliant transition will act, represents a specific design decision about the semantics of the current subsystem.

While it might be possible to justify such a design decision on requirements grounds when the variable belongs to the system being updated, the same design decision can seem very unnatural when the variable in question belongs to a *different* subsystem, in which its behaviour is being governed by a pliant transition that started in that subsystem earlier, and which demands some non-constant behaviour for the variable. Then, the idea that behaviour is suddenly overridden by a constant behaviour that "appears out of nowhere" (from the point of view of that other subsystem) is very counterintuitive. So it is highly preferable that such variables be allowed to continue with their pre-existing behaviour.

Taking this latter view complicates the semantic picture a little. On the face of it, the definition of the sequence of times t_i at which discontinuities take place becomes more problematic—the sequence that is "naturally" seen by one subsystem need not coincide with the sequence that is "naturally" seen by another subsystem. Additionally, specifying the moments at which mode transitions arise, and

---

[4] Dually, one can approach the same issue by decomposing simpler abstract systems into collections of smaller, more detailed subsystems, as happens in Event-B for instance.

Table 1. Notations utilised in the semantics

Notation	Explanation
$\mathcal{T}$	Time interval, duration of the dynamics
U^{var}	Type for variable var
$\mathcal{R}$	Set of rules
$\mathcal{S}$	Semantics of $\mathcal{R}$, a set of system traces
ζ_{var}	System trace of var, $\zeta_{var} : \mathcal{T} \to U^{var}$
$PlRl(pli, t)$	Pliant rule for variable pli that ζ_{pli} obeys at time t
$InitUDS$	Set of consistent update sets for the initial rules of $\mathcal{R}$
$PliRsEN$	Set of enabled pliant rules of $\mathcal{R}$ (at any execution of step [5] of the Semantics)
$PliRsCT$	Set of pliant rules that are to continue preceding execution (at any execution of step [6] of the semantics)
$PliREM$	Set of remaining pliant rules (not in $PliRsEN \cup PliRsCT$ at any execution of step [7] of the semantics)
$MoRs$	Set of non-INIT mode rules enabled at a preemption point (at any execution of step [12.2] of the semantics)
$MoRsUDS$	Set of consistent update sets for the rules in $MoRs$ (at any execution of step [12.3] of the semantics)

their scope, as well as determining the scope of pliant transitions, requires more care. Deciding what "subsystem" refers to, and how to handle it in the context of a rule system based formulation, also requires care.

Our semantics takes these considerations into account. It defines the behaviours of a set of rules $\mathcal{R}$, much as one would do for a monolithic system. However, we allow for the fact that $\mathcal{R}$ may itself be made of the union of one or more constituent sets of rules. We do this by: (i) allowing for several INITial rules (which must, of course, be consistent, originating from different constituent rule subsets), (ii) having a preemption mechanism that allows pliant rules to continue past a preemption point (when this is appropriate) as well to be preempted (when that is appropriate), using rule-variable dependencies to determine which course of action to apply after any mode transition. This gives a simple syntax-independent semantics for composition. With these thoughts in mind, the semantics is given in the following sections.

4.1 Semantic Context

We start with a number of contextual observations and definitions. Table 1 summarises the specialised notations introduced during the course of the technical details.

[A] Time, referred to as t, takes values in the real left-closed right-open set $[t_0 \ldots +\infty)$, where t_0 is an initial value for time. For every other system variable

var, there is a universe of values (or type) U^{var}. If var is pliant, then U^{var} is $\mathbb{R}$. (N. B. Earlier we were more lax concerning the types of pliant variables. Now we will be more specific, recognising that, in practice, more complex types that are of interest can be constructed from $\mathbb{R}$ anyway.)

[B] The semantics is given for $\mathcal{R}$ which is a set of rules. $\mathcal{R}$ contains one or more distinguished INITial rules. Each INIT rule has a guard which is either "true" or "$t = t_0$".

[C] Time is a distinguished variable (read-only, never assigned by rules). All other variables have interpretations which are functions of an interval of time starting at t_0. (See **[E]**.) As well as directly referring to the time variable, time may be handled indirectly by using clock variables. Their values may be assigned by mode rules, and their rates of change with respect to time may (during use in pliant rules) be specified directly, or defaulted to unity.

[D] $\mathcal{R}$ consists of mode rules and pliant rules. A mode rule (e.g. (4)), is *enabled* iff, under the current valuation of the system variables, the value of the *guard* of (4) lies in the topological closure of the true-set of the *guard*. A pliant rule (e.g. (5)), is *enabled* iff $IV \wedge guard$ evaluates to true under the current valuation of the system variables. A variable is *governed* by a mode rule iff it is assigned by that rule. A pliant variable is *governed* by a pliant rule iff it appears in the left hand side of the DE of the rule, or is directly assigned in the rule, or is constrained in the **with** clause.

[E] The semantics of $\mathcal{R}$ is a set of system traces $\mathcal{S}$. Each system trace $S \in \mathcal{S}$ is given by a time interval $\mathcal{T} = [t_0 \ldots t_{\text{FINAL}})$ (where t_{FINAL}, with $t_{\text{FINAL}} > t_0$, is finite or $+\infty$), and a set of time dependent variable interpretations $\zeta_{var} : \mathcal{T} \to U^{var}$, one for each variable var. If $\mathcal{S}$ is empty we say that the semantics of $\mathcal{R}$ is VOID.

[F] In order that the evolution of each pliant variables is suitably managed, an additional data structure is needed. For each pliant variable pli, the function $PlRl(pli, t)$ returns the pliant rule that the interpretation ζ_{pli} of variable pli is obeying at time t.

[G] The set of traces $\mathcal{S}$ is constructed by the step by step process below, which describes how individual system traces are incrementally constructed.[5] Whenever a CHOOSE is encountered, the current trace-so-far is replicated as many times as there are different possible choices, a different choice is allocated to each copy, and the procedure is continued for each resulting trace-so-far. Whenever a TERMINATE is encountered, the current trace-so-far is complete. Whenever an ABORT is encountered, the current trace-so-far is abandoned, and eliminated from the semantics $\mathcal{S}$, of $\mathcal{R}$.

4.2   Operational Semantics

In the context of the assumptions **[A]**–**[G]** above, the operational semantics of the Continuous ASM can be given as follows.

---

[5] N. B. The process is not intended to be executable. All traces-so-far are intended to be explored simultaneously.

[1] Let $i := 0$ (where i is a meta-level variable).

[2] Let $InitUDS$ be the set of consistent update sets for the collection of initial rules of $\mathcal{R}$. If $InitUDS$ is empty then VOID. Otherwise, CHOOSE an update set from $InitUDS$ and assign all variables accordingly, thereby interpreting their values at time t_0. (N. B. We assume that all system variables acquire an initial value in this manner.)

[3] If any non-INIT mode rule is enabled when the variables have the values at t_i then ABORT.

[4] If no pliant rule from $\mathcal{R}$ is enabled then ABORT.

[5] Let $PliRsEN$ be the set of enabled pliant rules from $\mathcal{R}$.

  [5.1] If any pliant variable occurs in the left hand side of the DE (or direct assignment) of more than one rule in $PliRsEN$ (or more than once in the left hand side of the DE in the same rule), then ABORT.

[6] If $i = 0$ let $PliRsCT = \varnothing$. Otherwise, let $PliRsCT$ be the set of pliant rules from $\mathcal{R}$, such that: $PliRsCT$ is maximal; no rule in $PliRsCT$ is in $PliRsEN$; no variable governed by any rule in $PliRsCT$ is governed by any rule in $PliRsEN$; for every rule PLIRLCT in $PliRsCT$, for every pliant variable pli governed by PLIRLCT, $\overrightarrow{PlRl(pli, t_i)} = $ PLIRLCT; for every rule PLIRLCT in $PliRsCT$, for every mode variable v which occurs in the $guard$ of PLIRLCT, $\overrightarrow{v(t_i)} = v(t_i)$.

[7] Let $PliREM$ consist of any pliant variables pli that are not governed by any rule in either $PliRsEN$ or $PliRsCT$. If $PliREM$ is nonempty, then ABORT.

[8] If there does not exist a $t_{\text{NEW}} > t_i$ such that there is a simultaneous solution of all the DEs and direct assignments in the rules in $PliRsEN \cup PliRsCT$ in the left-closed, right-open interval $[t_i \ldots t_{\text{NEW}})$, using as initial values the variable values and right limits of inputs and outputs at t_i, and such that the rel predicates also evaluate to true in the interval $[t_i \ldots t_{\text{NEW}})$, then ABORT.

[9] Otherwise, CHOOSE a simultaneous solution as in [8], and let t_{MAX} be maximal such that $t_{\text{MAX}} > t_i$ and this solution is defined in the interval $[t_i \ldots t_{\text{MAX}})$.

  [9.1] For all pliant variables pli, for all $t \in [t_i \ldots t_{\text{MAX}})$, let $PlRl(pli, t)$ be the rule governing the behaviour of pli. (N. B. This assignment is total by [7] and unambiguous by [5.1].)

  [9.2] For every mode variable, extend its value at t_i to a constant function in the interval $[t_i \ldots t_{\text{MAX}})$.

[10] If no non-INIT mode rule is enabled at any time t_{NEXT} in the open interval $(t_i \ldots t_{\text{MAX}})$, or no non-$Init$ mode rule is enabled by the left-limit values of the state variables at time t_{MAX} in the case that these left-limit values exist and are finite at t_{MAX}, then TERMINATE.

[11] Let $i := i + 1$.

[12] Let t_i be the smallest time t_{NEXT} at which some non-INIT mode rule is enabled in [10].

[**12.1**] Discard the interpretation of all variables, and the definition of $PlRl$, in the interval $[t_i \ldots t_{\text{MAX}})$.

[**12.2**] Let $MoRs$ be the set of non-INIT mode rules that are enabled when all variables var are interpreted as the left-limit values at t_i, i.e. as $\overrightarrow{var(t_i)}$.

[**12.3**] Let $MoRsUDS$ be the set of consistent update sets for the rules in $MoRs$. If $MoRsUDS$ is empty then ABORT. Otherwise, CHOOSE an update set from $MoRsUDS$ and assign all the updated variables accordingly, thereby interpreting their values at time t_i.

[**12.4**] For all other variables var, interpret their values at time t_i to be their left-limit values at t_i, i.e. to be $\overrightarrow{var(t_i)}$.

[**13**] Goto [**3**].

4.3   Mode-Pliant and Pliant-Mode Handovers

Before commenting further, we make some observations on the consistency of the above definition. As noted earlier, we can take certain things for granted, such as well definedness of mode transitions via ASM update semantics, and the existence of solutions to differential equations. The key remaining points then, are whether the handovers from pliant to mode transitions, and those from mode to pliant transitions, are well defined.

We observe that the handover from pliant to mode transitions is trouble-free as follows. Since the set of values at which any mode rule becomes enabled is closed (being given by the closure of the true-set of the *guard* of the rule, by [**D**]), and since the system trajectory is a continuous function during any interval in which a pliant rule is active, if the system trajectory meets the closure at all during such an interval, it first meets it at some specific time point. Since there are only finitely many rules, the minimum of these points is a unique well defined time point, and so t_i in [**12**] emerges as this minimum. Thus the earliest moment that a mode transition becomes enabled during a pliant transition is a well defined time point, and the time at which the pliant transition is preempted is well defined, from which a consistent set of mode updates is derived, by [**12**], [**12.1**], [**12.2**], [**12.3**].

We argue that the handover from mode to pliant transitions is also consistent. Firstly, upon completion of a mode transition, some pliant rules will (typically) be enabled, [**5**]; these are required to be unambiguous and consistent by [**5.1**]. Secondly, these rules need not govern *all* the pliant variables of the whole system. By [**6**], if there were pliant rules contributing to the pliant transition that was just preempted, which govern variables disjoint from those governed by the first case, they are permitted to continue—we might term this figurative interruption and resumption a "virtual skip". Thirdly, the former two measures may still not take care of *all* pliant variables, since there is no requirement for pliant rules and the sets of variables that they govern to dovetail neatly together. If there are any pliant variables left over, [**7**] ensures that the run is ABORTed.

With suitable attention to routine details, the above remarks can be turned into a formal proof of the consistency of the definition of system traces.

4.4   Multiple Subsystems

We return to the questions that were raised earlier concerning the definition of, and interaction between, subsystems that coexist within a single encompassing system. We examine how the formal semantics above helps to address these, and we tie up the semantic loose ends.

To start with, we would normally expect that a separate subsystem would control (i.e. have write access to) an exclusive set of variables. We therefore take that as a fundamental principle.[6]

The next basic insight comes from [13], which promotes a perspective in which a system's variables are either *monitored* or *controlled*. Controlled variables are written to by the system, whereas monitored variables are merely read. For the latter, it is assumed that the environment supplies the values that are read, but aside from the condition that the values of monitored variables should be stable when read, no further restriction is placed on them. Thus, there is nothing to prevent their values from being supplied by another ASM system, the original system and its environment thus becoming two subsystems of a larger system. In other words, the conventional definition of an ASM system is intended to enable it to play the role of subsystem, essentially without modification.

For our purposes, we add a couple of observations to the above picture to make it suit the Continuous ASM situation. Firstly, since pliant variables' values will change continuously in general, we can modify "stable when read" to "reliably readable when needed", to avoid any possible confusion. Secondly, we emphasise that in the context of a system comprising several subsystems (i,e, one constructed via the composition of the subsystems' rulesets), each writable variable is written to by the rules belonging to exactly one of the subsystems, and no rule (of the whole system) writes to the writable variables of more than one of the subsystems. With these simple structural restrictions in place, the semantics of a system consisting of the composition of multiple subsystems is simply given by aggregating all of the rules of all of the subsystems in the usual way, and processing them according to the single system semantics given above.

Of course, a non-VOID semantics for subsystem $A1$ which assigns a variable $x1$ while reading variable $x2$ which belongs to subsystem $A2$, and a non-VOID semantics for subsystem $A2$ which assigns a variable $x2$ while reading variable $x1$, does not guarantee a non-VOID semantics for the entire system consisting of $A1$ together with $A2$, since there may not be values for $x1$ and $x2$ that *simultaneously* satisfy all the constraints imposed by the two subsystems.

The second proviso above also acts in concert with the stipulations of the formal semantics to ensure that, in a multi-subsystem system, each preemption point is caused by an identifiable subset of the subsystems,[7] and upon completion

---

[6] We can imagine that write access to some variable might, exceptionally, be shared by more than one subsystem, but under such circumstances a suitable protocol will have be in place to prevent race conditions, such as in the case of familiar mutual exclusion protocols [31,40]. We do not consider such cases here.

[7] For simplicity, we permit simultaneous preemption by more than one subsystem, even if it would be a little impractical in reality.

of the preemption, an identifiable subset of the subsystems embarks on new pliant behaviour, with the remainder resuming the pliant behaviour they were executing previously. (N.B. The two subsets need not be the same.)

The last point brings us to the issue of the how the indexing of mode and pliant transitions works in a system conceptually divided into separate subsystems. We see that the semantics defines a global indexing, which is a strict sequentialisation of the mode transitions of the entire system, regardless of which subsystem they might arise from. From the vantage point of any given subsystem, only a subset of these mode transitions might be "visible", but this amounts to simply re-indexing the mode transitions if we want to describe the system dynamics from that subsystem's viewpoint. Allied to this is the fact that if a remote subsystem undergoes a mode transition of which a given subsystem is unaware, some values being read by the local subsystem might still undergo discontinuous change in the midst of a pliant transition (of the local subsystem). This discontinuity causes no discomfort, since we understand differential equations in the sense of Carathéodory. Provided that the right hand side of each ODE in the system has the uniformly bounded Lipschitz property in the system variables, and remains measurable over time, it is guaranteed that a solution exists and is absolutely continuous.

Lastly, a note on Zeno behaviour. Nothing in the semantics that we have discussed precludes it. Therefore the semantic model does not of itself guarantee the recommendation **I** of Sect. 2.2. As we remarked, Zeno-freeness normally depends on global reachability, so our view is that if a system model is capable of exhibiting Zeno behaviour, then there is potentially something wrong with the model, and, depending on circumstances, the model ought to be improved to remove or mitigate those aspects that lead to it.[8]

5   The Tennis Ball Revisited

What is interesting about the tennis ball example is to consider how the formal semantics of Sect. 4 views the behaviour of the tennis ball system. We start by noting that the guards of the various mode rules in the tennis ball system all featured strict inequalities. However, in order that in any given run, the times at which mode events occur are well defined, the runtime interpretation of mode events' guards is via the closure regions of their true-sets. In other words, the strict inequalities of guards are reinterpreted non-strictly. This gives rise to some interesting effects, which we comment on now.

One interesting effect concerns the constraint $py > 0$ in the guard of RACQUET. If this is replaced by $py \geq 0$, then the scenario is possible in which $py = 0$ becomes true at the precise time that the ball strikes the racquet. In this case both BOUNCE and RACQUET are enabled. If BOUNCE runs first, then RACQUET will be enabled immediately afterwards, and the run will be aborted by point **[3]** of the formal semantics. If RACQUET runs first, then BOUNCE will be enabled

---

[8] This could depend on requirements. Zeno behaviour may sometimes be tolerable and sometimes not.

immediately afterwards, and the run will also be aborted. These aborts are typical of the "cleaning up" that the semantics performs when the rules do not neatly conform to the requirements of the strucure of runs that we have demanded. This also supports the view that one should design *and reason about* systems using such guards etc. as most eloquently address the needed system requirements. Regarding behaviours at awkward boundary cases which arise because of the semantics, even if the semantics does not abort, behaviours may be forced that could justifiably be regarded as anomalous.

As an example of this consider the flight of the ball after the racquet strike, as it approaches the net. If the path of the ball is low, it will hit the net after having travelled a horizontal distance L from the point of impact with the racquet. Assuming the racquet strikes the ball so that it has an upward velocity component, let us consider increasing the velocity with which the racquet strikes the ball. As this increases, the ball will typically hit the net at points higher and higher up. Eventually, the top of the net will be reached. If the net is modelled as a vertical line, closed at the top (i.e. using a definition such as $0 \leq net\_y \leq N$), then the family of ball trajectories including these net impacts, generated by the above rules will behave smoothly as the limit of the top of the net is reached.

By contrast, keeping the same net definition, consider the case when the ball has a lot of energy on its return towards the net. Then it will fly over the net towards the right. Now consider reducing the ball's energy gradually. At the limit point, i.e. when the flight of the ball touches the net at height N, there will be a discontinuity in the family of ball trajectories. At the limit point, instead of the ball flying over the net, it will hit the net and drop leftwards. Assuming that the net is modelled as we said, and that we have a mode rule to model the impact with the net, this anomalous limiting behaviour of the family of trajectories in which the ball flies over the net will be generated by the semantics.

By contrast, if we model the net as $0 \leq net\_y < N$, then the anomaly would be generated the other way round, as the point of impact with the net rose.

Is the existence of either anomalous limit harmful? We argue that it is not. The anomaly exists at a single set of values for the system parameters. Viewed from the perspective of the system as a whole, this is a set of measure zero. In engineering terms, it is something which cannot be observed since the slightest departure from the specific parameter values causes a change in behaviour—only behaviours that are modelled by systems in which the behaviours are robust over parameter sets of non-zero measure can play a role in real life, so the existence of relatively isolated anomalous behaviours does no harm. These relatively isolated anomalous behaviours are the price that one sometimes has to pay for being able to model at an idealised level. And although such idealised models are clearly unrealistic to a degree, the clarity they can bring to high level system conceptualisation makes the price one that is worth paying.

A further set of interesting behaviours arises if we allow the coefficient of restitution parameter c, to vary. Under normal circumstances, one would expect a single bounce of the ball before the racquet returns the ball over the net, this being what is allowed by the rules of tennis. If, however, we consider a succession of further bounces, more interesting things can occur.

We assume that the racquet stays at a horizontal distance L away from the net. Then the number of bounces that the ball can experience depends on the value of c. Provided there is at least one bounce, then by adjusting the value of c we can increase the number of subsequent bounces arbitrarily. To see this we observe that the (vertical part of the) kinetic energy reduces by a factor of c^2 on every bounce. So the height of the parabolic flight segment after every bounce also reduces by a factor of c^2, and so does its width, which corresponds to the horizontal distance travelled during that parabolic flight segment. So, aside from the initial part in which the ball comes over the net, the total horizontal distance travelled is proportional to $\sum_{k=1}^{\infty} c^{2k} = c^2(1 - c^2)^{-1}$. As c reduces to zero, this approaches c^2, which also approaches zero.

So the total additional horizontal distance travelled after the first bounce can get arbitrarily small, and thus the ball may never reach the racquet while still in the air. What we see here is an example of a Zeno effect. To absorb all the initial vertical kinetic energy takes an infinite number of bounces.

What happens afterwards? Arguing physically (though still in a highly idealised way), since the vertical and horizontal elements of the kinetic energy are decoupled, after the vertical kinetic energy has been absorbed by the bounces, the horizontal kinetic energy remains, so the ball rolls along the ground with velocity vx_{in}. Arguing according to the semantics of Sect. 4, if c is small enough for this behaviour to ensue, then the single system run allowed by a fixed set of system parameters never gets past the limiting Zeno point of the behaviour described. This is because the semantic construction in Sect. 4 only allows for a number of steps that is indexable by the naturals. Going beyond the Zeno point would require a transfinite construction, which we have not explored in this paper. Such constructions, while possible, would always be unphysical to a greater or lesser extent, so are of limited value for application modelling.

What we have just been discussing, illustrates in a very clear way the difficulties inherent in the Zeno recommendation of Sect. 2. For one set of parameters, the Zeno effect is absent, and the model behaves in an exemplary way. For another set of parameters, looking not much different from the first, the behaviour is completely different, exhibiting Zeno effects. (And, of course, there is the boundary scenario, in which the limit of the Zeno behaviour reaches exactly the distance L, which we didn't explore.)

5.1 The Tennis Ball as a Multiple System

Let us contemplate our tennis ball example from the vantage point of multiple subsystems. Taking the rules we wrote at face value, there is no sensible possibility of partitioning the system, since its simplicity dictates that all the rules modify all variables, more or less.

However, we can consider enlarging the system, for example by adding a television camera that follows the flight of the ball.[9] This would have read access to the ball's dynamical variables, but not write access. In this case, the TV camera's variables and rules would reside in a separate subsystem from those of the tennis ball itself. Depending on the detail of its model, the ball's dynamical variables would, via read access, determine which pixels of the camera's CCD sensor changed in response to the ball's flight, etc.

An alternative approach might note that we have focused on building a system for the left hand side of the tennis court. We could thus contemplate building a complementary system to cover the right hand side of the court, subsequently composing the two subsystems to get a system covering the entire court.

However, there are problems with this approach, should we attempt to construct the system described. The ball would alternately be found, first in one half of the court, and next in the other. This means that the ball's behaviour would not be the responsibility of a single subsystem. This flies in the face of the restriction made at the beginning of Sect. 4.4, to ensure each variable is updated by a single subsystem.

The motivations for imposing such a restriction differ between mode and pliant variables. For mode variables, the fact that a mode variable can retain its value indefinitely, without special supervision, until an update changes the value, reduces the problem of its semantics' consistency to well understood questions of mutual exclusion, so that a single agent has authority to update the variable at any moment. The restriction of updates to a variable to a single subsystem is therefore just the simplest incarnation of this policy.

For pliant variables, the situation is different. A pliant variable's semantics demands that its value needs to be supervised at all times, by a differential equation for example. This corresponds to the physical reality that the laws of nature hold at all times, and therefore that any description of a physical process must adhere to the same principle. If responsibility for ensuring this is divided among a number of subsystems, the challenge of verifying that it is met becomes the harder. In particular, the description must be continuous and unbroken over time. This is easiest to ensure if all updates to the pliant variable are contained in one subsystem.

In [7] there is a much more extensive discussion of the implications of physical law for language systems intended for the definition and description of cyber-physical systems.

6   Continuous ASM Refinement

Now we develop our Continuous ASM framework to encompass refinement of Continuous ASM models. We start by describing the usual ASM refinement formulation, appropriate to pure mode transitions, and then show how to extend this to encompass the new kinds of transition.

---

[9] A sports programme that genuinely did this would make viewers dizzy, but we can tolerate the idea of it for the sake of the example.

6.1   The Discrete Case

In general, to prove a conventional ASM refinement, we verify so-called (m, n) diagrams, in which m abstract steps simulate n concrete ones in an appropriate way. This means that there is nothing that the n concrete steps can do that is not suitably reflected in m appropriately chosen abstract steps, where both m and n can be freely chosen to suit the application. It will be sufficient to focus on the refinement proof obligations (POs) which are the embodiment of this policy. The situation for refinement is illustrated in Fig. 2, in which we suppress input and output for clarity.

In Fig. 2 the refinement relation $R_{A,C}$ (also often referred to as the gluing relation) between abstract and concrete states, holds at the beginning and end of the (m, n) pair. This permits us to abut such (m, n) diagrams, by identifying the last (abstract and concrete) states of one (m, n) diagram, with the first (abstract and concrete respectively) states of the next, and thereby to create relationships between abstract and concrete runs in which $R_{A,C}$ is periodically reestablished. (N. B. In much of the ASM literature, the main focus is on an *equivalence*, usually written $\equiv$, between abstract and concrete states.

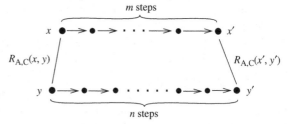

Fig. 2. An ASM (m, n) diagram, showing how m abstract steps, going from state x to state x' simulate n concrete steps, going from y to y'. The simulation is embodied in the refinement relation $R_{A,C}$, which holds for the before-states of the series of steps $R_{A,C}(x, y)$, and is re-established for the after-states of the series $R_{A,C}(x', y')$.

This is normally deemed to contain a "practically useful" subrelation $R_{A,C}$, chosen to be easier to work with. The approach via $R_{A,C}$ will be the focus of our treatment, and is also focus of the KIV [30] formalization in [43,44].)

The first PO is the initialization PO:

$$\forall y' \bullet CInit(y') \Rightarrow (\exists x' \bullet AInit(x') \land R_{A,C}(x', y')) \tag{6}$$

In (6), it is demanded that for each concrete initial state y', there is an abstract initial state x' such that $R_{A,C}(x', y')$ holds.

The second PO is correctness. The PO is concerned with the verification of (m, n) diagrams. For this, we have to have some way of deciding which (m, n) diagrams are sufficient for the application. In practice, this is part of the design process, so let us assume that this has been done. Let $CFrags$ be the set of fragments of concrete runs that we have previously determined will permit a covering of all the concrete runs of interest for the application. Using :: to denote concatenation, we write $y :: ys :: y' \in CFrags$ to denote an element of $CFrags$ starting with concrete state y, ending with concrete state y', and with intervening concrete state sequence ys. Likewise we write $x :: xs :: x' \in AFrags$ for abstract fragments. Let is, js, os, ps denote the sequences of abstract inputs, concrete inputs, abstract outputs, concrete outputs, respectively, belonging to $x::xs::x'$ and $y::ys::y'$ and let $In_{\text{AOPS,COPS}}(is, js)$ and $Out_{\text{AOPS,COPS}}(os, ps)$ denote suitable input and output relations. Then the correctness PO reads:

$$\forall x, is, y, ys, y', js, ps \bullet y::ys::y' \in CFrags \wedge$$
$$R_{\text{A,C}}(x, y) \wedge In_{\text{AOPS,COPS}}(is, js) \wedge \text{COPS}(y :: ys :: y', js, ps) \Rightarrow$$
$$(\exists xs, x', os \bullet x::xs::x' \in AFrags \wedge$$
$$\text{AOPS}(x::xs::x', is, os) \wedge R_{\text{A,C}}(x', y') \wedge Out_{\text{AOPS,COPS}}(os, ps)) \qquad (7)$$

In (7), it is demanded that whenever there is a concrete run fragment of the form $\text{COPS}(y :: ys :: y', js, ps)$, carried out by a sequence of concrete operations[10] COPS, with state sequence $y :: ys :: y'$, input sequence js and output sequence ps, such that the refinement and input relations $R_{\text{A,C}}(x, y) \wedge In_{\text{AOPS,COPS}}(is, js)$ hold between the concrete and abstract before-states and inputs, then an abstract run fragment $\text{AOPS}(x::xs::x', is, os)$ can be found to re-establish the refinement and output relations $R_{\text{A,C}}(x', y') \wedge Out_{\text{AOPS,COPS}}(os, ps)$.

The ASM refinement policy also demands that non-termination be preserved from concrete to abstract. We retain this in our extension of the formalism for when it is needed.

Assuming that (6) holds, and that we can prove enough instances of (7) to cater for the application of interest, then the concrete model is a **correct refinement** of the abstract model. In a correct refinement, all the properties of the concrete model (that are visible through the refinement and other relations), are suitably reflected in properties of the abstract model (because of the direction of the implication in (7)). If in addition, the abstract model is also a correct refinement of the concrete model (using the converses of the same relations), then the concrete model is a **complete refinement** of the abstract model. In a complete refinement, all relevant properties of the abstract model are also present in the concrete model (because of the direction of the implication in the modified version of (7)). Therefore, to ensure that the complete set of requirements of an intended system is faithfully preserved through a series of refinement steps, it is enough to express them all in a single abstract model, and then to ensure that each refinement step is a complete refinement.

6.2   The Continuous Case

The preceding was formulated for the discrete world. However, to extend it to the continuous world is not very hard. The essence of the approach is to reinterpret the run fragments $\text{AOPS}(x::xs::x', is, os)$ and $\text{COPS}(y::ys::y', js, ps)$ appearing in (7) in a way that yields a natural extension of the discrete case.

In the discrete context, such a notation refers to a sequence of states, i.e. a map from some natural number indexes to state values. In a context including

---

[10] We define an operation as a maximal enabled set of rules—provided its updates are consistent. Enabled inconsistent updates cause abortion of the run, as usual in ASM.

real time, the analogue of this is a function from an interval of time to state values, which is piecewise constant. More precisely, the interval of time in question will be a finite closed interval $[t_A \ldots t_B]$, where $t_A < t_B$. Such an interval corresponds to a typical left-closed right-open interval $[t_A \ldots t_B)$ on which the function is piecewise constant *plus* the right endpoint t_B. The interval $[t_A \ldots t_B)$ itself is partitioned into a finite sequence of left-closed right-open subintervals, on each piece of which the function is constant (as seen in the semantics of mode variables in Sect. 4).

The purpose of the right endpoint t_B, is to record the after-state of the last mode transition, so that it can be identified with the initial state of a successor function, when (m, n) diagrams are abutted. Referring to Fig. 2, we can view the rightward pointing arrows (both abstract and concrete) as the constant functions on non-empty left-closed right-open subintervals (with the blobs at their tails representing the leftmost values), and the final blob (both abstract and concrete) representing the isolated value at the right closure of the entire interval.

We allow an exception to this convention when $t_B = \infty$. In that case the last subinterval of $[t_A \ldots t_B)$ is of infinite length, corresponding to a nonterminating final transition, and there is no isolated right endpoint, and no abutting of an (m, n) diagram featuring this kind of final subinterval to any successor.

The obvious generalisation of this for the framework of Continuous ASM is to use piecewise absolutely continuous functions from intervals of time to state values. These would be defined on finite closed intervals $[t_A \ldots t_B]$, with $t_A < t_B$. As above, such an interval would partition into one or more left-closed right-open subintervals on each of which the state function is absolutely continuous and without internal discontinuities, *plus* an isolated state at the right endpoint t_B, included to allow identification with the initial state of a successor function.

In this context, $\mathrm{AOPS}(x :: xs :: x', is, os)$ at the abstract level and $\mathrm{COPS}(y :: ys :: y', js, ps)$ at the concrete level, each consist of an alternating sequence of pliant and mode transitions (starting with pliant and ending with mode—unless the $t_B = \infty$ exception applies, and the last transition is pliant too).

Similar principles apply to inputs and outputs. Mode inputs and outputs are mapped to the time instant at which the after-state is established, while pliant inputs and outputs are mapped to the (left- and right-) open interval during which the pliant transition runs. We thereby derive an interpretation for the notation used in (7) appropriate for the current context. Thus $R(x, y)$ becomes a predicate about the earliest abstract and concrete state values referred to by the state functions mentioned, while $R(x', y')$ refers to the latest state values.

In this way, (7) continues to define refinement in the Continuous ASM context. The piecing together of (m, n) diagrams to build an abstract simulation of a concrete run, now reduces to the identification of the latest (abstract and concrete) state values reached by one (m, n) diagram, with the earliest (abstract and concrete) state values of its successor (m, n) diagram, in the way indicated above for the discrete case.

Given the above, it is instructive to point out what the PO (7) *does not* demand. We have already said that the ASM PO does not mention states that are internal to the length m and length n fragments that occur in a given (m, n) diagram. This frequently simplifies the relations R, In, Out etc., that capture the relationship between abstract and concrete worlds—the policy is particularly useful when it is easy to predict that the systems are *guaranteed* to schedule their steps in the particular way exploited in a given (m, n) diagram.

Finally, there is no explicit mention of time in (7). In particular there is nothing in (7) that indicates, in any relative way, how time is expected to progress during the respective length m and length n fragments—the abstract and concrete systems are free to progress according to their own notions of time.

Such aspects give the ASM POs great flexibility. Designers can define relationships between systems in the most practically useful way, a perspective the ASM philosophy promotes. The appropriateness of the policy adopted for a given development becomes a matter for the wider requirements arena.

6.3 Continuous ASM Refinement and Multiple Subsystems

It is clear that refinement, as thus defined, suits a monolithic semantics—the correctness PO implicitly speaks of the state and I/O spaces in their entirety, and makes no concession to the subsystem issues debated in detail in Sect. 4.4. We comment on this now.

Suppose we have two subsystems $A1$ and $A2$. Taking $A1$ in isolation, its semantics is given in Sect. 4, on the understanding that any external values needed (e.g. from $A2$) appear as values of free variables of $A1$ that are "reliably readable when needed". On the other hand, viewing the system as a whole, forces the same Sect. 4 semantics to address the whole system, and to supply values of variables for both $A1$ and $A2$ simultaneously and consistently. Additionally, concerning the effect of a single rule, the ASM rule firing policy (namely that any enabled rule executes), also allows us to largely ignore whether the rule is being executed by a monolithic system, or by one of its subsystems.[11] We observed already though, that a non-VOID semantics for $A1$ and a non-VOID semantics $A2$, do not in themselves guarantee a non-VOID semantics for the combination of $A1$ and $A2$, since a globally consistent assignment of values to variables does not follow from individually consistent partial assignments.

The last observation makes clear that things get more complicated when refinement is considered. Suppose abstract subsystem $A1$, with variables $x1$, $is1$, $os1$, is refined by concrete subsystem $C1$, with variables $y1$, $js1$, $ps1$, using relations $R_{A1,C1}$, $In_{A1\text{OPS}_d, C1\text{OPS}_d}$ and $Out_{A1\text{OPS}_d, C1\text{OPS}_d}$, where d indexes over the (m, n) diagrams of the $A1$ to $C1$ refinement. Suppose abstract subsystem $A2$, with variables $x2$, $is2$, $os2$, is refined by concrete subsystem $C2$, with variables $y2$, $js2$, $ps2$, using relations $R_{A2,C2}$, $In_{A2\text{OPS}_d, C2\text{OPS}_d}$ and $Out_{A2\text{OPS}_e, C2\text{OPS}_e}$, where e indexes over the (m, n) diagrams of the $A2$ to $C2$ refinement.

---

[11] Contrast that with the case in which only one enabled rule is chosen to execute at a time. Then, whether a single rule executes, or a single rule *per subsystem* executes (and how this is reflected in observable effects), has a significant impact on the semantics and becomes very visible to the environment.

Now, even if there is a non-VOID semantics for the combination of $A1$ and $A2$, there is no guarantee of a non-VOID semantics for the combination of $C1$ and $C2$. Furthermore, even if there is a non-VOID semantics for the combination of $C1$ and $C2$, there is no *a priori* guarantee that the non-VOID semantics for the $C1$ and $C2$ combination is a refinement of the non-VOID semantics of the combination of $A1$ and $A2$.

The root cause of these problems is the presence of the existential quantifiers in the conclusion of (7), since a conjunction of existential quantifications does not imply the existential quantification of the conjunction, as would be needed if refinement of the combined system were to follow from the refinements of the subsystems individually.

Moreover, even contemplating the $C1$ and $C2$ combination as a refinement of the combination of $A1$ and $A2$ raises difficulties, since the lexical scope of (7) reaches beyond just the state variables of the abstract and concrete systems, to include any read-only input variables and write-only output variables (via $In_{A1OPS,C1OPS}$ and $Out_{A1OPS,C1OPS}$ respectively for $A1$, for example). If some of the read-only input variables or write-only output variables of one subsystem are identified with state variables of the other, the form of (7) itself would have to be adapted to reflect this, depending on the context.

Further difficulties in eliciting a refinement of $A1$ and $A2$ to $C1$ and $C2$ from individual refinements of $A1$ to $C1$ and $A2$ to $C2$ come from the fact that these individual refinements need not use (m, n) diagrams that are necessarily congruent. Thus the 'shape' of the diagram used at a particular point of the $A1/C1$ execution (in terms of the number of steps and their durations at the two levels of abstraction) need not coincide with the 'shape' of the diagram used at the corresponding point of the $A2/C2$ execution. All this notwithstanding the fact that the two separate refinements do not have to agree about the way that time itself progresses in their constituent models.

In the face of all the difficulties pointed out, there are two approaches that make sense. The first approach is to leave the resolution of all the issues that come up, case by case, to individual application developments. Most often, an individual application will be characterised by features that reduce most of the points raised to trivialities, and by other features that indicate the way to resolve those that remain in ways that are relatively convincing and evident from the structure of the application.

The second approach is to simplify matters drastically, until a point is reached at which the difficulties pointed out are sufficiently reduced that a relatively tractable generic formulation results. We illustrate what can be done using a simple example of this approach.

To start with, we make a number of restrictions, and we argue for their sufficiency as a scheme for combining two subsystems (having a restricted structure) below.[12] Thus let $A1$ be refined to $C1$ and $A2$ be refined to $C2$, and let us assume

---

[12] In the sequel, we refer to manipulations on relations via the logical operations on the logical definition of their bodies, for simplicity—e.g., (set theoretic) intersection of relations (over the same signature) is expressed via conjunction.

the other notations introduced above. We will refer to the combination of $A1$ and $A2$ as A, with state variables x, and the combination of $C1$, and $C2$ as C, with state variables y. The sought for refinement from A to C will be described by a refinement relation $R_{A,C}$, which we define in terms of the *per subsystem* ones already introduced.

In less technical terms, what follows can be seen as the opening of the lexical scopes of the separate name spaces of the $A1$ and $A2$ systems, and the creation of the name space of A via their union. This allows name capture of identical identifiers. Similarly for $C1$ and $C2$, yielding C. To then get a valid A to C refinement requires a number of additional compatibility properties to hold, so that the desired refinement can be proved.

(1) Time is deemed to progress at the same rate in all models of the construction.

(2) For simplicity, we assume that none of $A1$, $A2$, $C1$, $C2$ have any I/O.

(3) The state variables $x1$ of $A1$ partition into $x1_1$, $x1_2$. The state variables $C1$ $y1$ partition into $y1_1$, $y1_2$. The state variables $x2$ of $A2$ partition into $x2_1$, $x2_2$. The state variables $C2$ $y2$ partition into $y2_1$, $y2_2$.

(4) The following pairs of variables are identical: $x1_2 \equiv x2_1$ ($\equiv x_{12}$); $y1_2 \equiv y2_1$ ($\equiv y_{12}$). There are no other variable clashes.

(5) The refinement relations of the $A1/C1$ and $A2/C2$ refinements decompose as follows, being nontrivial on only the variables mentioned. $R_{A1,C1}(x1, y1) \equiv R_{A1,C1}^{11}(x1_1, y1_1) \wedge R_{A1,C1}^{12}(x1_2, y1_2)$; $R_{A2,C2}(x2, y2) \equiv R_{A2,C2}^{21}(x2_1, y2_1) \wedge R_{A2,C2}^{22}(x2_2, y2_2)$. $R_{A1,C1}^{12}(x1_2, y1_2) = R_{A2,C2}^{21}(x2_1, y2_1) \equiv R_{A2,C2}^{12}(x_{12}, y_{12})$. We define $R_{A,C}(x, y) \equiv R_{A1,C1}^{11}(x1_1, y1_1) \wedge R_{A2,C2}^{12}(x_{12}, y_{12}) \wedge R_{A2,C2}^{22}(x2_2, y2_2)$.

(6) There is a relation $\rho_{1,2}$ from the (m, n) diagrams of the $A1/C1$ refinement to the (m, n) diagrams of the $A2/C2$ refinement. It satisfies the following conditions. (i) For every (m, n) diagram of the $A1/C1$ refinement featuring a given behaviour of $x1_2$ in $A1$ and $y1_2$ in $C1$ (over the duration of the diagram), there is a (m, n) diagram of the $A2/C2$ refinement featuring an identical behaviour of $x2_1$ in $A2$ and $y2_1$ in $C2$ (over the identical duration), and the pair of (m, n) diagrams is in $\rho_{1,2}$. (ii) As for (i), but directed from the $A2/C2$ refinement to the $A1/C1$ refinement using the converse of $\rho_{1,2}$. (iii) $\rho_{1,2}$ is universal on all $A1/C1$ and $A2/C2$ (m, n) diagram pairs having a given common $x1_2/y1_2$ ($\equiv x2_1/y2_1$) behaviour.

(7) For each pair of $\rho_{1,2}$-related (m, n) diagrams, we construct an (m, n) diagram of the A/C refinement as follows. The execution fragment of A is the conjunction of: the execution fragment of $A1$ on $x1_1$, the execution fragment of $A1$ (or $A2$) on x_{12}, and the execution fragment of $A2$ on $x2_2$. The execution fragment of C is the conjunction of: the execution fragment of $C1$ on $y1_1$, the execution fragment of $C1$ (or $C2$) on y_{12}, and the execution fragment of $C2$ on $y2_2$. (And the refinement relation satisfied at the beginning and end of the constructed (m, n) diagram is $R_{A,C}$.

Theorem 1. *Let system A1 have refinement C1 and system A2 have refinement C2, as described. Let systems A and C be as constructed above, and suppose points (1)–(7) above hold. Then*

1. *The (m, n) diagrams constructed for A and C in (7) are valid (m, n) diagrams, in that the abstract and concrete execution fragments are related via the $R_{A,C}$ refinement relation.*
2. *The collection of (m, n) diagrams for A and C thereby constructed yields a Continuous ASM refinement from A to C.*

Proof: To show claim 1., we consider a typical constructed (m, n) diagram, created by fusing an $A1/C1$ (m, n) diagram with a corresponding $A2/C2$ (m, n) diagram. The $C1$ execution fragment starts in a $C1$ state that is related by $R_{A1,C1}$ to the starting state of the $A1$ execution fragment. Likewise for $C2$ and $A2$. Composing the two concrete execution fragments in parallel while fusing the two identical behaviours of $y1_2$ and $y2_1$, yields an execution fragment of C that starts in a state which is related by $R_{A,C}$ to the starting state of A, and because the two concrete execution fragments have the same duration as a consequence of being related by $\rho_{1,2}$, they end simultaneously, in states that are $R_{A1,C1}$ and $R_{A2,C2}$ related respectively to end states of the corresponding abstract execution fragments (whose identical $x1_2$ and $x2_1$ behaviours have also been fused), reestablishing $R_{A,C}$ for the after-state of the constructed (m, n) diagram.

To show claim 2., we work by induction on an arbitrary execution of C. The initial C state decomposes into a $C1$ initial state fused with a $C2$ initial state. The combination of these is related by $R_{A,C}$ to an A initial state, similarly decomposed and fused. For the inductive hypothesis we assume that the execution of C has been simulated, using a succession of the constructed (m, n) diagrams, reaching a concrete state $y \equiv (y1_1, y1_2, y2_2)$ which is related by $R_{A,C}(x, y)$ to an abstract state $x \equiv (x1_1, x1_2, x2)$.

Consider the concrete execution continuing from y. The $(y1_1, y1_2)$ part of the initial portion of it is a $C1$ execution fragment that is simulated by an $A1$ execution fragment via an (m, n) diagram of the $A1/C1$ refinement. The $y1_2 \equiv y_{12}$ part of the $C1$ execution fragment is common to the $(y2_1, y2_2)$ part of the concrete execution continuing from y, i.e. common to a $C2$ execution fragment that is simulated by an $A2$ execution fragment via an (m, n) diagram of the $A2/C2$ refinement. By (6).(i) and (6).(ii), we can choose the $A2$ execution fragment to have the same x_{12} behaviour exhibited by the $A1$ execution fragment. Therefore, by (6).(iii) we can fuse the two (m, n) diagrams to give an (m, n) diagram of the A to C refinement that simulates the concrete execution continuing from y. By (5) we easily derive that the refinement relation satisfied at the end of the (m, n) diagram is R.   □

The preceding constitutes a basic generic result of the kind being sought. We can imagine many variations on a result like this. For instance, we could involve inputs and outputs. Alternatively, we could insist that some of the R relations (and/or their I/O analogues) were *functions* from concrete to abstract. As another option we could relax the independence of the various relations on shared vs. unshared variables in various ways. And so on.

It is now evident that the all the mechanisms involved in combining the $A1$ to $C1$ refinement with the $A2$ to $C2$ refinement to get the A to C refinement —whether as described above, or via the generalisations suggested— centre on manipulation of the name spaces of the individual subsystems. In principle, these act as lexical binders, fixing the meaning of each identifier within the context of that subsystem. The objective of the manipulation is to then open these name spaces, in order to permit name capture of the free identifiers inside, which then become variables shared across the larger system. The ASM approach of allowing an individual subsystem's behaviour to be influenced by monitored variables —whose updates need not be defined within the subsystem— allows each subsystem to have enough available behaviours, that the behaviours required for shared variable cooperation are available, and can be specified within a larger system by the name capture technique.

7   Refinement and the Tennis Ball

We now expand our tennis ball example to illustrate the potential for our framework to express the inclusion of design detail via refinement. However, rather than developing a more elaborated version of the previous model —which typically would entail the introduction of copious quantities of technical detail— we develop an *abstraction* of the model of Sect. 3, and we argue that our original model arises as a refinement of the new one via the ASM refinement mechanism.

Thus, suppose the rally was taking place in a court surrounded by a fence of height H. One approach to the design might be that the fence is sufficiently high that no player can hit the ball out of the court. Of course, to enforce such a restriction absolutely might well be too demanding in a realistic setting, but we can pursue it in our idealised scenario anyway.

From this perspective, the only property of the ball that we need to care about is its total energy. That determines the maximum height it can reach via the conversion of all that energy into potential energy via the law of conservation of energy. In fact, since the ball's kinetic energy arises from the inner product of the velocity vector with itself, and the vertical and horizontal components of the velocity are orthogonal, we can identify the vertical energy (arising from the square of the vertical component of the velocity) as a separately conserved quantity, and it is only that energy that is available to be converted into potential energy as the ball flies upwards. Therefore, aside from an abstract INIT rule, we can model the path of the ball in abstract terms using the following abstract pliant event, in which, for clarity, the subscript A distinguishes abstract variables from their former (now concrete) counterparts:

FLIGHT$_A$ $\overset{c}{=}$

choose $px'_A(t), py'_A(t), vx'_A(t), vy'_A(t)$ **with** $\dfrac{1}{2}vy'^2_A(t) + g\,py_A(t) \leq E_{max}$

do $px_A(t), py_A(t), vx_A(t), vy_A(t) := px'_A(t), py'_A(t), vx'_A(t), vy'_A(t)$

In FLIGHT$_A$, we use the direct assignment form of a pliant rule to allow the dynamics of the abstract tennis ball to evolve arbitrarily, subject only to the constraint that its vertical energy, $\frac{1}{2}vy'^2_A(t) + g\,py_A(t)$ remains within E_{max}. In such direct assignment pliant rules, it is tacitly assumed that the behaviours of the variables are only ever assigned to (piecewise) absolutely continuous functions of time. This restriction implies that the derivatives of these functions exist in the Carathéodory sense, and thus the semantics of such direct assignment cases falls within the scope of the previously given differential equation semantics when we interpret a direct assignment $z := \Theta$ via differentiation, i.e., $\mathcal{D}z := \mathcal{D}\Theta$. (We observe that if Θ has discontinuities, then these can be handled via the "virtual skip" mechanism discussed earlier.) With the FLIGHT$_A$ rule in place, it is now easy to build an (m, n) diagram to show the refinement of FLIGHT$_A$ to some of the behaviours we discussed in the previous section.

Firstly, we model the passage of time in the same way in our two systems. Secondly, the m of our (m, n) diagram will be 2: i.e. an execution of the abstract INIT rule, followed by a single execution of FLIGHT$_A$ at the abstract level. Thirdly, the n of our (m, n) diagram will be covered by two broad cases: it will be 6 for the normal dynamics case discussed in Sect. 3; and it will be $6 + 2k$ (with $k \geq 1$) for the "approaching Zeno" cases.

For the normal dynamics case, the sequence of 6 steps consists of INIT, FLIGHT, BOUNCE, FLIGHT, RACQUET, FLIGHT. For the "approaching Zeno" cases, it consists of INIT, FLIGHT, then k repetitions of (BOUNCE, FLIGHT), and then RACQUET, FLIGHT. The Zeno case itself would correspond to FLIGHT, followed by an infinite number of repetitions of (BOUNCE, FLIGHT). But we do not regard that as a proper (m, n) diagram because of the infinite number of steps.

To qualify as ASM refinements, we need to make explicit the equivalence R that such (m, n) diagrams preserve. For this, we observe that provided that the parameters of the earlier model are confined (in the static algebra within which the dynamics takes place), to values that limit the vertical energy of the ball appropriately, then R can taken to be a partial identity relation between abstract and concrete states, being defined as the identity on those states which have a vertical energy that does not exceed the specified maximum. The fact that the equivalence is preserved is proved by the observation that the concrete dynamics permitted by the explicit model of Sect. 3 is simply one of the arbitrary behaviours allowed in the abstract model (provided that the energy of the concrete model remains suitably constrained).

8   Related Work

The framework we described above is similar to many ways of formulating hybrid systems present in the literature. We comment on some aspects of that here. Earlier work includes [4,5,27,32]. Shortly after these works were published, there appeared a spate of other papers, such as [20,21,32]. Much further activity ensued, too much to be surveyed comprehensively. A large proportion of it is described in the *Hybrid Systems: Computation and Control* series of international conferences. Slightly later formulations include [10,28,33]. Many of these earlier approaches, and especially the tools that support the relevant methodologies are surveyed in [14]. A less old theoretical overview is to be found in [48].

The majority of these works take an automata-theoretic view of hybrid systems. Thus, they have named states for the discrete control, within each of which, continuous behaviour evolves. This continues until the next preemption point arrives, triggered by the guard condition of the next discrete state becoming true. We achieve a similar effect via our mode and pliant operations. This relatively small degree of difference is in fact reassuring, since, in attempting to describe physical behaviour we have little leeway: the physical world is as it is and all descriptions must conform to it.

From our point of view, the capabilities of most of these systems are rather similar, except in those cases where the expressivity of the continuous part has been deliberately curtailed in order to get greater decidability, e.g. the pioneering [28] where continuous behaviour is linear in time. The focus on decidability is pursued vigorously in the literature. The survey [19] is a contemporary overview of reachability analysis of hybrid systems, and discusses many sublanguages of the general hybrid framework, restricted so that one or other variation of the notion of reachability is decidable for them.

The general hybrid framework is so expressive, that its undecidability is relatively self-evident, even if attention has to be paid to the details in order to model a two counter machine, which is the usual route to the result. The consequence of this is that unbounded state values are needed, or the state space will have accumulation points. While these are fine theoretically, both are unrealistic from an engineering standpoint, since engineering state spaces have both a finite size, and a limited accuracy.

The absence of the automata-theoretic structure in our approach simplifies the description of systems somewhat. All aspects become expressible in a relatively recognisable "program-like" syntax. The separation of discrete transitions from continuous ones also chimes with our other goal, of developing a hybrid formalism as a clean extension of an existing discrete formalism, syntactically and semantically. This also allows for different kinds of mathematical reasoning, relevant to the two worlds, to be cleanly separated on a *per rule* basis.

One difference between these approaches and ours, is the greater attention we have paid to the general semantics of differential equations. Issues of noise aside, classical physics is invariably defined in these terms, so we took that as basic. Many of the approaches above sidestep the issue by merely positing the existence of a *continuous flow* over the time interval between two discrete transitions. The equivalent of that for us would have been to take the criteria at the end of Sect. 2.2 as part of our formalism's *definition*, rather than as *properties* to be demonstrated on the basis of that definition. We argued for the truth of these on the basis of "off the shelf" mathematics in Sect. 4.

Properly controlling the continuous behaviour is just as important as properly defining the discrete, of course. Innocent looking conditions, such as merely requiring the right hand side of a DE to be continuous (c.f. [6]), can, strictly speaking, be unsound.[13]

The way to avoid problems, is to restrict the form of the allowed differential equations to cases whose properties are known. The results surveyed in [19] give many examples of this kind. Among these are several that incorporate the standard textbook results on linear and non-linear DEs, long included in computer algebra systems like Mathematica [35] or Maple [34]. A more general approach to DEs is taken in [37,38]. In our case, we have broadened the class of allowed differential equations quite a bit, to maximise expressivity, relying on the "off the shelf" mathematics mentioned above to suply solutions, where they are available. (Of course, only a tiny fraction of the DEs that one can write down have solutions that one can write down [39].)

9   Conclusions

In the preceding sections we first reviewed traditional discrete ASM, founding it on a discussion of basic ASM rules, and then we embarked on an extension that would allow a convincing description of the continuous phenomena inherent in hybrid and cyber-physical systems. Our strategy was based on deciding on a simple semantic domain first, centered on piecewise absolutely continuous functions of time that were solutions of well posed initial value problems of ordinary differential equations. We then arranged the syntax and its formal semantics to map cleanly onto it. The benefits of this included the fact that the behaviour of every variable could be fully described by a straightforward function: from a semi-infinite or finite interval of time to its type, and satisfying the properties mentioned. Of the many available ways of formulating continuous phenomena within applied mathematics, this semantic domain covers the vast majority of the problems that arise in practice, and is ultimately behind most formulations of hybrid and cyber-physical systems, which are so intensively studied today [1,8,15,27,36,37,46–48,51,53,55,56].

The formal semantics was then described, in sufficient detail that a fully rigorous technical definition could be elaborated from it if desired. We did not go into the full details however, since so much of that could be straightforwardly taken from quite standard sources. After that we considered refinement, and having reviewed discrete ASM refinement, we formulated continuous ASM refinement as a minimal extension of the discrete case. Our various discussions of semantics were complemented by discussions of issues surrounding compositionality and multi-subsystem systems, in the light of the formulation given. Accompanying this, we gave a simple illustration of the formalism in an example involving the flight of a tennis ball. Despite the apparent simplicity, this example nevertheless provided an opportunity to discuss further technical issues that arise when we

---

[13] The standard counterexample that mere continuity of the right hand side admits is $\mathcal{D} x = x^2$. This has a solution $x(t) = (a - t)^{-1}$ (for some constant of integration a), which explodes at $t = a$. Such counterexamples are very familiar in the differential equations literature, typically being surveyed in the opening pages of standard texts.

model hybrid systems in a clean way, for example as Zeno effects. We illustrated the formulation of Continuous ASM refinement by showing an abstraction, illustrating how very general properties could be specified in our formalism, and could then be refined to more specific behaviours. In future work, we intend to use our formulation to explore larger, more complex case studies.

A   The RAQUET Rule

We recall the RACQUET rule of Sect. 3.

RACQUET =

if $py > 0 \wedge py = rpy \wedge px < 0 \wedge px = rpx \wedge vx < 0 \wedge vx.rvx + vy.rvy < 0$ **then**

do

$$vx := -(vx - rvx)(\cos^2(\alpha) + c\sin^2(\alpha)) + (vy - rvy)(1 - c)\cos(\alpha)\sin(\alpha) + rvx,$$

$$vy := (vx - rvx)(1 - c)\cos(\alpha)\sin(\alpha) - (vy - rvy)(\sin^2(\alpha) + c\cos^2(\alpha)) + rvy$$

To understand the assignments to vx and vy in the above we use vector notation. So let $\mathbf{p}, \mathbf{v}, \mathbf{rp}, \mathbf{rv}$ be 2D vectors (in the plane of Fig. 1) corresponding to our earlier quantities. Let $\mathbf{r} = [-\cos(\alpha), \sin(\alpha)]$ be a unit vector pointing upwards along the line of the racquet, and let $\mathbf{r}^{\perp} = [\sin(\alpha), \cos(\alpha)]$ be a unit vector normal to the racquet, pointing towards the net.

We make a rigid Galilean transformation into the rest frame of the racquet, keeping the orientation of the racquet the same, but reducing its velocity to zero. In this frame of reference, the ball approaches the racquet with velocity $\mathbf{v} - \mathbf{rv}$. When the ball strikes the racquet, the tangential component of the ball's velocity remains the same, while the perpendicular component is reflected, and is reduced by the coefficient of restitution c. Resolving the velocity into these two components, the velocity before the collision is $[(\mathbf{v} - \mathbf{rv}) \cdot \mathbf{r}, (\mathbf{v} - \mathbf{rv}) \cdot \mathbf{r}^{\perp}]$, while the velocity after the collision is $[(\mathbf{v} - \mathbf{rv}) \cdot \mathbf{r}, -c(\mathbf{v} - \mathbf{rv}) \cdot \mathbf{r}^{\perp}] = [-(vx - rvx)\cos(\alpha) + (vy - rvy)\sin(\alpha), -c(vx - rvx)\sin(\alpha) - c(vy - rvy)\cos(\alpha)]$, where in the last expression, we have evaluated the dot products in the rectiliear frame of reference, since dot products are rotationally invariant. We can re-express this after-velocity in the rectilinear frame by applying a rotation matrix as follows:

$$\begin{bmatrix} \cos(\alpha) & \sin(\alpha) \\ -\sin(\alpha) & \cos(\alpha) \end{bmatrix} \begin{bmatrix} -(vx - rvx)\cos(\alpha) + (vy - rvy)\sin(\alpha) \\ -c(vx - rvx)\sin(\alpha) - c(vy - rvy)\cos(\alpha) \end{bmatrix}$$

$$= \begin{bmatrix} -(vx - rvx)\cos^2(\alpha) + (vy - rvy)\cos(\alpha)\sin(\alpha) - \\ c(vx - rvx)\sin^2(\alpha) - c(vy - rvy)\cos(\alpha)\sin(\alpha) \\ (vx - rvx)\cos(\alpha)\sin(\alpha) - (vy - rvy)\sin^2(\alpha) - \\ c(vx - rvx)\cos(\alpha)\sin(\alpha) - c(vy - rvy)\cos^2(\alpha) \end{bmatrix}$$

Now, reversing the rigid Galilean transformation, we get the assignment given in RACQUET.

References

1. Summit Report: Cyber-Physical Systems (2008). http://iccps2012.cse.wustl.edu/_doc/CPS_Summit_Report.pdf
2. Abrial, J.R.: The B-Book: Assigning Programs to Meanings. Cambridge University Press, Cambridge (1996)
3. Abrial, J.R.: Modeling in Event-B: System and Software Engineering. Cambridge University Press, Cambridge (2010)
4. Alur, R., Courcoubetis, C., Henzinger, T.A., Ho, P.-H.: Hybrid automata: an algorithmic approach to the specification and verification of hybrid systems. In: Grossman, R.L., Nerode, A., Ravn, A.P., Rischel, H. (eds.) HS 1991-1992. LNCS, vol. 736, pp. 209–229. Springer, Heidelberg (1993). https://doi.org/10.1007/3-540-57318-6_30
5. Alur, R., Dill, D.: A theory of timed automata. Theoret. Comp. Sci. **126**, 183–235 (1994)
6. Back, R.J., Petre, L., Porres, I.: Continuous action systems as a model for hybrid systems. Nordic J. Comp. **8**, 2–21, 202–213 (2001). Extended version of FTRTFT-00. LNCS 1926
7. Banach, R., Zhu, H.: Language evolution and healthiness for critical cyber-physical systems. J. Soft. Evol. Proc. 2020, e2301, 24 (2021)
8. Barolli, L., Takizawa, M., Hussain, F.: Special issue on emerging trends in cyber-physical systems. J. Amb. Intel. Hum. Comput. **2**, 249–250 (2011)
9. Beauquier, D., Slissenko, A.: On Semantics of Algorithms with Continuous Time. Technical report TR-LACL-1997-15, LACL, University of Paris-12 (1997)
10. Bender, K., Broy, M., Péter, I., Pretschner, A., Stauner, T.: Model based development of hybrid systems: specification, simulation, test case generation. In: Engell, S., Frehse, G., Schnieder, E. (eds.) Modelling Analysis, and Design of Hybrid Systems, LNCIS, vol. 279, pp. 37–51. Springer (2002). https://doi.org/10.1007/3-540-45426-8_3
11. Börger, E.: The ASM refinement method. FACJ **15**, 237–257 (2003)
12. Börger, E., Gurevich, Y., Rosenzweig, D.: The bakery algorithm: yet another specification and verification. In: Börger, E. (ed.) Specification and Validation Methods. Oxford University Press (1995)
13. Börger, E., Stärk, R.: Abstract State Machines. A Method for High Level System Design and Analysis. Springer, Heidelberg (2003). https://doi.org/10.1007/978-3-642-18216-7
14. Carloni, L., Passerone, R., Pinto, A., Sangiovanni-Vincentelli, A.: Languages and tools for hybrid systems design. Found. Trends Electron. Design Autom. **1**, 1–193 (2006)
15. Clarke, E.M., Zuliani, P.: Statistical model checking for cyber-physical systems. In: Bultan, T., Hsiung, P.-A. (eds.) ATVA 2011. LNCS, vol. 6996, pp. 1–12. Springer, Heidelberg (2011). https://doi.org/10.1007/978-3-642-24372-1_1
16. Cohen, J., Slissenko, A.: Implementation of timed abstract state machines with instantaneous actions by machines with delays. Technical report, TR-LACL-2008-2, LACL, University of Paris-12 (2008)
17. Derrick, J., Boiten, E.: Refinement in Z and Object-Z: Foundations and Advanced Applications. Springer, London (2001). https://doi.org/10.1007/978-1-4471-0257-1
18. ESW: Embedded Systems Week Conferences

19. Fränzle, M., Chen, M., Kröger, P.: In Memory of Oded Maler: Automatic Reachability Analysis of Hybrid-State Automata. ACM SIGLOG News **6**, 19–39 (2019)
20. Friesen, V., Nordwig, A., Weber, M.: Object-oriented specification of hybrid systems using UML^h and ZimOO. In: Bowen, J.P., Fett, A., Hinchey, M.G. (eds.) ZUM 1998. LNCS, vol. 1493, pp. 328–346. Springer, Heidelberg (1998). https://doi.org/10.1007/978-3-540-49676-2_22
21. Friesen, V., Nordwig, A., Weber, M.: Toward an object-oriented design methodology for hybrid systems. In: Object-Oriented Technology and Computing Systems Re-Engineering, pp. 1–15. Elsevier (1999)
22. Geisberger, E., Broy, M. (eds.): Living in a Networked World. Integrated Research Agenda Cyber-Physical Systems (agendaCPS) (2015). http://www.acatech.de/fileadmin/user_upload/Baumstruktur_nach_Website/Acatech/root/de/Publikationen/Projektberichte/acaetch_STUDIE_agendaCPS_eng_WEB.pdf
23. Graf, S., Prinz, A.: A framework for time in FDTs. In: Proceedings of FORTE-04. LNCS, vol. 1092, pp. 266–290. Springer (2004)
24. Graf, S., Prinz, A.: Time in abstract state machines. Fund. Inf. **77**, 143–174 (2007)
25. Gratzer, G.: Universal algebra. In: Abstract Algebra. Graduate Texts in Mathematics, vol. 242. Springer, New York (2007). https://doi.org/10.1007/978-0-387-71568-1_15
26. Gurevich, Y., Huggins, J.K.: The railroad crossing problem: an experiment with instantaneous actions and immediate reactions. In: Kleine Büning, H. (ed.) CSL 1995. LNCS, vol. 1092, pp. 266–290. Springer, Heidelberg (1996). https://doi.org/10.1007/3-540-61377-3_43
27. He, J.: From CSP to hybrid systems. In: Roscoe (ed.) A Classical Mind, Essays in Honour of C.A.R. Hoare, pp. 171–189. Prentice-Hall (1994)
28. Henzinger, T.: The Theory of hybrid automata. In: Proceedings of IEEE LICS-96, pp. 278–292. IEEE (1996). http://mtc.epfl.ch/~tah/Publications/the_theory_of_hybrid_automata.pdf
29. HSCC: Hybrid Systems: Command and Control Conferences
30. Karlsruhe Interactive Verifier. http://www.informatik.uni-augsburg.de/lehrstuehle/swt/se/kiv/
31. Lynch, N.: Distributed Algorithms. Morgan Kaufmann (1996)
32. Lynch, N., Segala, R., Vaandrager, F., Weinberg, H.B.: Hybrid I/O automata. In: Alur, R., Henzinger, T.A., Sontag, E.D. (eds.) HS 1995. LNCS, vol. 1066, pp. 496–510. Springer, Heidelberg (1996). https://doi.org/10.1007/BFb0020971
33. Lynch, N., Segala, R., Vaandrager, F.a.: Hybrid I/O automata. Inf. Comput. **185**, 105–157. mIT Technical Report MIT-LCS-TR-827d
34. Maple. http://www.maplesoft.com
35. Mathematica. http://www.wolfram.com
36. National Science and Technology Council: Trustworthy Cyberspace: Strategic Plan for the Federal Cybersecurity Research and Development Program (2011). http://www.whitehouse.gov/sites/default/files/microsites/ostp/fed_cybersecurity_rd_strategic_plan_2011.pdf
37. Platzer, A.: Logical Analysis of Hybrid Systems: Proving Theorems for Complex Dynamics. Springer, Heidelberg (2010). https://doi.org/10.1007/978-3-642-14509-4
38. Platzer, A.: Logical Foundations of Hybrid Systems. Springer (2018). https://doi.org/10.1007/978-3-319-63588-0
39. Polyanin, A., Zaitsev, V.: Handbook of Ordinary Differential Equations: Exact Solutions, Methods, and Problems. C.R.C. Press, Boca Raton (2018)

40. Raynal, M.: Concurrent Programming: Algorithms, Principles and Foundations. Springer, Heidelberg (2013). https://doi.org/10.1007/978-3-642-32027-9
41. Royden, H., Fitzpatrick, P.: Real Analysis. Pearson, London (2010)
42. Rust, H.: Hybrid abstract state machines: using the hyperreals for describing continuous changes in a discrete notation. Technical report. Proc. ASM-00, TIK Report 87, ETH Zurich (2000)
43. Schellhorn, G.: Verification of ASM refinements using generalized forward simulation. JUCS **7**, 952–979 (2001)
44. Schellhorn, G.: ASM refinement and generalizations of forward simulation in data refinement: a comparison. Theoret. Comp. Sci. **336**, 403–435 (2005)
45. Slissenko, A., Vasilyev, P.: Simulation of timed abstract state machines with predicate logic model checking. JUCS **14**, 1984–2006 (2008)
46. Stehr, M.-O., Kim, M., Talcott, C.: Toward distributed declarative control of networked cyber-physical systems. In: Yu, Z., Liscano, R., Chen, G., Zhang, D., Zhou, X. (eds.) UIC 2010. LNCS, vol. 6406, pp. 397–413. Springer, Heidelberg (2010). https://doi.org/10.1007/978-3-642-16355-5_32
47. Sztipanovits, J.: Model integration and cyber physical systems: a semantics perspective. In: Butler, M., Schulte, W. (eds.) FM 2011. LNCS, vol. 6664, p. 1. Springer, Heidelberg (2011). https://doi.org/10.1007/978-3-642-21437-0_1
48. Tabuada, P.: Verification and Control of Hybrid Systems: A Symbolic Approach. Springer, Boston (2009). https://doi.org/10.1007/978-1-4419-0224-5
49. Walter, W.: Ordinary Differential Equations. Springer, New York (1998). https://doi.org/10.1007/978-1-4612-0601-9
50. Wechler, W.: Universal Algebra for Computer Scientists. Springer, Heidelberg (1992). https://doi.org/10.1007/978-3-642-76771-5
51. White, J., Clarke, S., Groba, C., Dougherty, B., Thompson, C., Schmidt, D.: R&D challenges and solutions for mobile cyber-physical applications and supporting internet services. J. Internet Serv. Appl. **1**, 45–56 (2010)
52. Wikipedia: Absolute continuity
53. Willems, J.: Open Dynamical Systems: Their Aims and their Origins. Ruberti Lecture, Rome (2007). http://homes.esat.kuleuven.be/~jwillems/Lectures/2007/Rubertilecture.pdf
54. Woodcock, J., Davies, J.: Using Z, Specification, Refinement and Proof. Prentice Hall, Hoboken (1996)
55. Zhang, L., He, J.: A formal framework for aspect-oriented specification of cyber physical systems. In: Lee, G., Howard, D., Ślęzak, D. (eds.) ICHIT 2011. CCIS, vol. 206, pp. 391–398. Springer, Heidelberg (2011). https://doi.org/10.1007/978-3-642-24106-2_50
56. Zhlke, L., Ollinger, L.: Agile automaton systems based on cyber-physical systems and service oriented architectures. In: Lee, G. (ed.) Proceedings of ICAR-11, LNEE, vol. 122, pp. 567–574. Springer, Heidelberg (2011). https://doi.org/10.1007/978-3-642-25553-3_70

Product Optimization in Stepwise Design

Don Batory[1(✉)], Jeho Oh[1], Ruben Heradio[2], and David Benavides[3]

[1] The University of Texas at Austin, Austin, TX 78712, USA
{batory,jeho}@cs.utexas.edu
[2] Universidad Nacional de Educación a Distancia, Madrid, Spain
rheradio@issi.uned.es
[3] University of Seville, Seville, Spain
benavides@us.es

Abstract. Stepwise design of programs is a divide-and-conquer strategy to control complexity in program modularization and theorems. It has been studied extensively in the last 30 years and has worked well, although it is not yet commonplace. This paper explores a new area of research, finding efficient products in colossal product spaces, that builds upon past work.

1 To My Friend Egon

Egon and I (Batory) first met at the "Logic for System Engineering" Dagstuhl Seminar on March 3–7, 1997. Egon presented his recent work on *Abstract State Machines* (**ASMs**) entitled "An Industrial Use of ASMs for System Documentation Case Study: The Production Cell Control Program". I presented my work on *Database Management System* (**DBMS**) customization via feature/layer composition. I had not yet directed my sights beyond DBMS software to software development in general.

At the heart of our presentations was the use and scaling of Dijkstra's concepts of layering and software virtual machines [15] and Wirth's notions of stepwise refinement [53]. The connection between our presentations was evident to us but likely no others. I was not yet technically mature enough to have a productive conversation with Egon then to explore our technical commonalities in-depth.

Our next encounter was at a Stanford workshop on Hoare's "Verifying Compiler Grand Challenge" in Spring 2006. Egon would make a point in workshop discussions and I would think: That is exactly what I would say! And to my delight as I learned later, Egon reacted similarly about my discussion points. At the end of the workshop, we agreed to explore interests and exchanged visits – I to Pisa and he to Austin. We wrote a joint paper [6] and presented it as a keynote at the June 2007 Abstract State Machine Workshop in June 2007. In doing so, I learned about his pioneering JBook case study [44]. Our interactions

This work was partially funded by the EU FEDER program, the MINECO project OPHELIA (RTI2018-101204-B-C22), the TASOVA network (MCIU-AEI TIN2017-90644-REDT), and the Junta de Andalucia METAMORFOSIS project.

were a revelation to me as our thinking, although addressing related problems from very different perspectives, led us to similar world view.

In this paper, I explain our source of commonality and where these ideas have been taken recently in my community, **Software Product Lines** (SPLs).

2   Similarity of Thought in Scaling Stepwise Design

Central to the **Stepwise Design** (**SWD**) of large programs is the scaling of a step to an increment in program functionality. A program is a composition of such increments. To demonstrate that such technology is possible, one must necessarily focus on the SWD of a single application (as in JBook [44]) or a family of related applications (as in SPLs) where stereotypical increments in functionality can be reused in building similar programs. These increments are **features**; think of features as the legos [49] of domain-specific software construction.

The JBook [44] presented a SWD of a suite of programs: a parser, **ASTs** (**Abstract Syntax Trees**), an interpreter, a compiler and a JVM for Java 1.0. At each step, there is a proof that the interpretation of *any* Java 1.0 program P and the compilation and then JVM execution of P produced identical results. The divide-and-conquer strategy used in JBook centered on the Java 1.0 grammar. The base language was the sublanguage of Java imperative expressions (ExpI). For this sublanguage, its grammar, ASTs, interpreter, compiler and JVM were defined, along with a proof of their consistency, Fig. 1a. Then imperative statements (ΔStmI) were added to ExpI, lock-step extending its grammar, ASTs, interpreter, compiler, JVM and proof of their composite consistency, Fig. 1b.[1]

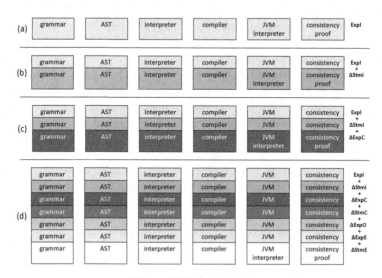

Fig. 1. SWD of JBook.

---

[1] All extensions were manually defined – this is normal.

And then static fields and expressions (ΔExpC) were added, Fig. 1c, and so on until the complete syntax of Java 1.0 was formed, with its complete AST definitions, a complete interpreter, compiler and JVM for Java 1.0 too, Fig. 1d.

The JBook was a masterful case study in SWD. It fit my SPL theory of features, where an application is defined by a set representations (programs, documents, property files, etc..). Features incrementally extend each representation so that they are consistent. Features could add new documents as well.

An SPL follows the JBook example, but with important differences. Some programs can have different numbers of features and different features can implement identical functionalities in different ways.[2] This enables a family of related programs to be built simply by composing features. Each program in an SPL is defined by a unique set of features. If there are n optional features, the size of the SPL's **product space** can be up to 2^n distinct programs/products.

It is well-known that features obey constraints: selecting one feature may demand the selection and/or exclusion of other features. And there is a preferred order in which features are composed. It was discovered that a context sensitive grammar could define the product space of an SPL whose sentences are legal sequences of features. Such a grammar is a **feature model** [7]. A partial feature model for JBook is below (given that each of the sublanguages in its design is useful); the first line is a context free grammar. Notation "[T]" denotes feature T is optional. Subsequent lines define propositional formulas as compositional constraints to make the grammar context sensitive:

```
JBook : Expl [ΔStml] [ΔExpC] [ΔStmC] ... ;
// constraints
ΔExpC ⇒ ΔStml;        // if ΔExpC then so too must ΔStml
ΔStmC ⇒ ΔExpC;        // if ΔStmC then so too must ΔExpC
```

These are the basics of SPLs [4]; a more advanced discussion is in [5].

3   SPL Feature Models and Product Spaces

A feature model can be translated to a propositional formula ϕ [2–4]. This is accomplished in two steps: (1) the context free grammar is translated to a propositional formula ϕ', and (2) composition constraints are conjoined with ϕ' to produce ϕ. For example, the lone production of the JBook context free grammar, defined above, is translated to:[3]

$$\phi' = \big(\ \texttt{JBook} \Leftrightarrow (\texttt{Expl})\ \big) \wedge \big(\ (\Delta\texttt{Stml} \vee \Delta\texttt{ExpC} \vee \Delta\texttt{StmC} \vee \ldots) \Rightarrow \texttt{JBook}\ \big)$$

where each term is a boolean variable. The complete propositional formula ϕ is:

$$\phi = \phi' \wedge (\Delta\texttt{ExpC} \Rightarrow \Delta\texttt{Stml}) \wedge (\Delta\texttt{StmC} \Rightarrow \Delta\texttt{ExpC})$$

---

[2] Much like different data structures implement the same container abstraction [8].

[3] More involved examples and explanations are given in [2–4].

Every solution of ϕ corresponds to a unique product (a unique set of features) in that SPL. ***Binary Decision Diagrams*** **(BDD)** and ***Sharp-SAT solvers*** **(#SAT)** can count the number of products of ϕ [12,21,32,40]. Industrial SPLs can have colossal product spaces. Consider the table below from [5,21]:

Model	#Variables	#SAT-solutions	Source
axTLS 1.5.3	64	10^{12}	http://axtls.sourceforge.net/
uClibc 201 50420	298	10^{50}	https://www.uclibc.org/
Toybox 0.7.5	316	10^{81}	http://landley.net/toybox/
BusyBox 1.23.2	613	10^{146}	https://busybox.net/
EmbToolkit 1.7.0	2331	10^{334}	https://www.embtoolkit.org
LargeAutomotive	17365	10^{1441}	[26]

272 is a magic number in SPLs. If an SPL has 272 optional features, it has 2^{272} unique products. $2^{272} \approx 10^{82}$ is a ***really big number***: 10^{82} is the current estimate of the number of atoms in the universe [46]. The **LargeAutomotive** SPL in the above table has a colossal space of 10^{1441} products. That makes the largest numbers theoretically possible in Modern Cosmology look really, really small ☺ [35].[4] And there are even *larger* known SPLs (e.g., the Linux Kernel), whose size exceeds the ability of state-of-the-art tools to compute.

Beyond admiring the size of these spaces, suppose you want to know which product in a space (or a user-defined subspace) has the best performance for a given a workload. Obviously, enumerating and benchmarking each product is infeasible. The immediate question is: How does one search colossal product spaces efficiently? A brief survey of current approaches is next.

4   Searching SPL Product Spaces

To predict the performance of SPL products, a mathematical performance model is created. Historically, such models are developed manually using domain-specific knowledge [1,13]. More recently, performance prediction models are learned from performance measurements of sampled products. In either case, a performance model is given to an optimizer, which can then find near-optimal products that observe user-imposed feature constraints (e.g., product predicates that exclude feature F and include feature G).

4.1   Prediction Models

Models can estimate the performance of any valid product [17,37,42,43,54]. The goal is to use as few samples as possible to learn a model that is 'accurate'. Finding a good set of training samples to use is one challenge; another is minimizing the variance in predictions.

---

[4] Still 10^{1441} does pale in comparison to 10^{284265}, the size of the space of texts a monkey can randomly type out, one text of which *is* Hamlet [34,36] or 10^{40000}, the size of the space of texts a monkey can type out, one text of which is this paper.

Let $\mathbb{C}$ be the set of all legal SPL products. 1st-order performance models have the following form: let $\$P$ be the estimated performance of an SPL product $P \in \mathbb{C}$, where f_P is the set of P's selected features and $\$F_i$ is the performance contribution of feature F_i:

$$\$P = \sum_{i \in f_P} \$F_i \tag{1}$$

$\$F_i$ might be as simple as a constant (c_i) or a constant-weighted expression [17]:

$$\$F_i = c_0 \tag{2}$$

$$\text{or}$$

$$= c_0 + c_1 \cdot n + c_2 \cdot n \cdot \log(n) + c_3 \cdot n^2 + \ldots \tag{3}$$

where n is a global variable that indicates a metric of product or application 'size'. The value for n is given; the values of constants (c_i) must be 'learned'.

1st-order performance models are linear regression equations without (feature) interaction terms. Such models are inaccurate. Let $\$F_{ij}$ denote the performance contribution of the interaction of features F_i and F_j, which requires both F_i and F_j to be present in a product; $\$F_{ij} = 0$ otherwise. 2nd-order models take into account 2-way interactions:

$$\$P = \left(\sum_{i \in f_P} \$F_i \right) + \left(\sum_{i \in f_P} \sum_{j \in f_P} \$F_{ij} \right) \tag{4}$$

Models with n-way interactions add even more nested-summations to (4) [42].

Manually-developed performance models [1,9,13] are different as they:

- Identify operations $[O_{1..}]$ invoked by system clients;
- Define a function $\$O_k$ to estimate the performance of each operation O_k;
- Encode system workloads by operation execution frequencies, where ν_k is the frequency of O_k; and
- Express performance $\$P$ of a program P as a weighted sum of frequency times operation cost:

$$\$P = \sum_k \nu_k \cdot \$O_k \tag{5}$$

Features complicate the cost function of each operation, where the set of features of product $P \in \mathbb{C}$ becomes an explicit parameter of each O_k:

$$\$P = \sum_k \nu_k \cdot \$O_k(f_P) \tag{6}$$

In summary, manual performance models include workload variances in their predictions, whereas current SPL performance models use a *fixed* workload. Workload variations play a significant role in SPL product performance. To include workloads in learned models requires relearning models from scratch or transfer learning which has its own set of issues [23].[5]

---

[5] Transfer learning is an automatic translation of one performance model to another.

4.2   Finding a Near-Optimal is NP-Hard

The simplest formulation of this problem, namely as linear regression equations, is NP-Hard [52]. Here's a reformulation of Eq. (1) as a 0–1 Integer Programming Problem. Let $\mathbb{1}_i(P)$ be a boolean indicator variable to designate if feature F_i is present ($\mathbb{1}_i(P) = 1$) or absent ($\mathbb{1}_i(P) = 0$) in P. Rewrite Eq. (1) as:

$$\$P = \sum_{i \in f_P} \$F_i = \sum_i \$F_i \cdot \mathbb{1}_i(P)$$

We want to find a configuration c_{near} to minimize $\$P$, Eq. (7). To do so, convert Eq. (7) into a inequality with a cost bound b, Eq. (8). By solving Eq. (8) a polynomial number of times (progressively reducing b) we can determine a near optimal performance $\$c_{near}$ and c_{near}'s features (the values of its indicator variables):

$$\min_{P \in \mathbb{C}} (\$P) = \min_{P \in \mathbb{C}} \left(\sum_i \$F_i \cdot \mathbb{1}_i(P) \right) \tag{7}$$

$$\min_{P \in \mathbb{C}} \left(\sum_{i \in f} \$F_i \cdot \mathbb{1}_i(P) \right) \leq b \tag{8}$$

Recall a feature model defines constraints among features, like those in the "**Prop Formula**" column of Fig. 2. There are well-known procedures to translate a propositional formula to a linear constraint, and then to $\leq$ inequalities [16,22].

Prop Formula	Linear Constraint	Linear Inequality
$x \wedge y$	$x + y = 2$	$-x - y \leq -2$
$x \vee y$	$x + y \geq 1$	$-x - y \leq -1$
$x \Rightarrow y$	$x - y \geq 0$	$x - y \leq 0$
$x \Rightarrow \neg y$	$x + y \leq 1$	$x + y \leq 1$
$x \Leftrightarrow y$	$x = y$	$(x - y \leq 0) \wedge (y - x \leq 0)$
$Choose1(x, y)$	$x + y = 1$	$(x + y \leq 1) \wedge (-x - y \leq -1)$

Fig. 2. Prop Formula to an integer inequality.

To optimize Eq. (8) correctly, feature model constraints must be observed. The general structure of the optimization problem described above is:

$$\text{find } \mathbf{x} \text{ such that} \quad \mathbf{c}^T \mathbf{x} \leq b \qquad\qquad \text{rewrite of Eq. (8)}$$
$$\text{subject to} \quad \mathbf{Ax} \leq \mathbf{d} \qquad\qquad \text{feature model constraints}$$

where $\mathbf{x} \in \mathbb{1}^n$ ($\mathbf{x}$ is an array of n booleans), $\mathbf{c} \in \mathbb{Z}^n$ and $b \in \mathbb{Z}$ ($\mathbf{c}$ is an array of n integers, b is an integer), $\mathbf{A} \in \mathbb{Z}^{m \times n}$ ($\mathbf{A}$ is an $m \times n$ array of integers), and $\mathbf{d} \in \mathbb{Z}^m$ ($\mathbf{d}$ is an array of n integers). This is the definition of 0–1 Linear Programming, which is NP-Complete [52]. The NP-hard version removes bound b and minimizes $\mathbf{c}^T \mathbf{x}$.

4.3 Uniform Random Sampling

Optimizers and prediction models [17–19,37–39,54] rely on 'random sampling', but the samples used are not provably uniform. **Uniform Random Sampling (URS)** conceptually enumerates all $\eta = |\mathbb{C}|$ legal products in an array $\mathbb{A}$. An integer $\mathtt{i} \in [\,1..\eta\,]$ is randomly selected (giving all elements in the space an equal chance) and $\mathbb{A}[\mathtt{i}]$ is returned. This simple approach is not used because η could be astronomically large. Interestingly, URS of large SPLs was considered infeasible as late as 2019 [24,33].

An alternative is to randomly select *features*. If the set of features is valid, a product was "randomly" selected. However, this approach creates far too many invalid feature combinations to be practical [17,18,37,39,54]. Another approach uses SAT solvers to generate valid products [19,38], but this produces products with similar features due to the way solvers enumerate solutions. Although Henard et al. [19] mitigated these issues by randomly permuting the parameter settings in SAT solvers, true URS was not demonstrated.

The top path of Fig. 3 summarizes prior work: the product space is non-uniformly sampled to derive a performance model; samples are interleaved with performance model learning until a model is sufficiently 'accurate'. That model is then used by an optimizer, along with user-imposed feature constraints, to find a near-optimal performing product.

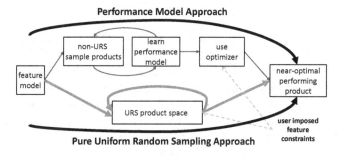

Fig. 3. Different ways to find near-optimal products.

In contrast, a pure URS approach (the bottom path of Fig. 3) uses neither performance models nor optimizers. Near-optimal products are found by uniformly probing the product space directly, and benchmarking the performance of sampled products using the required workload. User-imposed feature constraints simply reduce the space to probe. A benefit of URS is that it is a standard way to estimate properties accurately and efficiently of colossal spaces [14]. It replaces heuristics with no guarantees with mathematics with confidence guarantees.

> **Note ❮**For some, it may not evident that URS could be used for optimization. In fact, **Random Search** ($\mathbb{R_S}$) algorithms [10,50] do exactly this – find near-optimal solutions in a configuration space. We present evidence later that URS requires many fewer samples than existing performance model approaches [30].**❯**

5   URS Without Enumeration

Let $\eta = |\phi|$ be the size of an SPL product space whose propositional formula is ϕ. Let $\mathcal{F} = [\,F_1, F_2, ..F_\theta\,]$ be a list of optional SPL features. Randomly select an integer $i \in [1..\eta]$ and compute $s_1 = |\phi \wedge F_1|$, the number of products with feature F_1. If $i \leq s_1$, then F_1 belongs to the ith product and recurse on the subspace $\phi \wedge F_1$ using feature F_2. Otherwise $\neg F_1$ belongs to the ith product and recurse on subspace $\phi \wedge \neg F_1$ with $i = i - |\phi \wedge F_1|$ using feature F_2. Recursion continues until feature F_θ is processed, at that point every feature in the ith product is known.

> **Note** ❰Historically, Knuth first proposed this algorithm in 2012 [25]; Oh and Batory reinvented and implemented it in 2017 using classical BDDs [30]. Since then other SAT technologies were tried [11,32,40]. (A #SAT solver is a variant of a SAT solver: instead of finding a solution of ϕ efficiently, #SAT counts ϕ solutions efficiently.) The most scalable version today is by Heradio et al. and uses reduced BDDs [21], which in itself is surprising as for about a decade, SAT technologies have dominated feature model analysis.❱

Given the ability to URS a SPL colossal product space, how can a near-optimal product for a given workload be found? That's next.

6   Performance Configuration Space (PCS) Graphs

Let $\mathbb{C}$ denote the product space of ϕ, where $\eta = |\phi| = |\mathbb{C}|$. Imagine that for every product $P \in \mathbb{C}$ we predict or measure a **performance metric** $\$(P)$ for a given benchmark. By "performance metric", we mean any non-functional property of interest of P (response time, memory size, energy consumption, throughput, etc.). A small $\$$ value is good (efficient) and a large $\$$ value is bad (inefficient). An optimal product P_{best} in $\mathbb{C}$ has the smallest $\$$ metric:[6]

$$\exists P_{best} \in \mathbb{C} \; : \; \big(\forall P \in \mathbb{C} \; : \; \$(P_{best}) \leq \$(P)\big) \tag{9}$$

For large $\mathbb{C}$, creating all $(P, \$(P))$ pairs is impossible... **but imagine that we could do so.** Further, let's normalize the range of $\$$ values: Let $\$(P_{best}) = 0$ be the best performance metric and let $\$(P_{worst}) = 1$ be the worst. Now sort the $(P, \$(P))$ pairs in increasing $\$(P)$ order where $\$(P_{best}) = 0$ is first and $\$(P_{worst}) = 1$ is last, and plot them. The result is a **Performance Configuration/Product Space** (PCS) graph, Fig. 4a. This graph suggests that PCS graphs are continuous; they are not. PCS graphs are stair-stepped, discontinuous and non-differentiable [27] because consecutive products on the X-axis encode discrete decisions (features) that can make discontinuous jumps in performance, Fig. 4b.

---

[6] To maximize a metric, negate it.

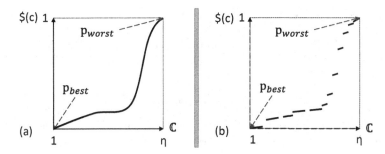

Fig. 4. Normalized PCS graphs.

Example ❮Suppose product P_i has feature F and F is replaced by G in P_{i+1}. If F increments performance by .01, say, and G increments performance by .20, there will be a discontinuity from $\$(P_i)$ and $\$(P_{i+1})$ in a PCS graph.❯

Note ❮Every PCS graph is monotonically non-decreasing. The latter means that consecutive products on the X-axis, like P_i and P_{i+1}, must satisfy $\$(P_i) \leq \$(P_{i+1})$. Many products in $\mathbb{C}$ may have indistinguishable performance values/metrics because their differing features have *no* impact on performance, leading to $\$(P_i) = \$(P_{i+1})$.❯

Random Search ($\mathbb{RS}$) is a family of numerical optimization algorithms that can be used on functions that are discontinuous and non-differentiable [10,50]. The simplest of all $\mathbb{RS}$ algorithms is the *Best-of-n-Samples* below. Here we use URS for sampling:

```
1. Initialize x with a random product in the search space.
2. Until a termination criterion is met (n − 1 samples) repeat:
   2.1 Sample a new product y in the search space.
   2.2 If $(y)<$(x) set x=y.
3. Return x.
```

Listing 1.1. Best-of-n-Samples

How accurate is the returned product? An answer can be derived by exploiting a PCS graph's monotonicity, next.

7   Analysis of Best-of-n-Samples

The X-axis of a PCS graph (i.e., the product space) can be approximated by the real unit interval $\mathbb{I} = [0..1]$ when $\eta > 2000$ [30]. $\mathbb{I}$ emerges from the limit:

$$\lim_{\eta \to \infty} \frac{1}{\eta} \cdot \left[1..\eta\right] = \lim_{\eta \to \infty} \left[\frac{1}{\eta}..\frac{\eta}{\eta}\right] = [0..1] = \mathbb{I} \tag{10}$$

Randomly select a product in $\mathbb{C}$, i.e., one point in $\mathbb{I}$. URS means each point in $\mathbb{I}$ is equally likely to be chosen. It follows that *on average* the selected product $p_{1,1}$ partitions $\mathbb{I}$ in half:

$$p_{1,1} = \int_0^1 x \cdot dx = \frac{1}{2} \tag{11}$$

Now randomly select n products from $\mathbb{C}$. *On average* n points partition $\mathbb{I}$ into $n+1$ equal-length regions. The kth-best product out of n, denoted $p_{k,n}$, has rank $\frac{k}{n+1}$, where the $k \cdot \binom{n}{k}$ term below is the normalization constant [5]:[7]

$$p_{k,n} = k \cdot \binom{n}{k} \cdot \int_0^1 x^k \cdot (1-x)^{n-k} \cdot dx = \frac{k}{n+1} \tag{12}$$

The left-most selected product, which is a near-optimal product $P_{nearOpt} = p_{1,n}$, is an average distance $\frac{1}{n+1}$ from the optimal $P_{best} = 0$ by Eq. (12):

$$p_{1,n} = \frac{1}{n+1} \tag{13}$$

Let's pause to understand this result. Look at Fig. 5. As the sample set size n increases (Fig. (a)$\to$Fig. (c)), $P_{nearOpt}$ progressively moves closer to P_{best} at $X = 0$. *If $n = 99$ samples are taken, $P_{nearOpt}$ on average will be 1% from P_{best} in the ranking along the X-axis.*

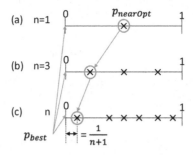

Fig. 5. Convergence to P_{best} by increasing Sample Size (n).

Note: *None* of the Eqs. (11–13) reference η or $|\mathbb{C}|$; both disappeared when we took the limit in (10). This means (11–13) predict sampled ranks (that is, X-axis

---

[7] Equation (12) is an example of the Beta function [47].

ranks) for an *infinite-sized space*. Taking n=99 samples on *any* colossal product space, on average $P_{nearOpt}$ will be 1% from P_{best} in the ranking. It is only for minuscule product spaces, $\eta \leq 2000$, where predictions by Eqs. (11–13) will be low [30]. Such small product spaces are enumerable anyway, and not really of interest to us.

Third, how accurate is the $p_{1,n} = \frac{1}{n+1}$ estimate? Answer: the standard deviation of $p_{1,n}$, namely $\sigma_{1,n}$ [5,29], can be computed from $v_{1,n}$, the second moment of $p_{1,n}$:

$$v_{1,n} = 1 \cdot \binom{n}{1} \cdot \int_0^1 x^2 \cdot (1-x)^{n-1} \cdot dx = \frac{2}{(n+1) \cdot (n+2)}$$

$$\sigma_{1,n} = \sqrt{v_{1,n} - p_{1,n}^2} = \sqrt{\frac{2}{(n+1) \cdot (n+2)} - \left(\frac{1}{n+1}\right)^2} \qquad (14)$$

Observe $p_{1,n}$ is almost equal to $\sigma_{1,n}$. Figure 6 plots the percentage difference between the two:

$$\%\texttt{diff} = 100 \cdot \left(\frac{p_{1,n}}{\sigma_{1,n}} - 1\right) \qquad (15)$$

For $n = 80$ samples $p_{1,n}$ is 1.2% higher than $\sigma_{1,n}$ which is itself small. *For $n \geq 100$ there is no practical difference between $p_{1,1}$ and $\sigma_{1,n}$.* Stated differently, URS offers remarkable good accuracy and variance with $n \geq 100$ [30].

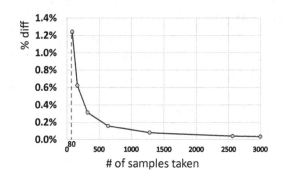

Fig. 6. %diff plot.

Bottom Line. To find a near-optimal product in colossal product space, take a uniform-random sample set of size n, predict or measure the performance of each product, and return the best performing product as it will be $\frac{100}{n+1}\%$ away, with variance approximately $\frac{100}{n+1}\%$, from the optimal product, P_{best}, in the space.

Percentiles. Readers may have noticed that our ranking is how 'close' $P_{nearOpt}$ is X-axis-based from P_{best}, where the more typical notion is Y-axis-based, i.e., the fraction $\$(P_{nearOpt})$ is from $\$(P_{best})$. The utility of PCS graphs is this: *optimizing X also optimizes Y.*

A common X-axis metric in statistics is a **percentile**, see Fig. 7 [28]. Candidates are lined up and the percentage of candidates that are "shorter" than You (the blue person) is computed.

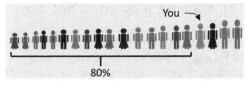

Fig. 7. Percentiles.

You are in the (top) 80% percentile. In performance optimization, we want to be in the lowest percentile, ideally <1% means "in the top <1 percentile".

8   Another Benefit of URS in Product Optimization

Prior to performance models for and URS of colossal product spaces, URS was compared with early results on small SPLs that used performance models [30]. The performance model contestants were **Sankar2015** [37] and **Siegmund2012** [42], which at the time were the best models to date.

Figure 8 shows two non-PCS graphs: the X-axis is the number **n** of samples taken to form a prediction model or a Best-of-n-Samples result and the Y-axis is the fraction distance of their returned $P_{nearOpt}$ to P_{best}.

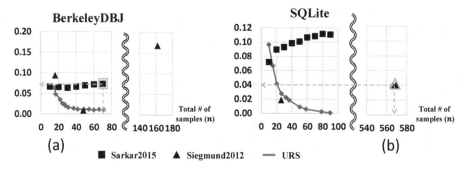

Fig. 8. Comparison of URS with existing performance models in 2017.

In short:

- The accuracy of both performance models did *not* improve with increasing N, unlike URS which progressively improves;
- URS obtained the same accuracy with less work (fewer samples) than both **Sarkar2015** (see ■ in Fig. 8a) and **Siegmund2012** (see ▲ in Fig. 8b); and
- URS obtained better results for the same work (see ■ and ▲ above).

Other similar results are reported in [30]. However, these results are outdated and need to be refreshed with the latest performance model technologies and URS technologies; at best they are provisional and suggest future work.

9   Choosing a Sample Set Size

An open problem with non-URS methods is: what sample set size is needed to find $P_{nearOpt}$ with a given accuracy? As there is no formal analysis of non-uniform sampling, it is not known how to answer this question. However, URS has an answer. The following elegant derivation is by Heradio [20], better than [31].

Confidence assertions are of the form: with 90% probability $P_{nearOpt}$ will be within the top ρ percentile. Let ρ be the desired percentile (e.g., top 1% has $\rho = .01$). In one random selection, we have probability ρ that a desired product was selected and $(1 - \rho)$ that it was not. After n selections, we have probability $(1 - \rho)^n$ that *no* selections were desirable and $1 - (1 - \rho)^n$ that at least one of them is. Let c denote the confidence (probability) that after n selections $P_{nearOpt}$ is in the top ρ percentile:

$$c = 1 - (1 - \rho)^n \tag{16}$$

Solving for n:

$$n = \frac{\ln(1 - c)}{\ln(1 - \rho)} \tag{17}$$

The table of Fig. 9 lists the sample set size to use for a given confidence (c) and accuracy (ρ) *no matter how colossal the space*. **Note**: 90% and 95% are common degrees of confidence; 99.7% is known as *near certainty* since it encompasses 3σ, virtually all values [51]:

n=sample set size	%confidence			
%accuracy	90.0%	95.0%	98.0%	99.7%
5.00%	45	58	76	113
4.00%	56	73	96	142
3.00%	76	98	128	191
2.00%	114	148	194	288
1.00%	229	298	389	578
0.50%	459	598	780	1159
0.30%	766	997	1302	1933
0.20%	1150	1496	1954	2902
0.10%	2301	2994	3910	5806

Fig. 9. Sample set size to achieve `%accuracy` with `%confidence`.

Example: A product:

- in the top 5% is returned in 45 samples with 90% confidence;
- in the top 2% is returned in 148 samples with 95% confidence; and
- in the top .20% is returned in 1954 samples with 98% confidence.

Equation (16) has three variables; given values of two, one can solve for the third. The previous discussion showed how to determine n given confidence c and accuracy ρ. Here are the two other possibilities:

Given c and n, what is the expected accuracy ρ? Solving (16) for ρ:

$$\rho = 1 - (1 - c)^{\frac{1}{n}} \tag{18}$$

The table of Fig. 10 has rows for confidence c values, columns are the number of samples taken n, and entries are the accuracy ρ of returned answers:

%ρ	n = sample set size						
%confidence	25	50	100	200	400	800	1600
90.0%	8.80%	4.50%	2.28%	1.14%	0.57%	0.29%	0.14%
95.0%	11.29%	5.82%	2.95%	1.49%	0.75%	0.37%	0.19%
98.0%	14.49%	7.53%	3.84%	1.94%	0.97%	0.49%	0.24%
99.7%	20.73%	10.97%	5.64%	2.86%	1.44%	0.72%	0.36%

Fig. 10. Expected accuracy ρ given c and n.

Example ❮Suppose a total of n = 100 samples are to be taken and 95% confidence is desired in an answer. The returned solution has accuracy in the top 2.95% of all solutions.❯

Given n and ρ, what is expected confidence c? Eq. (16) is already solved for c and is repeated below:

$$c = 1 - (1 - \rho)^n$$

The table of Fig. 11 has rows for accuracy values ρ, columns are the total number of samples taken n, and entries are the confidence c of returned answers:

%c = confidence	n = number of samples						
%ρ = accuracy	25	50	100	200	400	800	1600
4.000%	63.96%	87.01%	98.31%	99.97%	100.00%	100.00%	100.00%
2.000%	39.65%	63.58%	86.74%	98.24%	99.97%	100.00%	100.00%
1.000%	22.22%	39.50%	63.40%	86.60%	98.20%	99.97%	100.00%
0.500%	11.78%	22.17%	39.42%	63.30%	86.53%	98.19%	99.97%
0.250%	6.07%	11.76%	22.14%	39.38%	63.26%	86.50%	98.18%
0.125%	3.08%	6.06%	11.76%	22.13%	39.37%	63.24%	86.48%

Fig. 11. Expected confidence c given ρ and n.

Example ❮Taking n = 100 samples and wanting accuracy $\rho = 1\%$, the confidence of a returned answer is 63.4%. That is, there is a 63.4% chance that the returned answer is within the top 1% of all products.❯

There are other analyzes for the Best-of-n-Samples algorithm, like how many samples are needed to have two samples returned in the top $\rho\%$ with c% confidence, is much like the above. Also, a recursive search of a space is another interesting possibility – performing a search and using the data collected to reduce the size of the space to search in the next recursion. A noticeable improvement in $P_{nearOpt}$ was observed by recursive searching [30] – with two important caveats. We have not yet determined how to guarantee that P_{best} is not pruned in a space restriction, nor do we have mathematics to compute the confidence of returned results. These remain open problems.

10   Given a Limit of n Samples...

Benchmarking is *by far* the greatest cost in sampling. Suppose a client is willing to pay the cost of benchmarking 100 samples to find a near optimal product for a particular precision ρ and confidence c.

Question: Would it be better to conduct one experiment E of 100 samples for a given ρ accuracy, or two experiments E_1 and E_2 of 50 examples each again with the same ρ accuracy, and take the best result? In the latter case, would the confidence change by using two experiments?

Answer: There is no difference! It is not difficult to see that the $P_{nearOpt}$ answer in either case would be the same: $P_{nearOpt}$ would be the result of experiment E_1 or E_2, and would be the best-of-both result.

It is a bit harder to see is how the confidence of the two-experiment result is the same as a one-experiment result. Let c be the confidence of both E_1 and E_2. The confidence c2 we would have in the result of taking the best-of-both experiments is weighted. Namely, sum the product of confidences where at least one experiment succeeds:

$$c2 = c \cdot c \cdot 1 \ + \ c \cdot (1-c) \cdot 1 + (1-c) \cdot c \cdot 1 + (1-c) \cdot (1-c) \cdot 0$$
$$= 2 \cdot c - c^2 \tag{19}$$

Let $c(n, \rho) = 1 - (1 - \rho)^n$, Eq. (16), be the confidence of an experiment for a fixed n and ρ. It is easy to prove the following equality that shows the confidence of a single experiment with $2 \cdot n$ samples and ρ accuracy and the confidence of two smaller experiments with n samples and ρ accuracy, Eq. (19), are the same:

$$2 \cdot c(n, \rho) \ - \ c(n, \rho)^2 \ = \ c(2 \cdot n, \rho) \tag{20}$$

10.1   PCS Graphs of Real SPLs

What do real PCS graphs look like? This is not a fundamental question, but one asked out of curiosity. Several small SPLs were analyzed by Siegmund et al. [41,42] that took several months of benchmarking:

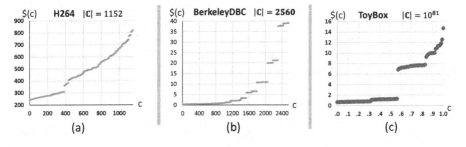

Fig. 12. H264, BerkeleyDBC, and ToyBox PCS graphs.

- **H264** is a video encoder library for H.264/MPEG-4 AVC format written in C. With 16 features and 1152 configurations, Sintel trailer encoding times were measured, see Fig. 12a and
- **BerkeleyDBC** is an embedded database system written in C. With 18 features and 2560 configurations, benchmark response times were measured. Note its multiple "*stairs*" or vertical leaps. See Fig. 12b.

Figure 12a–b are **Complete PCS graphs** – meaning all products are plotted. This is possible when configuration spaces are tiny. But what about SPLs with colossal spaces? What then? A number of techniques were tried, and the simplest performed best:

- Randomly select $n = 100$ or $n = 200$ configurations, as 100–200 points are sufficient resolution for a graph in a paper,
- Predict the performance or build-and-benchmark each sample,
- Sort the samples from best-performing to worst,
- Let p_i be the ith best performance. Plot a PCS graph using the n points $\{(x_i, y_i)\}_{i=1}^{n} = \{(\frac{i}{n+1}, p_i)\}_{i=1}^{n}$.

Example. ToyBox 0.7.5 provides Android systems with a collection of Linux command-line utilities within a single executable. It has 316 features and 10^{81} configurations [45]. Build size was measured. Its PCS graph, Fig. 12c, was produced with $n = 100$, although the graph for $n = 200$ was identical.

11   Future Work and Next Steps

There is a hunger in Software Engineering research for more scientific approaches to be used, where mathematics can help solve fundamental design problems. The use of mathematics is evident in the work of Börger et al. on ASMs and the JBook [44]; so too in the area of SPLs. Software design indeed has a mathematical foundation, but perhaps not how Dijkstra, Hoare, and Wirth initially envisioned. Science must deliver quite a lot before it can overcome Cowboy Programming [48]. The Science of Software Design will answer questions that were unanswerable previously.

This holds for finding near-optimal products in colossal SPL product spaces, a practical problem whose roots are found in early work on SWD. Given the ability to URS such spaces, an entire world of prior results on $\mathbb{R}_\mathbb{S}$ is now applicable. The simplest $\mathbb{R}_\mathbb{S}$ algorithm, Best-of-n-Samples, can answer scientific questions that prior approaches could not. Namely, given any two of (confidence of answer, accuracy of answer, and number of samples to take), the third can be computed. Perhaps other $\mathbb{R}_\mathbb{S}$ algorithms may be analyzable as well.

Software Engineering research is fad-driven – the latest is **Machine Learning** (ML). ML also can provide answers to questions that couldn't be answered before. We showed in Sect. 7 that near-optimal results can be accompanied with accuracy or confidence metrics – to give precision about returned results that could not be determined before. Or that the number of samples to take is no longer a guess – it can be precisely computed. And in Sect. 8, URS can also provide more accurate answers than performance models with less work (fewer samples), although these results need to be refreshed as they used small product spaces (what was available at that time). Today's open question is whether the provisional results in this paper scale to colossal spaces.

In this paper, URS may have been offered unintentionally as a tool to solve all analysis problems. Far from the truth, URS is but one in an ever-increasing sophisticated arsenal of techniques that can be used. Coupled with domain-specific knowledge, URS tools will be even better. URS will likely become a lower-bound on what can be accomplished and accepted (w.r.t. accuracy, confidence, and work) in future work. If so, we have indeed made progress.

To Egon. You and your work continue to inspire me and others. Thank you.

Acknowledgments. We thank the referees for their helpful comments on this paper.

References

1. Agrawal, S., Chaudhuri, S., Narasayya, V.R.: Automated selection of materialized views and indexes in SQL Databases. In: VLDB (2000)
2. Apel, S., Batory, D., Kästner, C., Saake, G.: Feature-Oriented Software Product Lines. Springer, Heidelberg (2013). https://doi.org/10.1007/978-3-642-37521-7
3. Batory, D.: Feature Models, Grammars and Propositional Formulas. In: SPLC (2005)
4. Batory, D.: Automated Software Design Volume 1. Lulu.com (2020)
5. Batory, D.: Automated Software Design Volume 2. in development (2022)
6. Batory, D., Börger, E.: Modularizing theorems for software product lines: the JBook case study. JUCS **14**(12) (2008)
7. Batory, D., O'Malley, S.: The Design and Implementation of Hierarchical Software Systems with Reusable Components. ACM TOSEM (1992)
8. Batory, D., Singhal, V., Thomas, J., Sirkin, M.: Scalable software libraries. In: ACM SIGSOFT (1993)
9. Batory, D.S., Gotlieb, C.C.: A unifying model of physical databases. In: ACM TODS (1982)
10. Bergstra, J., Bengio, Y.: Random search for hyper-parameter optimization. J. Mach. Learn. Res. (2012)
11. Chakraborty, S., Fremont, D.J., Meel, K.S., Seshia, S.A., Vardi, M.Y.: On parallel scalable uniform SAT witness generation. In: Baier, C., Tinelli, C. (eds.) TACAS 2015. LNCS, vol. 9035, pp. 304–319. Springer, Heidelberg (2015). https://doi.org/10.1007/978-3-662-46681-0_25
12. Chakraborty, S., Meel, K., Vardi, M.: A scalable and nearly uniform generator of SAT witnesses. In: CAV (2013)
13. Chaudhuri, S.: An overview of query optimization in relational systems. In: PODS (1998)
14. Devore, J.: Probability and Statistics for Engineering and the Sciences. Cengage Learning (2021)
15. Dijkstra, E.W.: The Structure of 'THE'-multiprogramming System. CACM (1968)

16. Emmanuel, J.: Integer Linear Programming - Binary (0–1) Variables (2021). https://www.youtube.com/watch?v=-3my1TkyFiM
17. Guo, J., Czarnecki, K., Apel, S., Siegmund, N., Wasowski, A.: Variability-aware performance prediction: a statistical learning approach. In: ASE (2013)
18. Guo, J., White, J., Wang, G., Li, J., Wang, Y.: A genetic algorithm for optimized feature selection with resource constraints in software product lines. J. Syst. Softw. **84**, 2208 (2011)
19. Henard, C., Papadakis, M., Harman, M., Le Traon, Y.: Combining multi-objective search and constraint solving for configuring large software product lines. In: ICSE (2015)
20. Heradio, R.: Derivation of Sample Set Size. Private Correspondence with Batory (2020)
21. Heradio, R., Fernandez-Amoros, D., Galindo, J., Benavides, D., Batory, D.: Uniform and Scalable Sampling of Highly Configurable Systems. In Submitted (2021)
22. Hodes, L.: Solving problems by formula manipulation in logic and linear inequalities. Artif. Intell. **3**, 165–174 (1972)
23. Jamshidi, P., et al.: Transfer learning for performance modeling of configurable systems: an exploratory analysis. In: ASE (2017)
24. Kaltenecker, C., Grebhahn, A., Siegmund, N., Guo, J., Apel, S.: Distance-based sampling of software configuration spaces. In: ICSE (2019)
25. Knuth, D.E.: The Art of Computer Programming, Fascicle 1: Bitwise Tricks and Techniques; Binary Decision Diagrams, vol. 4. Addison-Wesley Professional, New York (2009)
26. Krieter, S., Thüm, T., Schulze, S., Schröter, R., Saake, G.: Propagating configuration decisions with modal implication graphs. In: ICSE (2018)
27. Marker, B., Batory, D., van de Geijn, R.: Understanding performance stairs: elucidating heuristics. In: ASE (2014)
28. MathIsFun. Percentiles. https://www.mathsisfun.com/data/percentiles.html
29. MathIsFun. Standard Deviation Formulas (2019). https://www.mathsisfun.com/data/standard-deviation-formulas.html
30. Oh, J., Batory, D., Myers, M., Siegmund, N.: Finding near-optimal configurations in product lines by random sampling. In: FSE (2017)
31. Oh, J., Gazzillo, P., Batory, D., Heule, M., Myers, M.: Percentile Calculations for Randomly Searching Colossal Product Spaces. Technical Report TR-18-05, Dept. of Computer Science, University of Texas at Austin (2018)
32. Oh, J., Gazzillo, P., Batory, D., Heule, M., Myers, M.: Scalable Uniform Sampling for Real-World Software Product Lines. Technical Report TR-20-01, Dept. of Computer Science, University of Texas at Austin (2020)
33. Plazar, Q., Acher, N., Perrouins, G., Devroey, X., Cordy, M.: Uniform sampling of SAT solutions for configurable systems: are we there yet? In: Software Testing, Verification, and Validation (2019)
34. Saibian, S.: Sbiis Saibian's Large Number Site. https://sites.google.com/site/largenumbers/home
35. Saibian, S.: The Largest Numbers Theoretically Possible in Modern Cosmology. https://sites.google.com/site/largenumbers/home/2-1/Largest_Numbers_in_Science
36. Saibian, S.: Larger Numbers in Probability, Statistics, and Combinatorics (2021). https://sites.google.com/site/largenumbers/home/2-1/Large_Numbers_in_Probability
37. Sarkar, A., Guo, J., Siegmund, N., Apel, S., Czarnecki, K.: Cost-efficient sampling for performance prediction of configurable systems. In: ASE (2015)

38. Sayyad, A.S., Ingram, J., Menzies, T., Ammar, H.: Scalable product line configuration: a straw to break the camel's back. In: ASE (2013)
39. Sayyad, A.S., Menzies, T., Ammar, H.: On the value of user preferences in search-based software engineering: a case study in software product lines. In: ICSE (2013)
40. Sharma, S., Gupta, R., Roy, S., Meel, K.: Knowledge compilation meets uniform sampling. In: Logic for Programming, Artificial Intelligence, and Reasoning (LPAR) (2018)
41. Siegmund, N., et al.: Dataset for Siegmund (2012). http://fosd.de/SPLConqueror
42. Siegmund, N., et al.: Predicting performance via automated feature-interaction detection. In: ICSE (2012)
43. Siegmund, N., Grebhahn, A., Apel, S., Kästner, C.: Performance-influence models for highly configurable systems. In: FSE (2015)
44. Stärk, R., Schmid, J., Börger, E.: Java and the Java Virtual Machine: Definition, Verification, Validation. Springer-Verlag, Heidelberg (2001). https://doi.org/10.1007/978-3-642-59495-3
45. Toybox Website (2018). http://landley.net/toybox/
46. Villanueva, J.: How Many Atoms Are There in the Universe? https://www.universetoday.com/36302/atoms-in-the-universe/
47. Wikipedia. Beta Distribution. https://en.wikipedia.org/wiki/Beta_distribution
48. Wikipedia. Cowboy Coders. https://en.wikipedia.org/wiki/Cowboy_coding
49. Wikipedia. Lego. https://en.wikipedia.org/wiki/Lego
50. Random Search. https://en.wikipedia.org/wiki/Random_search
51. Wikipedia. 68–95-99.7 Rule (2019). https://en.wikipedia.org/wiki/68%E2%80%9395%E2%80%9399.7_rule
52. Wikipedia. Integer Programming (2021). https://en.wikipedia.org/wiki/Integer_programming
53. Wirth, N.: Program Development by Stepwise Refinement. CACM (1971)
54. Zhang, Y., Guo, J., Blais, E., Czarnecki, K.: Performance prediction of configurable software systems by fourier learning. In: ASE (2015)

Semantic Splitting of Conditional Belief Bases

Christoph Beierle[1(✉)], Jonas Haldimann[1], and Gabriele Kern-Isberner[2]

[1] FernUniversität in Hagen, 58084 Hagen, Germany
{christoph.beierle,jonas.haldimann}@fernuni-hagen.de
[2] Technische Universität Dortmund, 44227 Dortmund, Germany
gabriele.kern-isberner@cs.tu-dortmund.de

Abstract. An important concept for nonmonotonic reasoning in the context of a conditional belief base $\mathcal{R}$ is syntax splitting, essentially stating that taking only the syntactically relevant part of $\mathcal{R}$ into account should be sufficient. In this paper, for the semantics of ordinal conditional functions (OCF) of $\mathcal{R}$, we introduce the notion of semantic splitting of $\mathcal{R}$ where the combination of models of sub-belief bases of $\mathcal{R}$ corresponds to the models of $\mathcal{R}$. While this is not the case for all OCF models, we show that for c-representations which are a subclass of all OCFs governed by the principle of conditional preservation, every syntactic splitting is also a semantic splitting. Furthermore, for the semantics of c-representations, we introduce constraint splittings of a conditional belief base and show that they fully capture and go beyond syntax splittings, thus allowing for additional belief base splittings that enable the computation of models locally from sub-belief bases.

Keywords: Conditional · Conditional belief base · Ordinal conditional function · Ranking function · Syntax splitting · Semantic splitting

1 Introduction

In logic-based knowledge representation, conditionals of the form $(B|A)$ are used to express defeasible rules "*If A then usually B*". In contrast to a material implication, a conditional allows for exceptions. A propositional interpretation, also called (possible) world, ω can verify the conditional ($\omega \models A \wedge B$), it can falsify it ($\omega \models A \wedge \neg B$), or the conditional is not applicable ($\omega \models \neg A$). Many different semantics have been proposed for a conditional belief base $\mathcal{R}$ consisting of a set of conditionals, e.g. [1,11,15,16,18,22]. Among these semantics, ordinal conditionals functions (OCFs) [26], also called ranking functions, assign a natural number to each world representing the degree of surprise when seeing this world. There are various semantics that select a single ranking model or a class of ranking models to be taken into account for nonmonotonic reasoning in the context, e.g. [16,19,22,24].

The basic idea of syntax splitting [21,23,25] is that for entailments that involve only a part of the signature of $\mathcal{R}$ and $\mathcal{R}$ splits over this signature, only

© Springer Nature Switzerland AG 2021
A. Raschke et al. (Eds.): Börger Festschrift, LNCS 12750, pp. 82–95, 2021.
https://doi.org/10.1007/978-3-030-76020-5_5

the part of $\mathcal{R}$ corresponding to the subsignature should be relevant. Here, we study splittings of conditional belief bases together with splittings on the level of their models. Recently, it has been shown that skeptical c-inference taking all c-representations of $\mathcal{R}$ into account [5,6] is fully compatible with syntax splitting [20]. We introduce the concept of semantic splitting of a belief base $\mathcal{R}$ and show that for c-representations every syntax splitting of $\mathcal{R}$ is also a semantic splitting. Furthermore, we propose the notion of constraint splitting and show that every constraint splitting of $\mathcal{R}$ is also a semantic splitting, and that constraint splittings fully capture and properly exceed syntax splittings.

This paper relies on and extends our previous work on conditional belief bases. In [7–9], we employed Prof. Egon Börger's method for high-level system design and analysis [12–14], which is based an Gurevich's notion of abstract state machines (ASMs) [17], for specifying, refining, and verifying various belief management operations on conditional belief bases. Furthermore, one of the authors of this paper, Christoph Beierle, benefited very much from meeting Prof. Egon Börger when he gave a talk at IBM Germany in Stuttgart in April 1990 on a novel logic method for the definition of semantics of programming languages, and from joint work (e.g. [2–4]) when ASMs were still called evolving algebras. We are glad to contribute this paper on conditional belief bases to the Festschrift dedicated to Prof. Egon Börger on the occasion of his 75th birthday.

The rest of this paper is organized as follows. In Sect. 2, we summarize the background of conditional logics and ranking functions. In Sect. 3, syntax and semantic splittings of belief bases are presented, and in Sect. 4, we recall aspects from a compilation scheme for computing c-representations of a belief base. These aspects are used for establishing the concept of constraint splitting introduced and investigated in Sect. 5. In Sect. 6, we conclude and point out future work.

2 Background: Conditional Logic

Let $\mathcal{L}(\Sigma)$ be the propositional language over a finite signature Σ. The language may be denoted by $\mathcal{L}$ if the signature is clear from context. The formulas of $\mathcal{L}$ will be denoted by letters $A, B, C, \dots$. We write AB for $A \wedge B$ and $\overline{A}$ for $\neg A$. We identify the set of all complete conjunctions over Σ with the set Ω_Σ of possible worlds over $\mathcal{L}_\Sigma$. For $\omega \in \Omega$ and $A \in \mathcal{L}$, $\omega \models A$ means that A holds in ω. The set of worlds satisfying A is $\Omega_A = \{\omega \mid \omega \models A\}$. Two formulas A, B are $equivalent$, denoted as $A \equiv B$, if $\Omega_A = \Omega_B$.

By introducing a new binary operator $\mid$, we obtain the set $(\mathcal{L} \mid \mathcal{L})_\Sigma = \{(B|A) \mid A, B \in \mathcal{L}(\Sigma)\}$ of $conditionals$ over $\mathcal{L}(\Sigma)$. Again, Σ may be omitted. As semantics for conditionals, we use $ordinal\ conditional\ functions\ (OCF)$, also called ranking functions, first introduced (in a more general form) in [26]. An OCF (over Σ) is a function $\kappa : \Omega_\Sigma \to \mathbb{N}_0$ expressing degrees of (im-)plausibility of possible worlds where a lower degree denotes "less surprising". At least one world must be regarded as being normal; therefore, $\kappa(\omega) = 0$ for at least one $\omega \in \Omega$. Each κ uniquely extends to a function mapping formulas to $\mathbb{N}_0 \cup \{\infty\}$ given by

$\kappa(A) = \min\{\kappa(\omega) \mid \omega \models A\}$ where $\min \emptyset = \infty$. An OCF κ *accepts* a conditional $(B|A)$, written $\kappa \models (B|A)$, if the verification of the conditional is less surprising than its falsification, i.e., if $\kappa(AB) < \kappa(A\overline{B})$. A finite set $\mathcal{R} \subseteq (\mathcal{L} \mid \mathcal{L})_{\Sigma}$ of conditionals is called a *belief base* over Σ. An OCF κ accepts a belief base $\mathcal{R}$ if κ accepts all conditionals in $\mathcal{R}$, and $\mathcal{R}$ is *consistent* if an OCF accepting $\mathcal{R}$ exists. $Mod_{\Sigma}(\mathcal{R})$ denotes the set of all OCFs κ accepting $\mathcal{R}$.

3 Syntax Splitting and Semantic Splitting

Let us start with defining what a syntax splitting of a belief base is.

Definition 1 (syntax splitting of $\mathcal{R}$). *Let $\mathcal{R}$ be a belief base over a signature Σ. A partition $\{\mathcal{R}_1, \ldots, \mathcal{R}_n\}$ of $\mathcal{R}$ is a syntax splitting of $\mathcal{R}$ with respect to $\{\Sigma_1, \ldots, \Sigma_n\}$ if $\{\Sigma_1, \ldots, \Sigma_n\}$ is a partition of Σ such that $\mathcal{R}_i$ is a belief base over Σ_i, for $i = 1, \ldots, n$.*

In order to investigate whether a composition of models of sub-belief bases corresponding to the union $\mathcal{R} = \mathcal{R}_1 \cup \cdots \cup \mathcal{R}_n$ for syntax splitting exists, we introduce the notion of model combination for sets of ranking functions.

Definition 2 (model combination). *Let M_1, M_2 be sets of OCFs over Σ. The* model combination *of M_1 and M_2, denoted by $M_1 \oplus M_2$, is given by*

$$M_1 \oplus M_2 = \{\kappa \mid \kappa(\omega) = \kappa_1(\omega) + \kappa_2(\omega) \text{ for any } \omega \in \Omega_{\Sigma}, \kappa_1 \in M_1, \kappa_2 \in M_2\}.$$

Definition 3. *A (model based) semantics Sem for conditional belief bases is a function mapping every belief base $\mathcal{R}$ over Σ to a set of models $Mod_{\Sigma}^{Sem}(\mathcal{R}) \subseteq Mod_{\Sigma}(\mathcal{R})$.*

A semantic splitting of $\mathcal{R}$ depends on the combination of models given by an OCF-based semantics.

Definition 4 (semantic splitting of $\mathcal{R}$). *Let $\mathcal{R}$ be a belief base over a signature Σ. A partition $\{\mathcal{R}_1, \ldots, \mathcal{R}_n\}$ of $\mathcal{R}$ is a semantic splitting of $\mathcal{R}$ for a semantic Sem if*

$$Mod_{\Sigma}^{Sem}(\mathcal{R}) = Mod_{\Sigma}^{Sem}(\mathcal{R}_1) \oplus \cdots \oplus Mod_{\Sigma}^{Sem}(\mathcal{R}_n).$$

A semantic *Sem* may satisfy the following postulate:

Postulate (SynSem). *If $\{\mathcal{R}_1, \ldots, \mathcal{R}_n\}$ is a syntax splitting of $\mathcal{R}$, then it is also a semantic splitting of $\mathcal{R}$ for Sem.*

Example 1. System P is an axiom system stating desirable properties for non-monotonic reasoning with conditionals [1,22]. It also induces a semantic that maps a belief base $\mathcal{R}$ to all its models, i.e., $Mod_{\Sigma}^{System \ P}(\mathcal{R}) = Mod_{\Sigma}(\mathcal{R})$. This semantics does not satisfy (SynSem) which can be illustrated with $\mathcal{R} = \mathcal{R}_1 \cup \mathcal{R}_2$

and $\mathcal{R}_1 = \{(a|\top)\}$ and $\mathcal{R}_1 = \{(a|\top)\}$. Obviously, $\{\mathcal{R}_1, \mathcal{R}_2\}$ is a syntax splitting of $\mathcal{R}$. The ranking function

$$\kappa_1 = \{ab \mapsto 1, a\bar{b} \mapsto 0, \bar{a}b \mapsto 1, \bar{a}\bar{b} \mapsto 1\} \text{ accepts } \mathcal{R}_1, \text{ and}$$
$$\kappa_2 = \{ab \mapsto 1, a\bar{b} \mapsto 1, \bar{a}b \mapsto 0, \bar{a}\bar{b} \mapsto 1\} \text{ accepts } \mathcal{R}_2,$$

but $\kappa_1 + \kappa_2 = \{ab \mapsto 2, a\bar{b} \mapsto 1, \bar{a}b \mapsto 1, \bar{a}\bar{b} \mapsto 2\}$ is not even a ranking function and would also not model $\mathcal{R}$ if it were normalized by reducing all ranks by 1.

Among the OCF models of $\mathcal{R}$, c-representations are special models obtained by assigning an individual impact to each conditional and generating the world ranks as the sum of impacts of falsified conditionals. For an in-depth introduction to c-representations and their use of the principle of conditional preservation ensured by respecting conditional structures, we refer to [18,19]. The central definition is the following:

Definition 5 (c-representation [18]). *A c-representation of a belief base $\mathcal{R}$ over Σ is a ranking function $\kappa_{\vec{\eta}}$ constructed from $\vec{\eta} = (\eta_1, \dots, \eta_n)$ with integer impacts $\eta_i \in \mathbb{N}_0, i \in \{1, \dots, n\}$, assigned to each conditional $(B_i|A_i)$ such that κ accepts $\mathcal{R}$ and is given by:*

$$\kappa_{\vec{\eta}}(\omega) = \sum_{\substack{1 \leqslant i \leqslant n \\ \omega \models A_i \overline{B_i}}} \eta_i \tag{1}$$

We will denote the set of all c-representations of $\mathcal{R}$ by $Mod_{\Sigma}^C(\mathcal{R})$.

While for each consistent $\mathcal{R}$, the system Z ranking function κ^Z [16] is uniquely determined, there may be many different c-representations of $\mathcal{R}$. These can conveniently be characterized via the solutions of a constraint satisfaction problem (CSP). In [5], a modeling of c-representations as solutions of a constraint satisfaction problem $CR(\mathcal{R})$ is given.

Definition 6 ($CR_\Sigma(\mathcal{R})$ [5], $cr_i^{\mathcal{R}}$). *Let $\mathcal{R} = \{(B_1|A_1), \dots, (B_n|A_n)\}$ over Σ. The constraint satisfaction problem for c-representations of $\mathcal{R}$, denoted by $CR_\Sigma(\mathcal{R})$, on the constraint variables $\{\eta_1, \dots, \eta_n\}$ ranging over $\mathbb{N}_0$ is given by the constraints $cr_i^{\mathcal{R}}$, for all $i \in \{1, \dots, n\}$:*

$$(cr_i^{\mathcal{R}}) \qquad \eta_i > \underbrace{\min_{\substack{\omega \in \Omega_\Sigma \\ \omega \models A_i B_i}} \sum_{\substack{j \neq i \\ \omega \models A_j \overline{B_j}}} \eta_j}_{V_{min_i}} - \underbrace{\min_{\substack{\omega \in \Omega_\Sigma \\ \omega \models A_i \overline{B_i}}} \sum_{\substack{j \neq i \\ \omega \models A_j \overline{B_j}}} \eta_j}_{F_{min_i}} \tag{2}$$

The constraint $cr_i^{\mathcal{R}}$ in (2) is the constraint corresponding to the conditional $(B_i|A_i)$. The expressions V_{min_i} and F_{min_i} are induced by the worlds verifying and falsifying $(B_i|A_i)$, respectively. The constraint variable η_i corresponding to the conditional $r = (B_i|A_i)$ will sometimes also be denoted by η_r.

Example 2. Let $\Sigma = \{b, p, f, w\}$ represent birds, penguins, flying things and winged things, and let $\mathcal{R}_{bird} = \{r_1, r_2, r_3, r_4\}$ be the belief base with:

$$
\begin{array}{lll}
r_1 : (f|b) & \quad \text{birds usually fly} \\
r_2 : (\overline{f}|p) & \quad \text{penguins usually do no fly} \\
r_3 : (b|p) & \quad \text{penguins are usually birds} \\
r_4 : (w|b) & \quad \text{birds usually have wings}
\end{array}
$$

With Ω_Σ denoting the set of worlds over $\Sigma = \{b, p, f, w\}$ and $r_i = (B_i|A_i)$ for $i \in \{1, \ldots, 4\}$, the constraint system $CR_\Sigma(\mathcal{R}_{bird})$ is:

$$
(cr_1^{\mathcal{R}_{bird}}) \qquad \eta_1 > \min_{\substack{\omega \in \Omega_\Sigma \\ \omega \models bf}} \sum_{\substack{j \neq 1 \\ \omega \models A_j \overline{B_j}}} \eta_j \quad - \quad \min_{\substack{\omega \in \Omega_\Sigma \\ \omega \models b\overline{f}}} \sum_{\substack{j \neq 1 \\ \omega \models A_j \overline{B_j}}} \eta_j \tag{3}
$$

$$
(cr_2^{\mathcal{R}_{bird}}) \qquad \eta_2 > \min_{\substack{\omega \in \Omega_\Sigma \\ \omega \models p\overline{f}}} \sum_{\substack{j \neq 2 \\ \omega \models A_j \overline{B_j}}} \eta_j \quad - \quad \min_{\substack{\omega \in \Omega_\Sigma \\ \omega \models pf}} \sum_{\substack{j \neq 2 \\ \omega \models A_j \overline{B_j}}} \eta_j \tag{4}
$$

$$
(cr_3^{\mathcal{R}_{bird}}) \qquad \eta_3 > \min_{\substack{\omega \in \Omega_\Sigma \\ \omega \models pb}} \sum_{\substack{j \neq 3 \\ \omega \models A_j \overline{B_j}}} \eta_j \quad - \quad \min_{\substack{\omega \in \Omega_\Sigma \\ \omega \models p\overline{b}}} \sum_{\substack{j \neq 3 \\ \omega \models A_j \overline{B_j}}} \eta_j \tag{5}
$$

$$
(cr_4^{\mathcal{R}_{bird}}) \qquad \eta_4 > \min_{\substack{\omega \in \Omega_\Sigma \\ \omega \models bw}} \sum_{\substack{j \neq 4 \\ \omega \models A_j \overline{B_j}}} \eta_j \quad - \quad \min_{\substack{\omega \in \Omega_\Sigma \\ \omega \models b\overline{w}}} \sum_{\substack{j \neq 4 \\ \omega \models A_j \overline{B_j}}} \eta_j \tag{6}
$$

A solution of $CR_\Sigma(\mathcal{R})$ is an n-tuple $(\eta_1, \ldots, \eta_n) \in \mathbb{N}_0^n$. For a constraint satisfaction problem CSP, the set of solutions is denoted by $Sol(CSP)$. Thus, with $Sol(CR_\Sigma(\mathcal{R}))$ we denote the set of all solutions of $CR_\Sigma(\mathcal{R})$. Table 1 details how solutions of $CR_\Sigma(\mathcal{R}_{bird})$ translate to OCFs accepting $\mathcal{R}_{bird}$.

The following proposition is an immediate consequence of a theorem established in [18].

Proposition 1 (soundness and completeness of $CR_\Sigma(\mathcal{R})$ [5]). *Let $\mathcal{R} = \{(B_1|A_1), \ldots, (B_n|A_n)\}$ be a belief base over Σ. Then we have:*

$$
Mod_\Sigma^C(\mathcal{R}) = \{\kappa_{\vec{\eta}} \mid \vec{\eta} \in Sol(CR_\Sigma(\mathcal{R}))\} \tag{7}
$$

For showing that c-representations satisfy (SynSem), we employ this characterization of c-representations as solutions of a CSP and the following observation exploiting that c-representations satisfy the property of irrelevance regarding symbols not mentioned in $\mathcal{R}$ (cf. [6]).

Proposition 2. *Let $\mathcal{R}$ be a belief base over Σ' and $\Sigma' \subseteq \Sigma$. Then $\mathcal{R}$ is also a belief base over Σ and $Sol(CR_{\Sigma'}(\mathcal{R})) = Sol(CR_\Sigma(\mathcal{R}))$.*

In the sequel, we will use the following syntax splitting property of c-representations.

Table 1. Verification and falsification for the conditionals in $\mathcal{R}_{bird}$ from Example 2. $\vec{\eta}_1$, $\vec{\eta}_2$ and $\vec{\eta}_3$ are solutions of $CR_\Sigma(\mathcal{R}_{bird})$ and $\kappa_{\vec{\eta}_1}(\omega)$, $\kappa_{\vec{\eta}_2}(\omega)$, and $\kappa_{\vec{\eta}_3}(\omega)$ are their induced ranking functions according to Definition 5.

ω		$r_1:$ $(f\|b)$	$r_2:$ $(\overline{f}\|p)$	$r_3:$ $(b\|p)$	$r_4:$ $(w\|b)$	impact on ω	$\kappa_{\vec{\eta}_1}(\omega)$	$\kappa_{\vec{\eta}_2}(\omega)$	$\kappa_{\vec{\eta}_3}(\omega)$
ω_0	$bpfw$	v	f	v	v	η_2	2	4	5
ω_1	$bpf\overline{w}$	v	f	v	f	$\eta_2 + \eta_4$	3	7	12
ω_2	$bp\overline{f}w$	f	v	v	v	η_1	1	3	4
ω_3	$bp\overline{f}\,\overline{w}$	f	v	v	f	$\eta_1 + \eta_4$	2	6	11
ω_4	$b\overline{p}fw$	v	$-$	$-$	v	0	0	0	0
ω_5	$b\overline{p}f\overline{w}$	v	$-$	$-$	f	η_4	1	3	7
ω_6	$b\overline{p}\,\overline{f}w$	f	$-$	$-$	v	η_1	1	3	4
ω_7	$b\overline{p}\,\overline{f}\,\overline{w}$	f	$-$	$-$	f	$\eta_1 + \eta_4$	2	6	11
ω_8	$\overline{b}pfw$	$-$	f	f	$-$	$\eta_2 + \eta_3$	4	8	11
ω_9	$\overline{b}pf\overline{w}$	$-$	f	f	$-$	$\eta_2 + \eta_3$	4	8	11
ω_{10}	$\overline{b}p\overline{f}w$	$-$	v	f	$-$	η_3	2	4	6
ω_{11}	$\overline{b}p\overline{f}\,\overline{w}$	$-$	v	f	$-$	η_3	2	4	6
ω_{12}	$\overline{b}\,\overline{p}fw$	$-$	$-$	$-$	$-$	0	0	0	0
ω_{13}	$\overline{b}\,\overline{p}f\overline{w}$	$-$	$-$	$-$	$-$	0	0	0	0
ω_{14}	$\overline{b}\,\overline{p}\,\overline{f}w$	$-$	$-$	$-$	$-$	0	0	0	0
ω_{15}	$\overline{b}\,\overline{p}\,\overline{f}\,\overline{w}$	$-$	$-$	$-$	$-$	0	0	0	0
impacts:		η_1	η_2	η_3	η_4				
$\vec{\eta}_1$		1	2	2	1				
$\vec{\eta}_2$		3	4	4	3				
$\vec{\eta}_3$		4	5	6	7				

Proposition 3 ([20, **Prop. 8**]). *Let $\mathcal{R}$ be a belief base over Σ and $\{\mathcal{R}_1, \mathcal{R}_2\}$ be a syntax splitting of $\mathcal{R}$ with respect to $\{\Sigma_1, \Sigma_2\}$. Then $Sol(CR_\Sigma(\mathcal{R})) = \{(\vec{\eta}^1, \vec{\eta}^2) \mid \vec{\eta}^i \in Sol(CR_{\Sigma_i}(\mathcal{R}_i)), i = 1, 2\}$, i.e.:*

$$Sol(CR_\Sigma(\mathcal{R})) = Sol(CR_{\Sigma_1}(\mathcal{R}_1)) \times Sol(CR_{\Sigma_2}(\mathcal{R}_2))$$

For binary syntax splittings, we can now prove (SynSem) for c-representations.

Proposition 4. *Let $\mathcal{R}, \mathcal{R}_1, \mathcal{R}_2$ be belief bases over Σ such that $\{R_1, \mathcal{R}_2\}$ is a syntax splitting of $\mathcal{R}$ with respect to $\{\Sigma_1, \Sigma_2\}$. Then $Sol(CR_\Sigma(\mathcal{R})) = Sol(CR_\Sigma(\mathcal{R}_1)) \times Sol(CR_\Sigma(\mathcal{R}_2))$ implies*

$$Mod_\Sigma^C(\mathcal{R}) = Mod_\Sigma^C(\mathcal{R}_1) \oplus Mod_\Sigma^C(\mathcal{R}_2).$$

Proof. W.l.o.g. assume $\mathcal{R} = \{(B_1|A_1), \ldots, (B_n|A_n)\}$, $\mathcal{R}_1 = \{(B_1^1|A_1^1), \ldots, (B_k^1|A_k^1)\}$, $\mathcal{R}_2 = \{(B_1^2|A_1^2), \ldots, (B_l^2|A_l^2)\}$ and $(B_i|A_i) = (B_i^1|A_i^1)$ for $i = 1, \ldots, k$ and $(B_{i+k}|A_{i+k}) = (B_i^2|A_i^2)$ for $i = 1, \ldots, l$.

Direction "$\subseteq$":   Let $\kappa \in Mod_{\Sigma}^{C}(\mathcal{R})$. There must be $\vec{\eta} \in Sol(CR_{\Sigma}(\mathcal{R}))$ such that $\kappa = \kappa_{\vec{\eta}}$. Proposition 3 implies that there are $\vec{\eta}^{i} \in Sol(CR_{\Sigma_{i}}(\mathcal{R}_{i}))$, $i = 1, 2$ such that $\vec{\eta} = (\vec{\eta}^{1}, \vec{\eta}^{2})$. Due to Propositions 1 and 2, we have $\kappa_{\vec{\eta}^{1}} \in Mod_{\Sigma}^{C}(\mathcal{R}_{1})$ and $\kappa_{\vec{\eta}^{2}} \in Mod_{\Sigma}^{C}(\mathcal{R}_{2})$ and thus $\kappa_{\vec{\eta}^{1}} + \kappa_{\vec{\eta}^{2}} \in Mod_{\Sigma}^{C}(\mathcal{R}_{1}) \oplus Mod_{\Sigma}^{C}(\mathcal{R}_{2})$. Everything combined we have:

$$\kappa_{\vec{\eta}}(\omega) = \sum_{\substack{1 \leqslant i \leqslant n \\ \omega \models A_{i}\overline{B}_{i}}} \eta_{i} = \sum_{\substack{1 \leqslant i \leqslant k \\ \omega \models A_{i}^{1}\overline{B}_{i}^{1}}} \eta_{i}^{2} + \sum_{\substack{1 \leqslant i \leqslant l \\ \omega \models A_{i}^{2}\overline{B}_{i}^{2}}} \eta_{i}^{2} = \kappa_{\vec{\eta}^{1}} + \kappa_{\vec{\eta}^{2}} \tag{8}$$

and therefore $\kappa \in Mod_{\Sigma}^{C}(\mathcal{R}_{1}) \oplus Mod_{\Sigma}^{C}(\mathcal{R}_{2})$.

Direction "$\supseteq$":   Let $\kappa \in Mod_{\Sigma}^{C}(\mathcal{R}_{1}) \oplus Mod_{\Sigma}^{C}(\mathcal{R}_{2})$. By definition there are ranking functions $\kappa_{1} \in Mod_{\Sigma}^{C}(\mathcal{R}_{1}), \kappa_{2} \in Mod_{\Sigma}^{C}(\mathcal{R}_{2})$ such that $\kappa = \kappa_{1} + \kappa_{2}$. Due to Propositions 1 and 2, there must be $\vec{\eta}^{i} \in Sol(CR_{\Sigma}(\mathcal{R}_{i}))$ for $i = 1, 2$ such that $\kappa_{i} = \kappa_{\vec{\eta}^{i}}$ for $i = 1, 2$. Therefore, $\vec{\eta} = (\vec{\eta}^{1}, \vec{\eta}^{2}) \in Sol(CR_{\Sigma}(\mathcal{R}))$ according to Proposition 3 and thus $\kappa_{\vec{\eta}} \in Sol(CR_{\Sigma}(\mathcal{R}))$. With (8) it follows that $\kappa = \kappa_{1} + \kappa_{2} = \kappa_{\vec{\eta}^{1}} + \kappa_{\vec{\eta}^{2}} = \kappa_{\vec{\eta}} \in Mod_{\Sigma}^{C}(\mathcal{R})$. $\square$

The proof that c-representations fully satisfy (SynSem) is obtained by an induction on the number of partitions of $\mathcal{R}$.

Proposition 5. *C-representations satisfy (SynSem).*

Proof. Let $\{\mathcal{R}_{1}, \dots, \mathcal{R}_{n}\}$ be a syntax splitting of $\mathcal{R}$ over Σ. We show that $\{\mathcal{R}_{1}, \dots, \mathcal{R}_{n}\}$ is a semantic splitting for c-representations by induction over n. The base case $n = 1$ holds trivially.

Induction Step: Let $\{\mathcal{R}_{1}, \dots, \mathcal{R}_{n}, \mathcal{R}_{n+1}\}$ be a syntax splitting of $\mathcal{R}$ over Σ with respect to $\{\Sigma_{1}, \dots, \Sigma_{n}, \Sigma_{n+1}\}$. Then $\{(\mathcal{R}_{1} \cup \dots \cup \mathcal{R}_{n}), \mathcal{R}_{n+1}\}$ is a syntax splitting of $\mathcal{R}$ with respect to $\{\Sigma', \Sigma_{n+1}\}$ where $\Sigma' = \Sigma_{1} \cup \dots \cup \Sigma_{n}$. Proposition 3 implies $Sol(CR_{\Sigma}((\mathcal{R}_{1} \cup \dots \cup \mathcal{R}_{n}) \cup \mathcal{R}_{n+1})) = Sol(CR_{\Sigma'}(\mathcal{R}_{1} \cup \dots \cup \mathcal{R}_{n})) \times Sol(CR_{\Sigma_{n+1}}(\mathcal{R}_{n+1}))$. Proposition 2 allows us to lift $\mathcal{R}_{1} \cup \dots \cup \mathcal{R}_{n}$ and $\mathcal{R}_{n+1}$ from their sub-signatures to Σ with $Sol(CR_{\Sigma'}(\mathcal{R}_{1} \cup \dots \cup \mathcal{R}_{n})) = Sol(CR_{\Sigma}(\mathcal{R}_{1} \cup \dots \cup \mathcal{R}_{n}))$ and $Sol(CR_{\Sigma_{n+1}}(\mathcal{R}_{n+1})) = Sol(CR_{\Sigma}(\mathcal{R}_{n+1}))$, receptively. With Proposition 4, it follows that $Mod_{\Sigma}^{C}(\mathcal{R}_{1} \cup \dots \cup \mathcal{R}_{n} \cup \mathcal{R}_{n+1}) = Mod_{\Sigma}^{C}(\mathcal{R}_{1} \cup \dots \cup \mathcal{R}_{n}) \oplus Mod_{\Sigma}^{C}(\mathcal{R}_{n+1})$. Applying the induction hypothesis yields $Mod_{\Sigma}^{C}(\mathcal{R}_{1} \cup \dots \cup \mathcal{R}_{n} \cup \mathcal{R}_{n+1}) = Mod_{\Sigma}^{C}(\mathcal{R}_{1}) \oplus \dots \oplus Mod_{\Sigma}^{C}(\mathcal{R}_{n}) \oplus Mod_{\Sigma}^{C}(\mathcal{R}_{n+1})$, i.e., $\{\mathcal{R}_{1}, \dots, \mathcal{R}_{n}, \mathcal{R}_{n+1}\}$ is a semantic splitting of $\mathcal{R}$. $\square$

Thus, for c-representations, every syntax splitting is also a semantic splitting. In the next sections, we will show that we can go even further. We will develop a syntactic criterion for belief bases that captures all syntax splittings and allows for additional semantic splittings.

4   Optimizing Constraint Systems for C-Representations

The key ingredients determining the constraint $cr_{i}^{\mathcal{R}}$ in Definition 6 are the constraint variables occurring in the expressions $V_{min_{i}}$ und $F_{min_{i}}$. This observation is exploited in the powerset representation of $CR_{\Sigma}(\mathcal{R})$.

Definition 7 (powerset representation, $PSR(\mathcal{R})$ [10]). *Given the notation as in Definition 6, let*

$$\Pi(V_{min_i}) = \{\, \{\eta_j \mid j \neq i,\ \omega \models A_j \overline{B_j}\} \mid \omega \models A_i B_i \}\ \ \ (9)$$

$$\Pi(F_{min_i}) = \{\, \{\eta_j \mid j \neq i,\ \omega \models A_j \overline{B_j}\} \mid \omega \models A_i \overline{B_i} \}\ \ \ (10)$$

The powerset representation of $cr_i^{\mathcal{R}}$, also called the PSR term corresponding to $(B_i|A_i)$ relative to $\mathcal{R}$, is the pair $psr_i^{\mathcal{R}} = \langle \Pi(V_{min_i}), \Pi(F_{min_i}) \rangle$ and the power set representation of $\mathcal{R}$ is $PSR(\mathcal{R}) = \{psr_1^{\mathcal{R}}, \ldots, psr_n^{\mathcal{R}}\}$.

Thus, $\Pi(V_{min_i})$ and $\Pi(F_{min_i})$ are the sets of sets of constraint variables η_j occurring in the two sum expressions in (2).

Example 3. The PSR terms for the four constraints (3)–(6) are:

$$\langle \Pi(V_{min_1}),\ \Pi(F_{min_1}) \rangle\ =\ \langle \{\{\eta_2\}, \{\eta_2, \eta_4\}, \varnothing, \{\eta_4\}\},\ \{\varnothing, \{\eta_4\}\} \rangle\ \ \ (11)$$

$$\langle \Pi(V_{min_2}),\ \Pi(F_{min_2}) \rangle\ =\ \langle \{\{\eta_1\}, \{\eta_3\}, \{\eta_1, \eta_4\}\},\ \{\varnothing, \{\eta_4\}, \{\eta_3\}\} \rangle\ \ \ (12)$$

$$\langle \Pi(V_{min_3}),\ \Pi(F_{min_3}) \rangle\ =\ \langle \{\{\eta_2\}, \{\eta_2, \eta_4\}, \{\eta_1\}, \{\eta_1, \eta_4\}\},\ \{\varnothing, \{\eta_2\}\} \rangle\ \ \ (13)$$

$$\langle \Pi(V_{min_4}),\ \Pi(F_{min_4}) \rangle\ =\ \langle \{\{\eta_2\}, \{\eta_1\}, \varnothing\},\ \{\{\eta_2\}, \{\eta_1\}, \varnothing\} \rangle\ \ \ (14)$$

In general, a PSR term $\langle \mathcal{V},\ \mathcal{F} \rangle$ is a pair of sets of subsets of the involved constraint variables. The following definition assigns an arithmetic expression to any set of sets of constraint variables that will be used for recovering a CSP capturing $CR_\Sigma(\mathcal{R})$. For every set S, we will use $\mathcal{P}(S)$ to denote the power set of S.

Definition 8 (represented arithmetic term, ρ [10]). *Let $CV = \{\eta_1, \ldots, \eta_n\}$ be a set of constraint variables and let $M \subseteq \mathcal{P}(CV)$ be a set of subsets of CV. The arithmetic expression represented by $M = \{S_1, \ldots, S_r\}$, denoted by $\rho(M)$, is:*

$$\rho(M) = \min\Big\{ \sum_{\eta \in S_1} \eta,\ \ldots,\ \sum_{\eta \in S_r} \eta \Big\}\ \ \ (15)$$

Note that $\min \varnothing = \infty$, and if $S = \varnothing$ then $\sum_{\eta \in S} \eta = 0$. We extend the definition of ρ to PSR terms $\langle \mathcal{V},\ \mathcal{F} \rangle$:

$$\rho(\langle \mathcal{V},\ \mathcal{F} \rangle) = \begin{cases} \infty & \text{if } \rho(\mathcal{V}) = \infty \text{ and } \rho(\mathcal{F}) = \infty \\ \rho(\mathcal{V}) - \rho(\mathcal{F}) & \text{else} \end{cases}\ \ \ (16)$$

The first case in (16) catches the extreme case of a belief base $\mathcal{R}$ containing a conditional of the form $(B|\bot)$. Because such a conditional is never applicable, it is never verified or falsified. Thus, a belief base containing $(B|\bot)$ is inconsistent, since no ranking function accepting $(B|\bot)$ exists. The following proposition states that a subtraction expression occurring in $CR_\Sigma(\mathcal{R})$ that corresponds to a conditional that is applicable in at least one world can safely be replaced by the arithmetic expressions obtained from its PSR representation.

Proposition 6 ([10]). *Let* $\mathcal{R} = \{(B_1|A_1), \ldots, (B_n|A_n)\}$ *be a belief base, CV be the constraint variables occurring in* $CR_\Sigma(\mathcal{R})$, *and let* $\eta_i > V_{min_i} - F_{min_i}$ *be the constraint* $cr_i^\mathcal{R}$. *If* $A_i \not\equiv \bot$, *then for every variable assignment* $\alpha : CV \to \mathbb{N}_0$ *we have:*

$$\alpha(V_{min_i}) - \alpha(F_{min_i}) = \alpha(\rho(\langle \Pi(V_{min_i}), \Pi(F_{min_i})\rangle)) \qquad (17)$$

In order to be able to directly compare the different representations, we will use $\rho(PSR(\mathcal{R}))$ to denote the result of replacing every PSR term $psr_i^\mathcal{R}$ in $PSR(\mathcal{R})$ by the constraint $\eta_i > \rho(psr_i^\mathcal{R})$.

Proposition 7 (Soundness and completeness of $PSR(\mathcal{R})$**).** *For every belief base* $\mathcal{R}$, *we have* $Sol(CR_\Sigma(\mathcal{R})) = Sol(\rho(PSR(\mathcal{R})))$.

In order to ease our notation, we may omit the explicit distinction between a PSR term $\langle \mathcal{V}, \mathcal{F}\rangle$ and its represented subtraction expression in $\rho(\langle \mathcal{V}, \mathcal{F}\rangle)$. Likewise, we may omit the distinction between $PSR(\mathcal{R})$ and $\rho(PSR(\mathcal{R}))$.

The powerset representation $\langle \mathcal{V}, \mathcal{F}\rangle$ of a subtraction expression can be optimized. For instance, the minimum of two sums S_1 and S_2 of non-negative integers is S_1 if all summands of S_1 also occur in S_2. Figure 1 contains a set $\mathcal{T}$ of transformation rules that can be applied to pairs of elements of the powerset of constraint variables, and thus in particular to the PSR representation of a subtraction expression:

(**ss-V**) removes a set S' that is an element in the first component if it is a superset of another set S in the first component.

(**ss-F**) removes a set S' that is an element in the second component if it is a superset of another set S in the second component.

(**elem**) removes an element η that is in every set in both the first and the second component from all these sets.

$$(ss\text{-}V) \ subset\text{-}V : \qquad \frac{\langle \mathcal{V} \cup \{S, S'\}, \mathcal{F}\rangle}{\langle \mathcal{V} \cup \{S\}, \mathcal{F}\rangle} \qquad S \subsetneq S'$$

$$(ss\text{-}F) \ subset\text{-}F : \qquad \frac{\langle \mathcal{V}, \mathcal{F} \cup \{S, S'\}\rangle}{\langle \mathcal{V}, \mathcal{F} \cup \{S\}\rangle} \qquad S \subsetneq S'$$

$$(elem) \ element : \qquad \frac{\langle \{V_1 \cup \{\eta\}, \ldots, V_p \cup \{\eta\}\}, \{F_1 \cup \{\eta\}, \ldots, F_q \cup \{\eta\}\}\rangle}{\langle \{V_1, \ldots, V_p\}, \{F_1, \ldots, F_q\}\rangle}$$

Fig. 1. Transformation rules $\mathcal{T}$ for optimizing PSR terms of subtractions.

Note that $\mathcal{T}$ is functionally equivalent to the set of transformation rules given in [10]. The set given in Fig. 1 is a minimal set of transformation rules, and the two additional rules given in [10] can be replaced by a finite chain of applications of $(ss\text{-}V)$ and $(ss\text{-}F)$, respectively. The transformation system $\mathcal{T}$ is terminating and confluent, and it is sound in the sense that the result of applying $\mathcal{T}$ exhaustively to a PSR term $\langle \mathcal{V}, \mathcal{F} \rangle$ is a simplified PSR term that is equivalent to $\langle \mathcal{V}, \mathcal{F} \rangle$ with respect to all variable assignments.

Definition 9 ($CCR(\mathcal{R})$ [10]). *For a belief base $\mathcal{R}$, the* compilation *of $\mathcal{R}$, denoted by $CCR(\mathcal{R})$, is obtained from $PSR(\mathcal{R})$ by replacing every PSR term $psr_i^{\mathcal{R}}$ by its optimized normal form $ccr_i^{\mathcal{R}} = \mathcal{T}(psr_i^{\mathcal{R}})$.*

Example 4. $CCR(\mathcal{R}_{bird})$, the compilation of $\mathcal{R}_{bird}$, is given by:

$$
\begin{aligned}
\eta_1 &> \langle \{\varnothing\}, \{\varnothing\} \rangle \\
\eta_2 &> \langle \{\{\eta_1\}, \{\eta_3\}\}, \{\varnothing\} \rangle \\
\eta_3 &> \langle \{\{\eta_2\}, \{\eta_1\}\}, \{\varnothing\} \rangle \\
\eta_4 &> \langle \{\varnothing\}, \{\varnothing\} \rangle
\end{aligned}
$$

Similar as before, $\rho(CCR(\mathcal{R}))$ denotes the constraint system obtained from $CCR(\mathcal{R})$ by replacing every PSR term $\langle \mathcal{V}, \mathcal{F} \rangle$ by $\rho(\langle \mathcal{V}, \mathcal{F} \rangle)$. In [10], it is shown that the compilation is sound and complete, i.e., for every belief base $\mathcal{R}$,

$$\{\kappa_{\vec{\eta}} \mid \vec{\eta} \in Sol(\rho(CCR(\mathcal{R})))\} = \{\kappa \mid \kappa \text{ is a c-representation of } \mathcal{R}\} \tag{18}$$

and thus

$$Sol(\rho(CR_{\Sigma}(\mathcal{R}))) = Sol(\rho(CCR(\mathcal{R}))). \tag{19}$$

5 Constraint Splittings

For $CCR(\mathcal{R})$, we can extract the information which constraint variables are independent from each other.

Definition 10 ($\sim_{\mathcal{R}}^{dp}$, $\sim_{\mathcal{R}}^{*}$, $\perp\!\!\!\perp$). *Let $\mathcal{R}$ be a belief base and $(B_i|A_i), (B_j|A_j) \in \mathcal{R}$. Then $(B_i|A_i)$ directly depends on $(B_j|A_j)$ in $\mathcal{R}$, denoted by $(B_i|A_i) \sim_{\mathcal{R}}^{dp} (B_j|A_j)$, if $\eta_i \in Vars(ccr_j^{\mathcal{R}})$ where $Vars(ccr_j^{\mathcal{R}})$ is the set of constraint variables occurring in $ccr_j^{\mathcal{R}}$. The reflexive and transitive closure of $\sim_{\mathcal{R}}^{dp}$ is denoted by $\sim_{\mathcal{R}}^{*}$.*

For $\mathcal{R}', \mathcal{R}'' \subseteq \mathcal{R}$ we say that $\mathcal{R}'$ and $\mathcal{R}''$ are constraint independent, *denoted $\mathcal{R}' \perp\!\!\!\perp \mathcal{R}''$, if for all $r' \in \mathcal{R}'$, $r'' \in \mathcal{R}''$ we have $r' \not\sim_{\mathcal{R}}^{*} r''$.*

A constraint splitting of $\mathcal{R}$ must respect the independence relation $\perp\!\!\!\perp$ on the partitions.

Definition 11 (constraint splitting). *Let $\mathcal{R}$ be a belief base. A partition $\{\mathcal{R}_1, \ldots, \mathcal{R}_k\}$ of $\mathcal{R}$ is a* constraint splitting *of $\mathcal{R}$ if $\mathcal{R}_i \perp\!\!\!\perp \mathcal{R}_j$ for all $i, j \in \{1, \ldots, k\}$ with $i \neq j$.*

Example 5. The partition $\{\mathcal{R}_1, \mathcal{R}_2\}$ with $\mathcal{R}_1 = \{r_1, r_2, r_3\}$ and $\mathcal{R}_2 = \{r_4\}$ is a constraint partition of $\mathcal{R}_{bird}$.

Obviously, constraint variables belonging to different partitions of a syntax splitting are constraint independent, yielding the following observation.

Proposition 8. *For c-representations, every syntax splitting of $\mathcal{R}$ is also a constraint splitting of $\mathcal{R}$.*

While constraint splittings fully capture syntax splittings, they properly go beyond them. An example illustrating this is $\mathcal{R}_{bird}$ for which $\{\{r_1, r_2, r_3\}, \{r_4\}\}$ is a constraint splitting, but not a syntax splitting.

Proposition 9. *For c-representations, in general, a constraint splitting of $\mathcal{R}$ is not necessarily a syntax splitting of $\mathcal{R}$.*

For showing that for c-representations every constraint splitting is also a semantic splitting, we employ projections of impact vectors and of belief bases. If $\vec{\eta}$ is an impact vector with impacts corresponding to the conditionals in $\mathcal{R}$, then for $\mathcal{R}' \subseteq \mathcal{R}$, the subvector of $\vec{\eta}$ containing only the impacts related to the conditionals in $\mathcal{R}'$ is called the *projection* of $\vec{\eta}$ to $\mathcal{R}'$ and is denoted by $\vec{\eta}_{\mathcal{R}'}$. We extend the notion to a set M of impact vectors by writing $M_{\mathcal{R}'}$ The following definition extends the notion of projection to constraint satisfaction problems for c-representations.

Definition 12 (CSP projection $CR_\Sigma(\mathcal{R})_{\mathcal{R}'}$). *Let $\mathcal{R} = \{(B_1|A_1), \ldots, (B_n|A_n)\}$ be a set of conditionals, and let $\mathcal{R}' \subseteq \mathcal{R}$. The projection of $CR_\Sigma(\mathcal{R})$ to $\mathcal{R}'$, denoted by $CR_\Sigma(\mathcal{R})_{\mathcal{R}'}$, is the constraint satisfaction problem given by the set of constraints $\{cr_i^{\mathcal{R}} \mid (B_i|A_i) \in \mathcal{R}'\}$.*

W.l.o.g. let us assume that $\mathcal{R} = \{(B_1|A_1), \ldots, (B_n|A_n)\}$ and $\mathcal{R}' = \{(B_1|A_1), \ldots, (B_k|A_k)\}$ where $k \leqslant n$. Note the difference between

$$CR_\Sigma(\mathcal{R}') = \{cr_1^{\mathcal{R}'}, \ldots, cr_k^{\mathcal{R}'}\} \tag{20}$$

and

$$CR_\Sigma(\mathcal{R})_{\mathcal{R}'} = \{cr_1^{\mathcal{R}}, \ldots, cr_k^{\mathcal{R}}\}. \tag{21}$$

While $CR_\Sigma(\mathcal{R}')$ in (20) is a CSP over the constraint variables $\eta_1, \ldots, \eta_k$, $CR_\Sigma(\mathcal{R})_{\mathcal{R}'}$ in (21) is a CSP over the constraint variables $\eta_1, \ldots, \eta_k, \eta_{k+1}, \ldots, \eta_n$. Both CSP have $|\mathcal{R}'|$-many constraints, but in contrast to $cr_i^{\mathcal{R}}$, any of the constraint variables $\eta_{k+1}, \ldots, \eta_n$ in the sum in the minimizations terms given in Eq. (2) are removed in $cr_i^{\mathcal{R}'}$.

Using projections both for $CR_\Sigma(\mathcal{R})$ and for the solutions of the projection of $CR_\Sigma(\mathcal{R})$, we obtain a proposition for constraint splittings of $\mathcal{R}$ that corresponds to Proposition 3 for syntax splittings.

Proposition 10. *If $\{\mathcal{R}_1, \mathcal{R}_2\}$ is a constraint splitting of $\mathcal{R}$ with respect to $\Sigma = \{\Sigma_1, \Sigma_2\}$ then*

$$Sol(CR_\Sigma(\mathcal{R})) = Sol(CR_\Sigma(\mathcal{R})_{\mathcal{R}_1})_{\mathcal{R}_1} \times Sol(CR_\Sigma(\mathcal{R})_{\mathcal{R}_2})_{\mathcal{R}_2}.$$

Example 6. For the constraint splitting $\{\mathcal{R}_1, \mathcal{R}_2\}$ with $\mathcal{R}_1 = \{r_1, r_2, r_3\}$ and $\mathcal{R}_2 = \{r_4\}$ of $\mathcal{R}_{bird}$, the vector $\vec{\eta}^1 = (1, 2, 2, 0)$ is a solution of $CR_\Sigma(\mathcal{R}_{bird})_{\mathcal{R}_1}$, and the vector $\vec{\eta}^2 = (4, 3, 2, 1)$ is a solution of $CR_\Sigma(\mathcal{R}_{bird})_{\mathcal{R}_2}$. Note that neither $\vec{\eta}^1$ nor $\vec{\eta}^2$ is a solution of $CR_\Sigma(\mathcal{R}_{bird})$. However, when composing the projections of $\vec{\eta}^1$ and $\vec{\eta}^2$ to $\mathcal{R}_1$ and $\mathcal{R}_2$, respectively, we get $\vec{\eta}^1_{\mathcal{R}_1} = (1, 2, 2)$ and $\vec{\eta}^2_{\mathcal{R}_2} = (1)$ and thus $\vec{\eta} = (\vec{\eta}^1_{\mathcal{R}_1}, \vec{\eta}^2_{\mathcal{R}_2}) = (1, 2, 2, 1)$, with $\vec{\eta}$ being a solution of $CR_\Sigma(\mathcal{R}_{bird})$ as stated in Proposition 10.

The next proposition states that a binary constraint splitting is also a semantic splitting.

Proposition 11. *If $\{\mathcal{R}_1, \mathcal{R}_2\}$ is a constraint splitting of $\mathcal{R}$ with respect to $\Sigma = \{\Sigma_1, \Sigma_2\}$ then*

$$Mod_\Sigma^C(\mathcal{R}) = Mod_\Sigma^C(\mathcal{R}_1) \oplus Mod_\Sigma^C(\mathcal{R}_2).$$

Proof. The proof is obtained similar as in the proof of Proposition 4 by employing the relationship established in Proposition 10. □

We now have all ingredients for establishing the semantic splitting property of constraint splittings.

Proposition 12. *For c-representations, every constraint splitting of $\mathcal{R}$ is a semantic splitting of $\mathcal{R}$.*

Proof. Let $\{\mathcal{R}_1, \ldots, \mathcal{R}_n\}$ be a constraint splitting of $\mathcal{R}$. Using Propositions 10 and 11 the claim follows by induction over n analogously to the induction proof in Proposition 5. □

Thus, also for the syntax splitting exceeding constraint splittings, ranking models of $\mathcal{R}$ can be determined by simply combining models of the respective sub-belief bases of $\mathcal{R}$.

6 Conclusions and Future Work

Inspired by the notion of syntax splitting, we formulated the notion of semantic splitting under which the ranking models of a conditional belief base $\mathcal{R}$ can be easily computed from the models ob sub-belief bases of $\mathcal{R}$. For c-representations, every syntax splitting is a semantic splitting and also a constraint splitting. Furthermore, also every constraint splitting is a semantic splitting, while constraint splittings cover and extend syntax splittings. Our current work includes transferring the concepts developed here from ranking models to total preorders of worlds and to other semantics of conditional belief bases.

References

1. Adams, E.W.: The Logic of Conditionals: An Application of Probability to Deductive Logic. Synthese Library. Springer, Dordrecht (1975). https://doi.org/10.1007/978-94-015-7622-2

2. Beierle, C., Börger, E.: Correctness proof for the WAM with types. In: Börger, E., Jäger, G., Kleine Büning, H., Richter, M.M. (eds.) CSL 1991. LNCS, vol. 626, pp. 15–34. Springer, Heidelberg (1992). https://doi.org/10.1007/BFb0023755

3. Beierle, C., Börger, E.: Refinement of a typed WAM extension by polymorphic order-sorted types. Formal Aspects Comput. **8**(5), 539–564 (1996). https://doi.org/10.1007/BF01211908

4. Beierle, C., Börger, E.: Specification and correctness proof of a WAM extension with abstract type constraints. Formal Aspects of Comput. **8**(4), 428–462 (1996). https://doi.org/10.1007/BF01213533

5. Beierle, C., Eichhorn, C., Kern-Isberner, G., Kutsch, S.: Properties of skeptical c-inference for conditional knowledge bases and its realization as a constraint satisfaction problem. Ann. Math. Artif. Intell. **83**(3–4), 247–275 (2018)

6. Beierle, C., Eichhorn, C., Kern-Isberner, G., Kutsch, S.: Properties and interrelationships of skeptical, weakly skeptical, and credulous inference induced by classes of minimal models. Artif. Intell. **297** (2021). (in press). Online 2 Mar 2021

7. Beierle, C., Kern-Isberner, G.: Modelling conditional knowledge discovery and belief revision by abstract state machines. In: Börger, E., Gargantini, A., Riccobene, E. (eds.) ASM 2003. LNCS, vol. 2589, pp. 186–203. Springer, Heidelberg (2003). https://doi.org/10.1007/3-540-36498-6_10

8. Beierle, C., Kern-Isberner, G.: A Verified AsmL implementation of belief revision. In: Börger, E., Butler, M., Bowen, J.P., Boca, P. (eds.) ABZ 2008. LNCS, vol. 5238, pp. 98–111. Springer, Heidelberg (2008). https://doi.org/10.1007/978-3-540-87603-8_9

9. Beierle, C., Kutsch, S., Kern-Isberner, G.: From concepts in non-monotonic reasoning to high-level implementations using abstract state machines and functional programming. In: Mashkoor, A., Wang, Q., Thalheim, B. (eds.) Models: Concepts, Theory, Logic, Reasoning and Semantics - Essays Dedicated to Klaus-Dieter Schewe on the Occasion of his 60th Birthday, pp. 286–310. College Publications (2018)

10. Beierle, C., Kutsch, S., Sauerwald, K.: Compilation of static and evolving conditional knowledge bases for computing induced nonmonotonic inference relations. Ann. Math. Artif. Intell. **87**(1–2), 5–41 (2019)

11. Benferhat, S., Dubois, D., Prade, H.: Possibilistic and standard probabilistic semantics of conditional knowledge bases. J. Logic Comput. **9**(6), 873–895 (1999)

12. Börger, E.: The ASM refinement method. Formal Aspects Comput. **15**(2–3), 237–257 (2003)

13. Börger, E., Raschke, A.: Modeling Companion for Software Practitioners. Springer, Heidelberg (2018). https://doi.org/10.1007/978-3-662-56641-1

14. Börger, E., Stärk, R.: Abstract State Machines: A Method for High-Level System Design and Analysis. Springer, Heidelberg (2003). https://doi.org/10.1007/978-3-642-18216-7

15. Dubois, D., Prade, H.: Conditional objects as nonmonotonic consequence relationships. Spec. Issue Conditional Event Algebra, IEEE Trans. Syst. Man Cybern. **24**(12), 1724–1740 (1994)

16. Goldszmidt, M., Pearl, J.: Qualitative probabilities for default reasoning, belief revision, and causal modeling. Artif. Intell. **84**(1–2), 57–112 (1996)
17. Gurevich, Y.: Evolving algebras 1993: Lipari guide. In: Börger, E. (ed.) Specification and Validation Methods, pp. 9–36. Oxford University Press (1995)
18. Kern-Isberner, G.: Conditionals in Nonmonotonic Reasoning and Belief Revision. LNCS (LNAI), vol. 2087. Springer, Heidelberg (2001). https://doi.org/10.1007/3-540-44600-1
19. Kern-Isberner, G.: A thorough axiomatization of a principle of conditional preservation in belief revision. Ann. Math. Artif. Intell. **40**(1–2), 127–164 (2004)
20. Kern-Isberner, G., Beierle, C., Brewka, G.: Syntax splitting = relevance + independence: New postulates for nonmonotonic reasoning from conditional belief bases. In: KR-2020, pp. 560–571 (2020)
21. Kern-Isberner, G., Brewka, G.: Strong syntax splitting for iterated belief revision. In: Proceedings of the Twenty-Sixth International Joint Conference on Artificial Intelligence, IJCAI 2017, Melbourne, Australia, 19–25 August 2017, pp. 1131–1137 (2017)
22. Kraus, S., Lehmann, D., Magidor, M.: Nonmonotonic reasoning, preferential models and cumulative logics. Artif. Intell. **44**, 167–207 (1990)
23. Parikh, R.: Beliefs, belief revision, and splitting languages. Logic Lang. Computat. **2**, 266–278 (1999)
24. Pearl, J.: System Z: A natural ordering of defaults with tractable applications to nonmonotonic reasoning. In: Parikh, R. (ed) Proceedings of the 3rd Conference on Theoretical Aspects of Reasoning About Knowledge (TARK1990), San Francisco, CA, USA, pp. 121–135, Morgan Kaufmann Publishers Inc. (1990)
25. Peppas, P., Williams, M., Chopra, S., Foo, N.Y.: Relevance in belief revision. Artif. Intell. **229**, 126–138 (2015)
26. Spohn, W.: Ordinal conditional functions: a dynamic theory of epistemic states. In: Harper, W., Skyrms, B. (eds.) Causation in Decision, Belief Change, and Statistics, II, pp. 105–134. Kluwer Academic Publishers (1988)

Communities and Ancestors Associated with Egon Börger and ASM

Jonathan P. Bowen[1,2(✉)] (iD)

[1] School of Engineering, London South Bank University, Borough Road,
London SE1 1AA, UK
`jonathan.bowen@lsbu.ac.uk`
[2] Centre for Research and Innovation in Software Engineering (RISE),
Faculty of Computer and Information Science, Southwest University,
Chongqing 400715, China
`http://www.jpbowen.com`

Abstract. In this paper, we discuss the community associated with Abstract State Machines (ASM), especially in the context of a Community of Practice (CoP), a social science concept, considering the development of ASM by its community of researchers and practitioners over time. We also consider the long-term historical context of the advisor tree of Egon Börger, the main promulgator of the ASM approach, which can be considered as multiple interrelated CoPs, distributed over several centuries. This includes notable mathematicians and philosophers among its number with some interesting links between the people involved. Despite most being active well before the inception of computer science, a number have been influential on the field.

1 Background

> *There are two kinds of truths: those of reasoning and those of fact. The truths of reasoning are necessary and their opposite is impossible; the truths of fact are contingent and their opposites are possible.*
>
> – Gottfried Leibniz

This paper has been inspired by the work of the computer scientist Egon Börger [11,12] and is presented in celebration of his 75th birthday. The author has been involved in building and investigating communities [38], both in the area of formal methods, especially the Z notation [49], and also in museum-related [8,59] and arts-related [44,50] contexts. Börger has been central to building the community [10] around the Abstract State Machines (ASM) approach to modelling computer-based systems in a formal mathematical manner. This paper considers aspects of this community, especially with respect to Egon Börger's role, and also in the context of the Community of Practice (CoP) approach to considering the evolution of communities based around an area of developing knowledge [94,95].

© Springer Nature Switzerland AG 2021
A. Raschke et al. (Eds.): Börger Festschrift, LNCS 12750, pp. 96–120, 2021.
https://doi.org/10.1007/978-3-030-76020-5_6

1.1 Personal Appreciation

All our knowledge begins with the senses, proceeds then to the understanding, and ends with reason. There is nothing higher than reason.

– Immanuel Kant

I first met Egon Börger when he visited Tony Hoare [34] at the Oxford University Computing Laboratory's Programming Research Group (PRG) in September 1993, including a talk by him entitled *The methodology of evolving algebras for correctness proofs of compilation schemes: the case of OCCAM and TRANSPUTER* [12,19]. I was a Research Officer working on formal methods [68] and specifically the Z notation [33,67] at the time. I also became involved with the European ESPRIT **ProCoS** I and II projects on *Provably Correct Systems*, led by Tony Hoare at Oxford, Dines Bjørner at DTH in Denmark, and others in the early 1990s [9,31,47].

The subsequent **ProCoS-WG** Working Group of 25 partners around Europe existed to organize meetings and workshops in the late 1990s [32]. Egon Börger gave a talk to the group on *Proof of correctness of a scheme for compilation of Occam programs on the Transputer* at a January 1995 workshop in Oxford [12,19]. The **ProCoS-WG** final report in 1998 [48] included the following:

Prof. Egon Börger of the University of Pisa, Italy, has participated at many **ProCoS-WG** meetings, largely at his own expense. He was an invited speaker at ZUM'97 [46] and organized, with Prof. Hans Langmaack of the University of Kiel, an important set of case studies formalizing a Steam Boiler problem in a variety of notations [1], including a number of contributions by **ProCoS-WG** members. He reports:

The **ProCoS** meetings have been for me a very useful occasion for fruitful exchange of ideas and methods related to the application of formal methods. In particular I appreciate the occasion I had to present my work on the correctness theorem for a general compilation scheme for compiling Occam programs to Transputer code. This work appeared in [23]. Furthermore I appreciated the chance to present the Abstract State Machine (ASM) method to **ProCoS-WG** members.

See Fig. 1 for a group photograph of participants at the final **ProCoS-WG** workshop held at the University of Reading in the UK during 1997, including Egon Börger and myself. Egon Börger was also an invited speaker [27] at the co-located ZUM'97 conference [46] that I co-chaired, introducing the Z community to ASM.

Much later, in 2006, we both contributed to a book on *Software Specification Methods* [64] based around a common case study, including use of the Z notation [37] and ASM [24]. In the same year, I also attended Egon's 60th birthday *Festkolloquium* at Schloss Dagstuhl, Germany [3], later contributing to the associated *Festschift* volume [45].

Fig. 1. ProCoS-WG meeting at the University of Reading, 1997. Egon Börger is 4th from the right; I am 10th from the right.

From the early 2000s, I was Chair of the UK BCS-FACS (Formal Aspects of Computing Science) Specialist Group. In December 2003, Egon presented on *Teaching ASMs to Practice-Oriented Students with Limited Mathematical Background* at the BCS-FACS Workshop *Teaching Formal Methods: Practice and Experience*, held at Oxford Brookes University. Subsequently, I invited Egon to speak to the group at the BCS London office in March 2007, and a chapter appeared in a book of selected talks [18].

In 2008, we were co-chairs of the newly formed *Abstract State Machines, B and Z: First International Conference, ABZ 2008* in London, UK, edited by Egon Börger (ASM), Michael Butler (B), myself (Z), and Paul Boca as a local organizer [21,22], including an ASM-based paper co-authored by Egon Börger [30]. This was an extension of the previous ZB conferences, that were a combination of previously separate B and Z conferences. In 2011, we collaborated on a special issue of selected and extended papers from the ABZ 2008 conference in the *Formal Aspects of Computing* journal, edited by Egon Börger, myself, Michael Butler, and Mike Poppleton [20]. Most recently, I reviewed his 2018 book on *Modeling Companion for Software Practitioners* using the ASM approach, co-authored with Alexander Raschke [42].

2   Communities of Practice

Reason is purposive activity.   – Georg Hegel

A *Community of Practice* (CoP) [94] is a social science concept for modelling the collaborative activities of professional communities [10] with a common goal over time [95,96]. It can be used in various scenarios, for example, formal methods communities [39,41,49]. A CoP consists of:

1. A **domain** of knowledge and interest. In the case of ASM, this is the application of a mathematical approach to computer-based specification modelling and development.

2. **A community** around this domain. For ASM, this includes the ABZ conference organizers and programme committee members that are interested in ASM at its core, conference presenters and delegates, as well as other researchers and practitioners involved with ASM.
3. The **practice** undertaken by the community in this domain, developing its knowledge. The ASM community is encouraging the transfer of research ideas into practical use. The ASM approach has been used to model industrial-scale programming and specification languages, and a recent book has been produced to encourage use by software practitioners [28].

There are various stages in the development of a CoP:

1. **Potential:** There needs to be an existing network of people to initiate a CoP. In the case of ASM, researchers interested in theoretical computer science and formal methods were the starting point, especially the original progenitors, Yuri Gurevich and Egon Börger.
2. **Coalescing:** The community needs to establish a rhythm to ensure its continuation. In the case of ASM, a regular ASM workshop was established.
3. **Maturing:** The community must become more enduring. The ASM workshop combined with the ZB conference, already a conference for the Z notation and B-Method, to become the ABZ conference in 2008 [21].
4. **Stewardship:** The community needs to respond to its environment and develop appropriately. The ASM community has interacted with related organizations such as the Z User Group (ZUG), the B User Group (BUG), etc., and has fostered these relationships especially through the regular ABZ conference.
5. **Legacy:** All communities end eventually; if successful they morph into further communities. ASM continues as a community, although it has combined with other state-based approach communities such as those around B, VDM, Z, etc. Exactly how these related communities will continue is something that is worth considering and planning for at the appropriate time.

It remains to be seen precisely what legacy ASM leaves in the future. For the moment, the ASM community continues through the ABZ conference and other more informal and individual interactions.

3   The Development of ASM

The model should not dictate but reflect the problem.

– Egon Börger [28]

In the 1980s, the American computer scientist and mathematician Yuri Gurevich (originally from the Soviet Union) conceived of the idea of "evolving algebras" [62], based on the Church-Turing thesis, with algorithms being simulated by a suitable Turing machine [90]. He suggested the *ASM thesis*, that every algorithm, however abstract, can be emulated step-for-step with an appropriate ASM. In

2000, he axiomatized the notion of sequential algorithms, proving the ASM thesis for them [61]. Essentially, the axioms consist of state structures, state transitions on parts of the state, with everything invariant under isomorphisms of the structures. The structures can be considered as algebras; hence the original name of evolving algebras. However, later the term Abstract State Machine (ASM) was generally adopted and Yuri Gurevich's colleague Egon Börger became the leading figure in the ASM community. The axiomatization of sequential algorithms has subsequently been extended to interactive and parallel algorithms.

During the 1990s, a research community built up and an ASM method was developed, allowing ASMs to be used for the formal specification and development of computer-based software and hardware [14]. ASM models for widely used programming languages such as C, Java, and Prolog, as well as specification languages such as SDL, UML, and VHDL, were created. A more detailed historical presentation of ASM's early developments has been produced by Egon Börger in 2002 [15].

Subsequently, two ASM books have appears in 2003 [29] and 2018 [28], both with Egon Börger at the lead author. As we have seen, the original ASM workshops have been combined with the B-Method, Z notation, and other state-based formal approaches to form the ABZ conference, started in 2008 [21], and continuing to this day.

3.1 Publications

Some key evolving algebra and ASM publications are shown in Fig. 2. An ASM tutorial introduction as also available [17]. A 1996 Steam Boiler Control case study competition book for different formal methods [1, 2] included an "Abstract Machine" specification and refinement [7]. An annotated ASM bibliography is available, covering 1988–1998 [26].

Some of the main author influences of Egon Börger can be seen in Fig. 3, both in terms of who has influenced him and who he has influenced. Figure 4 shows mentions of Egon Börger in the Google corpus of books, from the late 1960s onwards. It is interesting to note the peak of "Egon Boerger" with no umlaut in the mid-1990s and the peak of "Egon Börger" with an umlaut in the mid-2000s, perhaps illustrating improvements in computer typesetting around the end of the 20th century. Similarly, Fig. 5 shows mentions of Abstract State Machines in books from the 1980s onwards. This indicates a peak of interest in the early 2000s, although there may be a slight revival in the late 2010s, perhaps due in part to the 2018 book on the subject [28].

4 A Longterm Historical View

What is reasonable is real; that which is real is reasonable.

– Georg Hegel

1995: Yuri Gurevich, *Evolving Algebras 1993: Lipari Guide* [62].

1995: Yuri Gurevich and Egon Börger, *Evolving Algebras: Mini-Course* [63].

1995: Egon Börger, *Why use Evolving Algebras for Hardware and Software Engineering?* [14].

2000: Yuri Gurevich, *Sequential Abstract-State Machines capture Sequential Algorithms* [61].

2002: Egon Börger, *The Origins and Development of the ASM Method for High-Level System Design and Analysis* [15].

2002: Wolfgang Grieskamp, Yuri Gurevich, Wolfram Schulte & Margus Veanes, *Generating Finite State Machines from Abstract State Machines* [60].

2003: Egon Börger & Robert Stärk, *Abstract State Machines: A Method for High-Level System Design and Analysis* [29].

2003: Egon Börger, *The ASM Refinement Method* [16].

2008: Egon Börger, Michael Butler, Jonathan Bowen & Paul Boca, *Abstract State Machines, B and Z: First International Conference, ABZ 2008* [21,30].

2010: Egon Börger, *The Abstract State Machines Method for High-Level System Design and Analysis* [18].

2018: Egon Börger & Alexander Raschke, *Modeling Companion for Software Practitioners* [28,42].

Fig. 2. Some key ASM publications.

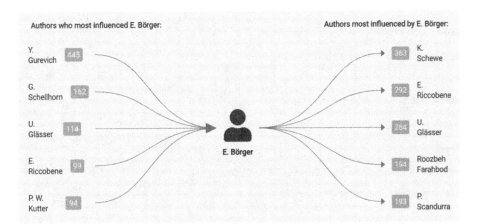

Fig. 3. Author influences of Egon Börger. (Semantic Scholar: http://www. semanticscholar.org.)

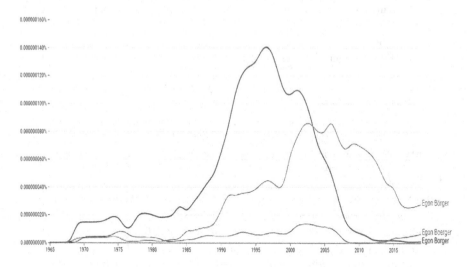

Fig. 4. Graph of mentions of Egon Börger in books. (Ngram Viewer, Google Books: http://books.google.com/ngrams.)

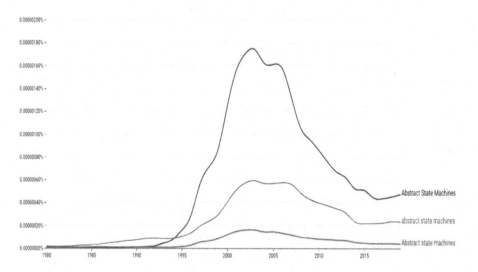

Fig. 5. Graph of mentions of Abstract State Machines in books. (Ngram Viewer, Google Books: http://books.google.com/ngrams.)

It is interesting to study the historical ancestry of ASM through the advisor tree of its leading promulgator, Egon Börger [97]. This itself forms a fascinating inter-related set of communities of related researchers through the centuries. Although Egon Börger has been based at the University of Pisa in Italy for much of his career, he was born in Germany and most of those in his advisor tree are of Germanic origin. This is assumed for those mentioned in the rest of this section, except where indicated otherwise.

We are lucky in the mathematical sciences such as computer science to have the excellent *Mathematics Genealogy Project* (MGP) resource available online (https://www.mathgenealogy.org), providing over a quarter of a million records of mathematicians (including many computer scientists), giving details of their degree, university, advisor(s), etc. This provides the foremost online resource for discovering degree and advisor information of people included in this database. In particular, records for advisors can be followed on the web through hyperlinks.

4.1   Logicians

Information on Egon Börger's doctoral thesis in 1971 is available on MGP [74]. It was entitled *Reduktionstypen in Krom- und Hornformeln* (in English, "Reduc-tion types in Krom and Horn formulas"), from Westfälische Wilhelms-Universität (now the University of Münster) in Germany and was supervised by Dieter Röd-ding [13]. Rödding (1937–1984) was a mathematical logician who made con-tributions to the classification of recursive functions and on recursive types in classical predicate logic. He was one of the first researchers to use a machine-oriented approach to complexity in his investigation of recursive functions and logical decision problems, before computer science had been established as an academic field.

Rödding's advisor, also at Münster, was Gisbert Hasenjaeger (1919–2006), another Germany mathematical logician. In 1949, Hasenjaeger developed a new proof for the completeness theorem of Kurt Gödel (1906–1978) for predicate logic. He worked as an assistant to the logician Heinrich Scholz (1884–1956) at the Cipher Department of the High Command of the Wehrmacht and was responsible for the security of the Enigma machine, used for encrypting German messages in World War II. The Enigma code was broken by Alan Turing (1912–1954) and his team at Bletchley Park in England [52]. Hasejaeger constructed a universal Turing machine (UTM) from telephone relays in 1963, now in the collection of the Heinz Nixdorf Museum in Paderborn. In the 1970s, Hasenjaeger learned about the breaking of the Enigma machine and he was impressed that Turing had worked successfully on this [81].

As well as working together, Heinrich Scholz was also Gisbert Hasenjaeger's doctoral advisor at Münster. Scholz was a philosopher and theologian, in addi-tion to being a logician. Alan Turing mentioned him regarding the reception of his ground-breaking paper *On Computable Numbers, with an Application to the Entscheidungsproblem*, read in 1936 and published in 1937 [90]. Turing received a preprint request from Scholz and was impressed by the German interest in the paper. Perhaps the use of a German term in the title helped! The *Entschei-dungsproblem* (German for "decision problem" [25]) was a challenge posed by the mathematicians David Hilbert (1862–1943) and Wilhelm Ackermann (1896–1962) in 1928. The origin of the *Entscheidungsproblem* goes back to Gottfried Leibniz (1646–1716) [83,84], of which more later. Scholz established the Institute of Mathematical Logic and Fundamental Research at Münster in 1936 and it was later led by Dieter Rödding.

Much later, during the early 21st century, Achim Clausing inspected Scholz's papers at Münster. He discovered two original preprints from Alan Turing, missing since 1945. The first paper *On Computable Numbers* [90] was with a postcard from Turing. In a letter from Turing to his mother while he was studying for his doctorate under the supervision of Alonzo Church (1903–1995) at Princeton University in the USA, there is indication that Scholz not only read Turing's paper, but also presented it in a seminar [51]. This could arguably have been the first theoretical computer science seminar. The second preprint found was Turing's foundational 1950 article on machine intelligence [91], foreseeing the subsequent development of artificial intelligence (AI). Turing noted by hand on the first page "This is probably my last copy" [51]!

Heinrich Scholz studied for two advanced degrees. The first was for a Licentiate theology degree at Humboldt-Universität zu Berlin (awarded in 1909) under the theologian Carl Gustav Adolf von Harnack (1851–1930) and Alois Adolf Riehl (1844–1924), an Austrian neo-Kantian philosopher. We shall hear more of Kant [5] later. The second was a Doctor of Philosophy degree at the Friedrich-Alexander-Universität Erlangen-Nürnberg (awarded in 1913) under Friedrich Otto Richard Falckenberg (1851–1920), entitled in Germany *Schleiermacher und Goethe; Ein Beitrag zur Geschichte des deutschen Geistes* (in ENglish, "Schleiermacher and Goethe; A contribution to the history of the German spirit"). This covered the theologian and philosopher Friedrich Schleiermacher (1768–1834), together with the renowned polymath Johann Wolfgang von Goethe (1749–1832).

Here we will follow each of these lines separately to an interesting denouement at the end. Those not interested in the history of mathematics [80] or philosophy [6,89] can safely skip to Sect. 4.4. The first line of advisors includes an eclectic mix of academics. The second line includes some of the leading philosophers and mathematicians of all time. Börger's advisor lineage as described in this section is illustrated in Fig. 6.

4.2   Polymaths: Astronomers, Geometrists, Mathematicians, Philosophers, Physicists, and Theologians

> *If others would but reflect on mathematical truths as deeply and continu-ously as I have, they would make my discoveries.*
>
> – Carl Gauss

Adolf von Harnack (sometimes known as just Adolf Harnack) was a Lutheran theologian and church historian. He gained his doctorate at the Universität Leipzig

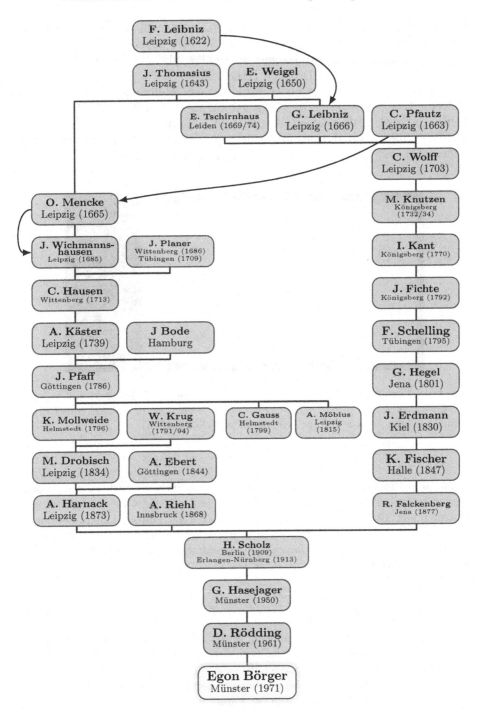

Fig. 6. Academic advisor tree for Egon Börger [74]. (Family relationships are indicated with additional curved arrow links, as discussed in the text.)

in 1873 under Moritz Wilhelm Drobisch (1802–1896) and Georg Karl Wilhelm Adolf Ebert (1820–1890), a philologist and literary historian. Drobisch was a mathematician, logician, psychologist, and philosopher. He was the brother of the composer Karl Ludwig Drobisch (1803–1854).

Moritz Drobisch studied under Karl Brandan Mollweide (1774–1825) and Wilhelm Traugott Krug (1770–1842), gaining his doctorate at the Universität Leipzig in 1824, with a dissertation entitled in Latin *Theoriae analyseos geometricae prolusio* on theories for analysis in geometry. Wilhelm Krug was a philosopher and writer who followed the Kantian school of logic.

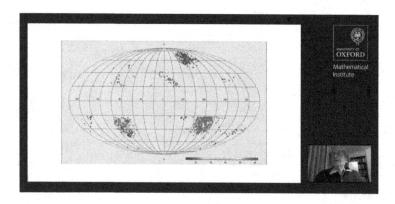

Fig. 7. A cosmological Mollweide projection, used in an online 2021 talk by the Oxford mathematical physicist Roger Penrose, in celebration of his 2020 Nobel Prize. (YouTube: http://www.youtube.com/watch?v=1zXC51o3Efl.)

Karl Mollweide was a mathematician and astronomer. He invented the *Mollweide projection* for maps, giving equal areas for different parts of a map of a curved surface like the spherical world. This is useful for wide-area global and cosmic maps, giving a projection in the form of a flat ellipse. This projection is used to the present day, as illustrated in Fig. 7. Mollweide also discovered what is now known as *Mollweide's formula* in trigonometry, useful in finding solutions relating to triangles:

$$\frac{a+b}{c} = \frac{cos(\frac{\alpha-\beta}{3})}{sin(\frac{\gamma}{2})} \qquad \frac{a-b}{c} = \frac{sin(\frac{\alpha-\beta}{3})}{cos(\frac{\gamma}{2})}$$

where a, b, c are the lengths of sides of a triangle and α, β, γ are the opposite angles. These pleasingly symmetrically matching equations both include all the important constituent parts (side lengths and angles) of a triangle.

Mollweide studied under the mathematician Johann Friedrich Pfaff (1765–1825) at the Universität Helmstedt, attaining his award in 1796. Pfaff was the advisor for two other famous mathematicians, Carl Friedrich Gauss (1777–1855), also a physicist, and August Ferdinand Möbius (1790–1888), also a theoretical astronomer [56]. Gauss has many mathematical and scientific concepts named after him as the leading scientist of his generation. Möbius has a number of mathematical, especially geometrical, ideas named after him too, the most well-known of which is the *Möbius strip*, a surface in three-dimensions with only one side, which he discovered in 1858.

Johann Pfaff achieved his doctorate in 1786 at the Georg-August-Universität Göttingen under Abraham Gotthelf Kästner (1719–1800) and the astronomer Johann Elert Bode (1747–1826), known for the Titus-Bode law for predicting the space between planets in a solar system. Kästner was a mathematician and also an epigrammatist. He wrote mathematical textbooks, compiled encyclopaedias, translated scientific proceedings, and even wrote poetry in an epigrammatic style. In 1789, he was elected a Fellow of the Royal Society, the leading scientific society based in London, England. The moon crater *Kästner*, 49 km in diameter, is named after him.

Abraham Kästner studied under the mathematician Christian August Hausen (1693–1743) Universität Leipzig, achieving his doctorate in 1739, with a dissertation entitled in Latin *Theoria radicum in aequationibus* (in English, "The theory of the roots of equations"). Hausen is also known for his research on electricity, using a *triboelectric* generator. The triboelectric effect is a type of electricity where some materials become electrostatically charged after being separated from another material that they were touching previously. Rubbing the two materials can increase their surface contact, and thus increase the triboelectric effect. Combing hair with a plastic comb is a common way of creating triboelectricity.

Christian Hausen gained his doctorate in 1713 at the Universität Wittenberg (now merged with Halle to later become the Martin-Luther-Universität Halle-Wittenberg), under the guidance of the philologist Johann Christoph Wichmannshausen (1663–1727) and the mathematician Johannes Andreas Planer (1665–1714). Wichmannshausen studied under the direction of his father-in-law, the philosopher and scientist Otto Mencke (1644–1707), gaining his doctorate at the Universität Leipzig in 1685, on issues concerning the ethical nature of divorces.

Otto Mencke also studied at Leipzig and his doctorate was awarded in 1665. His advisor was the philosopher and jurist Jakob Thomasius (1622–1684), and his dissertation was on theology. In 1682, he founded the first German scientific journal in Germany, entitled *Acta Eruditorum*. He was a professor of moral philosophy at Leipzig.

Jakob Thomasius conducted his studies at Leipzig as well, under Friedrich Leibniz (aka Leibnütz, 1597–1652), gaining his degree in 1643. He was an important foundational scholar in the history of philosophy [88]. Friedrich Leibniz was a Lutheran lawyer, a notary, registrar, and professor of moral philosophy at Leipzig. He was the father of the notable mathematician and polymath Gottfried Wilhelm (von) Leibniz (1646–1716), to whom we shall return [84].

4.3   Philosophers and Mathematicians

To comprehend what is, is the task of philosophy: and what is is Reason.
                                                                    – Georg Hegel

We now return to the advisor for Heinrich Scholz's second dissertation in 1913, under Richard Falckenberg (1851–1920) at the Friedrich-Alexander-Universität Erlangen-Nürnberg, at the bottom of the right-hand lineage in Fig. 6. Fackenberg was a historian of philosophy. He wrote the book *History of Modern Philosophy*, originally published in 1886, and still available in English translation through *Project Gutenberg* online as an open-access resource [55], scanned from the original editions [54]. Falckenberg received his doctorate from the Friedrich-Schiller-Universität Jena in 1877, having studied under the philosopher Ernst Kuno Berthold Fischer (1824–1907). His dissertation was entitled in German *Über den intelligiblen Charakter. Zur Kritik der Kantischen Freiheitslehre* (in English, "About the intelligent character. On the criticism of the Kantian doctrine of freedom").

Kuno Fischer was also a historian of philosophy and a critic, known for his lecturing skills. One of Fischer's philosophical contributions was to categorize philosophers into followers of empiricism and rationalism, including Gottfried Leibniz as a rationalist. He was a follower of Hegelianism and interpreted the works of Kant. He published a six-volume set of monographs entitled *Geschichte der neuern Philosophie* (in English, "History of modern philosophy") [57], which influenced the philosopher Friedrich Wilhelm Nietzsche (1844–1900). Fischer also taught the philosopher, logician, and mathematician Friedrich Ludwig Gottlob Frege (1848–1925) at Jena and, more unusually, the English playwright, novelist, and short-story writer William Somerset Maugham (1874–1965) at Heidelberg. Fischer's 80th birthday in 1904 was celebrated with a Festschrift, published three years later [73]. Fischer studied at the Universität Halle (now merged with Wittenberg to be the Martin-Luther-Universität Halle-Wittenberg) under Johann Eduard Erdmann (1805–1892). He was awarded a doctorate in 1847 for a dissertation on the ancient Athenian philosopher Plato.

Johann Erdmann was a pastor, historian of philosophy, and philosopher of religion. He wrote a three-volume set of books entitled *A History of Philosophy*, available in an English translation [53]. Erdmann studied for his doctorate under the leading philosopher Georg Wilhelm Friedrich Hegel (1770–1831) [76] at Christian-Albrechts-Universität zu Kiel. He received his doctoral degree in 1830 with a dissertation entitled in Latin *Quid intersit inter Philosophiam et Theologiam* (in English, "What is the difference between Philosophy and Theology").

Hegel was a leading figure in German idealism [85], developed from Kant's ideas [87], linked with Romanticism and the Enlightenment, where reality is seen depending on human perception or understanding. Hegel's ideas continue to be highly influential on contemporary Western philosophical issues, in areas relating to aesthetics, ontology, politics, both in analytic philosophy (mainly in English-speaking countries) and continental philosophy (largely in mainland Europe).

Hegel's philosophical ideas are now termed *Hegelianism*, summarized in the title of the preface in *Elements of the Philosophy of Right* as "What is rational is real; And what is real is rational" [66].

Hegel's *Science of Logic* [65] presented his idea of logic as a system of dialectics, later dubbed *Hegelian dialectic*. This is normally presented in a three-stage developmental style, as provided by Heinrich Moritz Chalybäus (1796–1862): 1) a thesis or problem; 2) an antithesis or reaction, contradicting this thesis; and 3) their resolution through a synthesis or solution. Although named after Hegel, this is a different formulation to his. Hegel was influenced in these ideas by Kant, with Johann Gottlieb Fichte (1762–1814, see later) elaborating and popularizing the approach. As well as science, Hegel has also been influential in artistic circles with respect to aesthetics to this day [77].

Hegel's dissertation was defended at Friedrich-Schiller-Universität Jena while with the philosopher Friedrich Wilhelm Joseph von Schelling (1775–1812) [86]. Hegel and Schelling shared a room at university. Schelling was an "Extraordinary Professor" (a professor without a chair in Germany) at Jena and encouraged his friend Hegel to come to Jena in 1801. Hegel became a *Privatdozent* (unsalaried lecturer) at Jena. His inaugural dissertation that year, in Latin, was entitled *De orbitis planetarum* (in English "The orbits of the planets"), interestingly in the field of astronomy, which then was considered as natural philosophy [4]. The dissertation is inscribed in Latin *Socio Assumto Carolo Schelling* (in English "an assumed ally Karl Schelling") and Schelling was present at Hegel's defence on 27 August 1801, Hegel's 31st birthday. At the time, Schelling was still only 26, several years younger than Hegel.

Karl Schelling was part of the German idealism philosophical movement. He and Hegel were early friends, but later became rivals, and Schelling became rather eclipsed by Hegel. Schelling undertook his doctorate at the Eberhard-Karls-Universität Tübingen, completing his dissertation in 1795, and working with the philosopher Johann Gottlieb Fichte (1762–1814) [86].

Johann Fichte's ideas were criticized by both Schelling and Hegel [92]. He studied for his doctorate with the well-known philosopher Immanuel Kant (1724–1804) [5] at the Universität Königsberg. His dissertation, produced in 1792, was entitled in German *Versuch einer Kritik aller Offenbarungen* (in English, "Attempt to criticize all revelations").

Immanuel Kant was an important thinker in the Age of Enlightenment (aka the Age of Reason). He produced works covering aesthetics, epistemology, ethics, and metaphysics. Subsequently, his ideas have made him one of the most influential historical figures within modern Western philosophy. He founded transcendental idealism, as expounded in his *Critique of Pure Reason* [70].

Kant enrolled at the University of Königsberg (now Kaliningrad) in 1740, at the age of 16, and remained there for his entire career. He studied the philosophy of Gottfried Leibniz and Christian Freiherr von Wolff (1679–1754) under the direction of Martin Knutzen (1713–1751), the Extraordinary Professor of Logic and Metaphysics there until his early death, aged 37. Knutzen was a rationalist who had an interest in British philosophy, especially empiricism, and science. He introduced Kant to the then relatively recent mathematical physics of Isaac Newton (1642–1727). Knutzen discouraged Kant from following Leibniz's theory of pre-established harmony [88], where substances only affect themselves despite apparent interactions, and idealism, the concept that reality is mental, which many 18th-century philosophers regarded negatively.

In 1755, Kant received a license to lecture at Königsberg, and produced a master's thesis in that year under the Prussian physicist and philosopher Johann Gottfried Teske (1704–1772). His doctoral dissertation was produced in 1770, entitled in Latin *De mundi sensibilis atque intelligibilis forma et principiis* (in English, "The form and principles of the sensible and the intelligible world").

Martin Knutzen studied mathematics, philosophy, and physics, at Königsberg, producing his inaugural dissertation in 1732, his Master of Arts dissertation in 1733, and his doctoral dissertation in 1734, resulting in him becoming a professor at the early age of 21 shortly after. Knutzen was a follower of Christian von Wolff at Königsberg and the rationalist school of thinkers. He also had an interest in natural sciences, teaching astronomy, mathematics, and physics, as well as philosophy. His interest in Newton's ideas led to him question the theory of pre-established harmony, as espoused by Leibniz and Wolff. He defended the idea of mechanical causality in moving physical objects and his lessons at Königsgberg influenced Kant, especially in his work on the *Critique of Judgement*, which attempted reconcile spiritual autonomy with respect to mechanical reality [71].

Christian von Wolff enrolled at Jena but moved to Leipzig in 1702, producing a *Habilitationsschrift* dissertation in 1703, written in Latin and entitled *Philosophia practica universalis, methodo mathematica conscripta* (in English, "On Universal Practical Philosophy, composed according to the Mathematical Method"). Wolff's main advisor was Ehrenfried Walther von Tschirnhaus (1651–1708). Otto Mencke, whom we have met early on the other branch of Egon Börger's advisor tree, served as an examiner for Wolff's dissertation. Mencke was impressed enough to send Wolff's dissertation to Gottfried Leibniz. Wolff and Leibniz remained in correspondence together until Leibniz died in 1716. The astronomer, geographer, librarian, and mathematician Christoph Pfautz (1645–1711) was also an advisor. Wolff was a mathematician, philosopher, and scientist during the German Enlightenment and is regarded by many to be the most influential and important German philosopher between Leibniz and Kant, two giants in the field.

Pfautz helped to co-found the first German scientific journal *Acta Eruditorum* (as mentioned earlier) with his brother-in-law Otto Mencke in 1682, in Egon Börger's other advisor line. To raise the journal's profile and encourage submissions, Pfautz took Otto Mencke to Holland and England in 1680, via Amsterdam, Antwerp, Delft, Leiden, Utrecht, London, and Oxford. Pfautz met leading scientists, including Isaac Newton, whose views he introduced to Germany scholars in the journal. Pfautz was a regular correspondent with Gottfried Leibniz from 1679 and was one of the early Enlightenment proponents at Leipzig.

Ehrenfried von Tschirnhaus was a mathematician, physician, physicist, and philosopher who originally studied at Leiden University in Holland. He developed the *Tschirnhaus transformation* to remove intermediate terms from an algebraic equation, published in the *Acta Eruditorum* journal in 1683. He is also considered by some to have invented European porcelain.

Gottfried Leibniz was a leading polymath of his age and one of the most important logicians, mathematicians, and philosophers during the Enlightenment. He followed the 17th-century philosophical tradition of rationalism. His most important mathematical contribution was the development of differential and integral calculus, at the same time that Isaac Newton developed these ideas independently too in England. They used different notations and Leibniz's more general notation is the one that has endured. Indeed, Newton's notation held back subsequent mathematical advances in England compared to continental Europe for centuries.

Leibniz introduced heuristic ideas of the law of continuity, allowing finite reasoning to be generalized to the infinite case (e.g., when considering a circle as being an infinite-sided polygon), and the transcendental law of homogeneity, allowing terms tending to the infinitesimal to be ignored (e.g., $a + dx = a$). Much later in the 1960s, these ideas became important in non-standard analysis, reformulating calculus using a logically rigorous notion of infinitesimal numbers, illustrating how long it can take for mathematical ideas to have a useful application.

Leibniz was also inventive in the development of mechanical calculators. While considering the inclusion of automatic multiplication and division in Pascal's calculator of the French mathematician Blaise Pascal (1623–1662), he originated a pinwheel calculator in 1685 with adjustable numbers of teeth (normally 0 to 9 in the decimal system). He also invented what became known as the Leibniz wheel, a cylindrical drum with stepped teeth, as used in the *arithmometer*, the first mechanical calculator to be mass-produced, introduced in 1820. This interest in mechanical reasoning can be seen as a precursor to later consideration of the nature of computation in a logical framework, including issues concerning the *Entscheidungsproblem* ("decision problem"), as previously mentioned [83].

Gottfried Leibniz studied for his doctorate under Jakob Thomasius (whose student Otto Mencke is also on Egon Börger's other line of advisors on the left-hand side of Fig. 6) and the astronomer, mathematician, and philosopher, Erhard Weigel (1625–1699) at the Universität Leipzig, producing his dissertation in Latin entitled *Disputatio arithmetica de complexionibus* on arithmetic in 1666.

Leibniz also studied with the legal scholar Bartholomäus Leonhard von Schwendendörffer (1631–1705) at the Universitt Altdorf and later with his mentor, the Dutch astronomer, inventor, mathematician, and physicist, Christiaan Huygens (1629–1695) through the French Académie Royale des Sciences (Royal Academy of Sciences) in Paris, after visiting the city from 1672. Hugygens was a major figure in the European Scientific Revolution that marked the emergence of modern science and was influential in the Age of Enlightenment. He invented the Huygens eyepiece with two lenses for telescopes.

As we saw earlier, the father and son Friedrich and Gottfried Leibniz are also related via Jakob Thomasius academically. Both are part of Egon Böger's academic lineage, Friedrich Leinbiz via both his major lines, as illustrated in Fig. 6.

4.4   The Origins of Binary Computing

There are 10 types of people: those that can count in binary and those that can't.   – Anon.

Gottfried Leibniz, 13 generations back in Egon Börger's academic genealogical tree (see Fig. 6), studied the binary numbering system in 1679, later published in an 1703 French article *Explication de l'Arithmétique Binaire* (in English, "Explanation of Binary Arithmetic", see Fig. 8) [72]. In 1854, the English mathematician George Boole (1815–1864), based at Queen's College (now University College), Cork, in southern Ireland, published a foundational book, *The Laws of Thought*, detailing an algebraic system of binary logic, later dubbed Boolean algebra, important in the design of digital electronics.

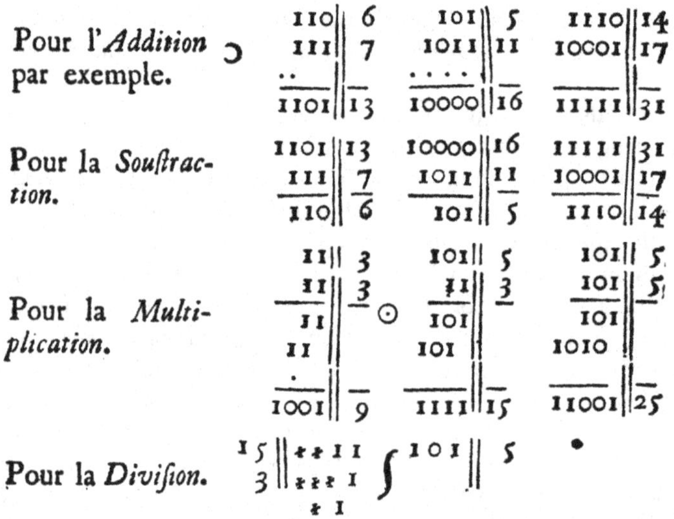

Fig. 8. Leibniz's examples of binary arithmetic, published in 1703 [72].

In 1937, the American engineer and mathematician Claude Shannon (1916–2001) [58] worked on his novel master's thesis (issued later in 1940) at the Massachusetts Institute of Technology (MIT) that implemented Boolean algebra and binary arithmetic using electronic relays and switches, entitled *A Symbolic Analysis of Relay and Switching Circuits* [82], foundational in digital circuit design.

Also in 1937, George Stibitz (1904–1995), while working at Bell Labs in the USA, created a relay-based computer called the *Model K* (for "Kitchen", where it was assembled!), which implemented binary addition [35]. In the same year, Alan Turing's foundational paper based on what became known as a *Turing machine*, a mathematical model for a digital computational device, appeared [90].

In 1945, the Hungarian-American mathematician and polymath John von Neumann (1903–1957) produced a draft report on the EDVAC binary computer design, that became dubbed *von Neumann architecture*, a standard style of digital computer design [93]. Thus, with all these subsequent developments, Leibniz's ideas on the binary number system were foundational for modern digital computer design. The discipline of computer science has developed especially from the second half of the 20th century [40], with the first academic computer science departments opening in the 1960s [36]. Without all these developments, there would be no need for formal methods in general and ASM in particular.

4.5   Further Academic Advisor Relationships

We have seen (as illustrated in Fig. 6) that Egon Börger's academic lineage goes back to the leading mathematician Gottfried Leibniz and his father. His immediate "ancestors" are logicians. Then there is a split into two major lineages with Heinrich Scholz through his two separate degrees in 1909 and 1913. The first line (on the left in Fig. 6) includes an eclectic mix of scientists and philosophers, including a relationship with Gauss and Möbius, leading to Gottfried Leibniz's father Friedrich Leibniz (also his "grandfather" in the academic tree of advisors via Jakob Thomasius). The second line (on the right in Fig. 6) of mainly philosophers includes two of the most important philosophers of all time, Hegel and Kant, as well as Gottfried Leibniz himself, and then links to the first line via Jakob Thomasius. These can be seen as a historical community of academics, each passing on their knowledge to the next generation, eventually to Egon Börger.

Egon Börger is a distant academic "relative" of the 19th/20th century mathematician David Hilbert, via Johann Pfaff and his student Gauss. Hilbert was influential on theoretical computer science through the likes of Kurt Gödel and then Alan Turing. Börger is also related to Turing, who's academic advisor line goes back to Gottfried Leibniz as well, via another follower of Leibniz in Paris, the rationalist philosopher Nicolas Malebranche (1638–1715) [43]. Even Yuri Gurevich and Egon Börger are distantly related by advisor. Following Gurevich's advisor tree back in time on the Mathematics Genealogy Project [75], we find the important Russian mathematician Pafnuty Lvovich Chebyshev (1821–1894) several generations back, eventually leading to Johann Pfaff in Börger's advisor tree via another of Pfaff's students, the mathematician Johann Martin Christian Bartelsi (1789–1836), who also tutored Gauss.

5   Conclusion

There is nothing without reason.   – Gottfried Leibniz

The Abstract State Machines (ASM) approach is one of a number of competing state-based formal methods for modelling computer-based systems. It has been used in this role for industrial-strength computer-based languages such as programming languages and specification notations. The community associated with ASM has developed since the 1980s and continues in tandem with other state-based approaches such as the Z notation, the B-Method, and Event-B. Each has their own advantages and disadvantages, which are beyond the scope of this paper. Each have their own community of adherents, that have now somewhat merged with the establishment of the ABZ conference in 2008 [21].

The 2003 book on ASM [29] is a general introduction to ASM. The subsequent 2018 ASM book [28] is entitled *Modeling Companion for Software Practitioners*. It can be used for self-study, as a reference book, and for teaching, but its title indicates the intention of being a practical book for potential industrial use. A third book with industrial case studies in due course could complete these books as a trilogy [42].

The ASM community is an example of a Community in Practice (CoP) in action. Other related CoPs are based around state-based specification and development approaches such as the Z notation, B-Method/Event-B, etc. CoPs can potentially merge and create new CoPs. For example, the B-Method and then Event-B were developed after the Z notation largely by the same progenitor, Jean-Raymond Abrial, and with some in the Z community becoming part of the B community.

A Community of Practice depends on people with different skills for success, be it for ideas, vision, organization, etc. Yuri Guevich and Egon Börger were both key for the success of ASM, just as Steve Wozniak and Steve Jobs [69] were both crucial for the initial launch of Apple. I will leave it for the reader to decide who has taken on which roles.

We have also considered Egon Börger's advisor tree historically, which started mainly in the fields of mathematics and philosophy, and more recently has included several logicians. All this knowledge has helped to lead to the development of the ASM approach. Members of the advisor tree have themselves participated in CoPs, such as the rationalist movement, German idealism, etc. Some have been eminent enough to be leaders of CoPs, like Kant and Hegel. They have inspired their own eponymous schools of thought, such as Kantian ethics and Hegelism.

Predicting the future is always difficult, but the community around ASM has been successful enough to leave its mark on the formal methods community as a whole. What is clear is that without Egon Börger, it is unlikely that the ASM community would have developed to the extent that it has.

6   Postscript

Genius is the ability to independently arrive at and understand concepts that would normally have to be taught by another person.

– Immanuel Kant

What it means to be a genius and how long it takes to become a genius are matters for debate [78]. However, Kant's definition above is perhaps a good one. Egon Börger completed his doctoral thesis in 1971 [13] and his first two decades of papers were mainly on logic and decision problems [11]. Subsequently his research centred increasingly around Abstract State Machines [15]. As the main leader of the ASM community, he has to this day produced papers developing ideas around ASM, advancing its use. He has been the teacher of ASM and co-author the two main books on the subject [28,29]. Thus, by Kant's definition above, Egon Börger is a genius.

Acknowledgments. The author is grateful to Egon Börger, for inspiration and collaboration over the years, and to Museophile Limited for financial support. Tim Denvir, Cliff Jones, and an anonymous reviewer provided helpful comments on earlier drafts of this paper.

References

1. Abrial, J.-R., Börger, E., Langmaack, H. (eds.): Formal Methods for Industrial Applications. LNCS, vol. 1165. Springer, Heidelberg (1996). https://doi.org/10.1007/BFb0027227
2. Abrial, J.-R., Börger, E., Langmaack, H.: The steam boiler case study: competition of formal program specification and development methods. In: Abrial, J.-R., Börger, E., Langmaack, H. (eds.) Formal Methods for Industrial Applications. LNCS, vol. 1165, pp. 1–12. Springer, Heidelberg (1996). https://doi.org/10.1007/BFb0027228
3. Abrial, J.R., Glässer, U. (eds.): 06191 Summary – Rigorous Methods for Software Construction and Analysis. No. 06191 in Dagstuhl Seminar Proceedings, Internationales Begegnungs- und Forschungszentrum für Informatik (IBFI), Schloss Dagstuhl, Germany, May 2006. http://drops.dagstuhl.de/opus/volltexte/2006/665
4. Adler, P., Hegel, G.W.F.: Philosophical dissertation on the orbits of the planets (1801). Graduate Faculty Philos. J. **12**(1&2), 269–309 (1987). Translated with Foreword and Notes
5. Ameriks, K.: Kant, Immanuel. In: Audi [6], pp. 398–404
6. Audi, R. (ed.): The Cambridge Dictionary of Philosophy. Cambridge University Press, Cambridge (1995)
7. Beierle, C., Börger, E., Đurđanović, I., Glässer, U., Riccobene, E.: Refining abstract machine specifications of the steam boiler control to well documented executable code. In: Abrial, J.-R., Börger, E., Langmaack, H. (eds.) Formal Methods for Industrial Applications. LNCS, vol. 1165, pp. 52–78. Springer, Heidelberg (1996). https://doi.org/10.1007/BFb0027231

8. Beler, A., Borda, A., Bowen, J.P., Filippini-Fantoni, S.: The building of online communities: an approach for learning organizations, with a particular focus on the museum sector. In: Hemsley, J., Cappellini, V., Stanke, G. (eds.) EVA 2004 London Conference Proceedings, EVA Conferences International, University College London, UK, pp. 2.1–2.15 (2004). https://arxiv.org/abs/cs/0409055

9. Bjørner, D., et al.: A ProCoS project description: ESPRIT BRA 3104. Bull. Eur. Assoc. Theoret. Comput. Sci. **39**, 60–73 (1989). http://researchgate.net/publication/256643262

10. Borda, A., Bowen, J.P.: Virtual collaboration and community. In: Information Resources Management Association Virtual Communities: Concepts, Methodologies, Tools and Applications, chap. 8.9, pp. 2600–2611. IGI Global (2011)

11. Börger, E.: Egon Boerger. Google Scholar. https://scholar.google.com/citations?user=j2lxsK0AAAAJ

12. Börger, E.: Prof. Dr. Egon Börger. Dipartimento di Informatica, Università di Pisa, Italy. http://pages.di.unipi.it/boerger/

13. Börger, E.: Reduktionstypen in Krom- und Hornformeln. Ph.D. thesis, Westfälische Wilhelms-Universität Münster, Germany (1971). Translation: "Reduction types in Krom and Horn formulas"

14. Börger, E.: Why use evolving algebras for hardware and software engineering? In: Bartosek, M., Staudek, J., Wiedermann, J. (eds.) SOFSEM 1995. LNCS, vol. 1012, pp. 236–271. Springer, Heidelberg (1995). https://doi.org/10.1007/3-540-60609-2_12

15. Börger, E.: The origins and development of the ASM method for high-level system design and analysis. J. Univ. Comput. Sci. **8**(1), 2–74 (2002). https://doi.org/10.3217/jucs-008-01-0002

16. Börger, E.: The ASM refinement method. Formal Aspects Comput. **15**, 237–257 (2003). https://doi.org/10.1007/s00165-003-0012-7

17. Börger, E.: The ASM method for system design and analysis. a tutorial introduction. In: Gramlich, B. (ed.) FroCoS 2005. LNCS (LNAI), vol. 3717, pp. 264–283. Springer, Heidelberg (2005). https://doi.org/10.1007/11559306_15

18. Börger, E.: The Abstract State Machines method for high-level system design and analysis. In: Boca, P.P., Bowen, J.P., Siddiqi, J.I. (eds.) Formal Methods: State of the Art and New Directions, chap. 3, pp. 79–116. Springer, Heidelberg (2010). https://doi.org/10.1007/978-1-84882-736-3_3

19. Börger, E.: Private communication, March 2021

20. Börger, E., Bowen, J.P., Butler, M.J., Poppleton, M.: Editorial. Formal Aspects Comput. **23**(1), 1–2 (2011). https://doi.org/10.1007/s00165-010-0168-x

21. Börger, E., Butler, M., Bowen, J.P., Boca, P. (eds.): ABZ 2008. LNCS, vol. 5238. Springer, Heidelberg (2008). https://doi.org/10.1007/978-3-540-87603-8

22. Börger, E., Butler, M.J., Bowen, J.P., Boca, P.P. (eds.): ABZ 2008 Conference: Short papers. BCS, London, UK (2008)

23. Börger, E., Durdanovic, I.: Correctness of compiling Occam to transputer code. Comput. J. **39**(1), 52–92 (1996)

24. Börger, E., Gargantini, A., Riccobene, E.: ASM. In: Habrias and Frappier [64], chap. 6, pp. 103–119. https://doi.org/10.1002/9780470612514

25. Börger, E., Grädel, E., Gurevich, Y.: The Classical Decision Problem. Springer, Heidelberg (1997/2001)

26. Börger, E., Huggins, J.K.: Abstract state machines 1988–1998: commented ASM bibliography. Bull. EATCS **64**, 105–127 (1998). https://arxiv.org/pdf/cs/9811014

27. Börger, E., Mazzanti, S.: A practical method for rigorously controllable hardware design. In: Bowen, J.P., Hinchey, M.G., Till, D. (eds.) ZUM 1997. LNCS, vol. 1212, pp. 149–187. Springer, Heidelberg (1997). https://doi.org/10.1007/BFb0027289

28. Börger, E., Raschke, R.: Modeling Companion for Software Practitioners (2018). https://doi.org/10.1007/978-3-662-56641-1

29. Börger, E., Stärk, R.: Abstract State Machines: A Method for High-Level System Design and Analysis. Springer, Heidelberg (2003). https://doi.org/10.1007/978-3-642-18216-7

30. Börger, E., Thalheim, B.: Modeling workflows, interaction patterns, web services and business processes: the ASM-based approach. In: Börger, E., Butler, M., Bowen, J.P., Boca, P. (eds.) ABZ 2008. LNCS, vol. 5238, pp. 24–38. Springer, Heidelberg (2008). https://doi.org/10.1007/978-3-540-87603-8_3

31. Bowen, J.P.: A ProCoS II project description: ESPRIT Basic Research project 7071. Bull. Eur. Assoc. Theoret. Comput. Sci. **50**, 128–137 (1993). http://researchgate.net/publication/2521581

32. Bowen, J.P.: A ProCoS-WG Working Group description: ESPRIT Basic Research 8694. Bull. Euro. Assoc. Theoret. Comput. Sci. **53**, 136–145 (1994)

33. Bowen, J.P.: Formal Specification and Documentation Using Z: A Case Study Approach. International Thomson Computer Press (1996). http://researchgate.net/publication/2480325

34. Bowen, J.P., Hoare, C., Antony R.: Rojas [79], pp. 368–370

35. Bowen, J.P.: Stibitz, George. In: Rojas [79], pp. 732–734

36. Bowen, J.P.: Computer science. In: Heilbron, J.L. (ed.) The Oxford Companion to the History of Modern Science, pp. 171–174. Oxford University Press (2003). https://doi.org/10.1093/acref/9780195112290.001.0001

37. Bowen, J.P.: Z. In: Habrias and Frappier [64], chap. 1, pp. 3–20. https://doi.org/10.1002/9780470612514, http://researchgate.net/publication/319019328

38. Bowen, J.P.: Online communities: visualization and formalization. In: Blackwell, C. (ed.) Cyberpatterns 2013: Second International Workshop on Cyberpatterns - Unifying Design Patterns with Security, Attack and Forensic Patterns, pp. 53–61. Oxford Brookes University, Abingdon (2013). http://arxiv.org/abs/1307.6145

39. Bowen, J.P.: A relational approach to an algebraic community: from Paul Erdős to He Jifeng. In: Liu, Z., Woodcock, J., Zhu, H. (eds.) Theories of Programming and Formal Methods. LNCS, vol. 8051, pp. 54–66. Springer, Heidelberg (2013). https://doi.org/10.1007/978-3-642-39698-4_4

40. Bowen, J.P.: It began with Babbage: the genesis of computer science, by Subrata Dasgupta: Oxford University Press, 2014, 334 p. £22.99 (hardback), 263–265, ISBN 978-0-19-930941-2. BSHM Bulletin: Journal of the British Society for the History of Mathematics 30(3) (2015). https://doi.org/10.1080/17498430.2015.1036336

41. Bowen, J.P.: Provably correct systems: community, connections, and citations. In: Hinchey, M.G., Bowen, J.P., Olderog, E.-R. (eds.) Provably Correct Systems. NMSSE, pp. 313–328. Springer, Cham (2017). https://doi.org/10.1007/978-3-319-48628-4_13

42. Bowen, J.P., Börger, E., Raschke, A.: Modeling companion for software practitioners. Formal Aspects Comput. **30**(6), 761–762 (2018). https://doi.org/10.1007/s00165-018-0472-4

43. Bowen, J.P.: The impact of Alan Turing: formal methods and beyond. In: Bowen, J.P., Liu, Z., Zhang, Z. (eds.) SETSS 2018. LNCS, vol. 11430, pp. 202–235. Springer, Cham (2019). https://doi.org/10.1007/978-3-030-17601-3_5

44. Bowen, J.P.: A personal view of EVA London: past, present, future. In: Weinel, J., Bowen, J.P., Diprose, G., Lambert, N. (eds.) EVA London 2020: Electronic Visualisation and the Arts, Electronic Workshops in Computing (eWiC), BCS, London, UK, pp. 8–15 (2020). https://doi.org/10.14236/ewic/EVA2020.2

45. Bowen, J.P., Hinchey, M.G.: Ten commandments ten years on: lessons for ASM, B, Z and VSR-Net. In: Abrial, J.-R., Glässer, U. (eds.) Rigorous Methods for Software Construction and Analysis. LNCS, vol. 5115, pp. 219–233. Springer, Heidelberg (2009). https://doi.org/10.1007/978-3-642-11447-2_14

46. Bowen, J.P., Hinchey, M.G., Till, D. (eds.): ZUM 1997. LNCS, vol. 1212. Springer, Heidelberg (1997). https://doi.org/10.1007/BFb0027279

47. Bowen, J.P., Hoare, C.A.R., Langmaack, H., Olderog, E.R., Ravn, A.P.: A ProCoS II project final report: ESPRIT Basic Research project 7071. Bull. Euro. Assoc. Theoret. Comput. Sci. **59**, 76–99 (1996). http://researchgate.net/publication/2255515

48. Bowen, J.P., Hoare, C.A.R., Langmaack, H., Olderog, E.R., Ravn, A.P.: A ProCoS-WG Working Group final report: ESPRIT Working Group 8694. Bull. Euro. Assoc. Theoret. Comput. Sci. **64**, 63–72 (1998). http://researchgate.net/publication/2527052

49. Bowen, J.P., Reeves, S.: From a community of practice to a body of knowledge: a case study of the formal methods community. In: Butler, M., Schulte, W. (eds.) FM 2011. LNCS, vol. 6664, pp. 308–322. Springer, Heidelberg (2011). https://doi.org/10.1007/978-3-642-21437-0_24

50. Bowen, J.P., Wilson, R.J.: Visualising virtual communities: from Erdős to the arts. In: Dunn, S., Bowen, J.P., Ng, K.C. (eds.) EVA London 2012: Electronic Visualisation and the Arts, Electronic Workshops in Computing (eWiC), BCS, pp. 238–244 (2012). http://arxiv.org/abs/1207.3420

51. Clausing, A.: Prof. Dr. Achim Clausing. University of Münster, Germany. https://ivv5hpp.uni-muenster.de/u/cl/

52. Copeland, B.J., Bowen, J.P., Sprevak, M., Wilson, R.J. (eds.): The Turing Guide. Oxford University Press, Oxford (2017)

53. Erdmann, J.E.: A History of Philosophy. Swan Sonnenschein & Co., London (1890). English translation by W. S. Hough

54. Falckenberg, R.: History of Modern Philosophy: From Nicolas of Cusa to the Present Time. H. Holt (1893). https://www.google.co.uk/books/edition/_/BZwvAAAAYAAJ

55. Falckenberg, R.: History of Modern Philosophy: From Nicolas of Cusa to the Present Time. Project Guttenberg (2004). http://www.gutenberg.org/ebooks/11100

56. Fauvel, J., Flood, R., Wilson, R. (eds.): Möbius and his Band: Mathematics and Astronomy in Nineteenth-century Germany. Oxford University Press, Oxford (1993)

57. Fischer, K.: Geschichte der neuern Philosophie. Stuttgart-Mannheim-Heidelberg (1854–77), New edn, Heidelberg (1897–1901)

58. Giannini, T., Bowen, J.P.: Life in code and digits: when Shannon met Turing. In: Bowen, J.P., Diprose, G., Lambert, N. (eds.) EVA London 2017: Electronic Visualisation and the Arts, Electronic Workshops in Computing (eWiC), BCS, pp. 51–58 (2017). https://doi.org/10.14236/ewic/EVA2017.9

59. Giannini, T., Bowen, J.P. (eds.): Museums and Digital Culture. SSCC. Springer, Cham (2019). https://doi.org/10.1007/978-3-319-97457-6

60. Grieskamp, W., Gurevich, Y., Schulte, W., Veanes, M.: Generating finite state machines from abstract state machines. ACM SIGSOFT Softw. Eng. Notes **27**(4), 112–122 (2002). https://doi.org/10.1145/566171.566190

61. Gurevich, Y.: Sequential abstract-state machines capture sequential algorithms. ACM Trans. Comput. Logic 1(1), 77–111 (2000). https://doi.org/10.1145/343369. 343384

62. Gurevich, Y., Börger, E.: Evolving algebras 1993: Lipari guide. In: Börger, E. (ed.) Specification and Validation Methods, pp. 231–243. Oxford University Press (1995). https://arxiv.org/abs/1808.06255

63. Gurevich, Y., Börger, E.: Evolving algebras: mini-course, Notes Series, vol. NS-95-4. BRICS: Basic Resarch in Computer Science, July 1995. https://www.researchgate.net/publication/221329427

64. Habrias, H., Frappier, M. (eds.): Software Specification Methods. ISTE (2006). https://doi.org/10.1002/9780470612514

65. Hegel, G.W.F.: Wissenschaft der Logik. Johann Leonhard Schrag, Rürnberg (1812–16). http://www.gutenberg.org/ebooks/55108

66. Hegel, G.W.F.: Grundlinien der Philosophie des Rechts. Nicolaischen Buchhandlung, Berlin (1821)

67. Henson, M.C., Reeves, S., Bowen, J.P.: Z logic and its consequences. Comput. Inf. 22(3–4), 381–415 (2003). https://doi.org/10289/1571

68. Hinchey, M.G., Bowen, J.P., Rouff, C.: Introduction to formal methods. In: Hinchey, M.G., Rash, J., Truszkowski, W., Gordon-Spears, D.F. (eds.) Agent Technology from a Formal Perspective, pp. 25–64. Springer, Cham (2006). https://doi.org/10.1007/1-84628-271-3_2

69. Isaacson, I.: Steve Jobs. Little, Brown, London (2011)

70. Kant, I.: Kritik der reinen Vernunft. Johann Friedrich Hartknoch, Riga, 2nd edn. (1781). https://www.gutenberg.org/ebooks/4280

71. Kant, I.: Kritik der Urteilskraft. Lagarde & Friedrich, Berlin & Libau (1790). http://www.gutenberg.org/ebooks/48433

72. Leibniz, G.W.: Explication de l'arithmétique binaire. Mémoires de Mathématique et de Physique de l'Académie Royale des Sciences, pp. 85–89 (1703). https://hal.archives-ouvertes.fr/ads-00104781

73. Liebmann, O., Wundt, W., Lipps, T., et al.: Die Philosophie im Beginn des 20. Jahrhunderts. Festschrift für Kuno Fischer, Heidelberg (1907)

74. MGP: Egon Börger. Mathematics Genealogy Project. https://www.mathgenealogy.org/id.php?id=155832

75. MGP: Yuri Gurevich. Mathematics Genealogy Project. https://www.mathgenealogy.org/id.php?id=7906

76. Pippin, R.B.: Hegel, Georg Wilhelm Friedrich. In: Audi [6], pp. 311–317

77. Polmeer, G.: Historical questions on being and digital culture. In: Giannini and Bowen [59], chap. 3, pp. 49–62. https://doi.org/10.1007/978-3-319-97457-6_3

78. Robinson, A.: Sudden Genius?. Oxford University Press, Oxford (2010)

79. Rojas, R. (ed.): Encyclopedia of Computers and Computer History. Fitzroy Dearborn Publishers, Chicago (2001)

80. Rooney, D. (ed.): Mathematics: How It Shaped Our World. Science Museum, London (2016)

81. Schmeh, K.: Enigma-schwachstellen auf der spur. Telepolis, Heise Online, Germany, August 2005. https://www.heise.de/tp/features/Enigma-Schwachstellen-auf-der-Spur-3402290.html. (in German)

82. Shannon, C.E.: A symbolic analysis of relay and switching circuits. Master of science, Massachusetts Institute of Technology, USA (1940). https://doi.org/1721.1/11173

83. Siekmann, J., Davis, M., Gabbay, D.M., et al.: Computational logic. In: Gabbay, D.M., Siekmann, J., Woods, J. (eds.) Handbook of the History of Logic, vol. 9. North Holland (2014)

84. Sleigh, R.C.: Liebniz, Gottfried Wilhelm. In: Audi [6], pp. 425–429

85. Smart, N.: Hegel, the giant of nineteenth-century German philosophy. In: Smart [89], pp. 243–244

86. Smart, N.: Idealism: Fichte and Schelling on the road to Hegel. In: Smart [89], pp. 241–243

87. Smart, N.: Immanuel Kant and the critical philosophy. In: Smart [89], pp. 238–241

88. Smart, N.: Leibniz and the idea of universal harmony. In: Smart [89], pp. 231–233

89. Smart, N.: World Philosophies. Routledge, London (1999)

90. Turing, A.M.: On computable numbers, with an application to the Entscheidungsproblem. In: Proceedings of the London Mathematical Society s2-42, pp. 230–265 (1937). https://doi.org/10.1112/plms/s2-42.1.230

91. Turing, A.M.: Computing machinery and intelligence. Mind **LIX**, 433–460 (1950). https://doi.org/10.1093/mind/LIX.236.433

92. Vater, M.G., Wood, D.W. (eds.): The philosophical rupture between Fichte and Schelling: selected texts and correspondence (1800–1802). Suny Press, Albany (2012)

93. von Neumann, J.: First draft of a report on the EDVAC. Moore School of Electrical Engineering, University of Pennsylvania, June 1945. http://web.mit.edu/STS.035/www/PDFs/edvac.pdf

94. Wenger, E.: Communities of Practice: Learning, Meaning, and Identity. Cambridge University Press, Cambridge (1998)

95. Wenger, E., McDermott, R.A., Snyder, W.: Cultivating Communities of Practice: A Guide to Managing Knowledge. Harvard Business School Press, Brighton (2002)

96. Wenger-Trayner, E., Wenger-Trayner, B.: Introduction to communities of practice: a brief overview of the concept and its uses (2015). https://wenger-trayner.com/introduction-to-communities-of-practice/

97. Wikipedia: Egon Börger: The Free Encyclopedia. https://en.wikipedia.org/wiki/Egon_Borger

Language and Communication Problems in Formalization: A Natural Language Approach

Alessandro Fantechi[1,2(✉)], Stefania Gnesi[2], and Laura Semini[2,3]

[1] Dipartimento di Ingegneria dell'Informazione, Università di Firenze, Florence, Italy
`alessandro.fantechi@unifi.it`
[2] Istituto di Scienza e Tecnologie dell'Informazione "A.Faedo",
Consiglio Nazionale delle Ricerche, ISTI–CNR, Pisa, Italy
`stefania.gnesi@isti.cnr.it`
[3] Dipartimento di Informatica, Università di Pisa, Pisa, Italy
`laura.semini@unipi.it`

Abstract. "The bride is dressed in red and the groom in white." Sometimes someone cannot believe their own ears, thinking they have misunderstood, and instead the communication is clear and exact, Egon actually got married in white and Donatella was in red. Some other times someone believe they have understood and instead the message is ambiguous and unclear.

When considering software requirements, one way to eliminate inaccuracies is to build a ground model and give it a formalization. In this paper, we propose an approach that begins by searching for terms and constructs that may cause communication problems and suggests a systematic way to disambiguate them.

1 Introduction

In the initial step of the classic software development process requirements are defined. This process aims to obtain a general description of functional (what the program should do) and non-functional (such as resources, performance, etc.) requirements. Although this process is somewhat systematic, requirements identification is usually intuitive and informal, and requirements are usually expressed informally through natural language phrases. Requirements are typically the basis on which implementations are built. However, natural language is inherently ambiguous, and ambiguity is seen as a possible (if not certain) source of problems in the subsequent interpretation of requirements.

Among the others, Egon Boerger advocated to start from an unambiguous, formal, *ground* model and proceed through a rigorous formal process that preserves correctness to develop the implementation [4,5]. Defining ground models is one of the three activities that constitute the formal method for engineering computer-based systems known as Abstract State Machine (ASM) method [6].

However, this introduces a necessary formalization step that should bridge the informal understanding of what is expected from the system to be built into its formal, rigorous model. Quoting Egon:

© Springer Nature Switzerland AG 2021
A. Raschke et al. (Eds.): Börger Festschrift, LNCS 12750, pp. 121–134, 2021.
https://doi.org/10.1007/978-3-030-76020-5_7

The notoriously difficult and error prone elicitation of requirements is largely a formalization task in the sense of an accurate task formulation, namely to realize the transition from usually natural-language problem descriptions to a sufficiently precise, unambiguous, consistent, complete and minimal formulation of ... "the conceptual construct" ... of a computer-based system, as distinguished from a software representation, e.g. by code.

This is mainly a language and communication problem between the domain expert or customer and the software designer who prior to coding have to come to a common understanding of "what to build", to be documented in a contract containing a model which can be inspected by the involved parties.

This, together with

the fact that there are no mathematical means to prove the correctness of the passage from an informal to a precise description

makes the formalization step quite challenging, especially for complex systems. Egon's approach has been that of expressing "the conceptual construct" as a so called "ground" model:

ground models must be apt to mediate between the application domain, where the task originates which is to be accomplished by the system to be built, and the world of models, where the relevant piece of reality has to be represented.

Our research has instead focused on the informal nature of requirement themselves. Pioneering a popular stream of research in the Requirements Engineering discipline, we have addressed the said communication mismatch problem by focusing on the identification and analysis of *ambiguity* in requirements documents. In particular, in previous works we have focused on the automated analysis of requirements documents by means of Natural Language Processing (NLP) tools [7]. This kind of analysis is aimed at identifying typical natural language defects, especially focusing on ambiguity sources [3].

In this paper we exemplify the principles of this research stream by applying the QuARS NLP tool to identify ambiguities in the informal description of the known Steam Boiler case study [2], showing how also that natural language description (considered as if it were a requirements document for the proposed case study) suffered from ambiguity: this ambiguity has necessarily to be solved when formalising the case study (and we show examples of solved ambiguities).

2   Sources of Ambiguity in Requirements

Natural language (NL) is the most commonly used means of expressing software requirements despite the fact that it is inherently ambiguous, and despite the fact that ambiguity is a known source of problems in the subsequent interpretation of requirements.

Natural language processing techniques (NLP) are frequently used today to help system engineers and project managers analyze and correct requirements in natural language and to provide the user with a quality assessment of each requirement analyzed, significantly automating and speeding up the task of finding possible errors in NL requirements documents. However this does not mean that human review of the requirements is unnecessary or unimportant. Rather, these NLP techniques and tools will provide a help in this task. Domain experts shall review the results the tools provide and use their expertise to correct the deficiencies they find.

There are many different sources of ambiguity that can cause communication problems between stakeholders, some are due to the use of ambiguous words (*lexical* ambiguities), others (the *syntactical* ambiguities) to particular syntactical structures or to the position of a word in a requirement (when used as an adjective or pronoun for example). We list the most common sources of ambiguity, with a classification inspired by [3].

Analytical, *attachment*, and *coordination* ambiguities, are different forms of syntactic ambiguities. They occur when a sentence admits more than one grammatical structure, and different structures have different meanings. We provide some examples, for a more in-depth discussion we refer to [3]: "software fault tolerance", can be tolerance to software faults or fault tolerance achieved by software (*analytical*); "I go to visit a friend with a red car" means either that I go with the red car or that the friend I'm visiting has a red car (*attachment*); "She likes red cars and bikes", where it is not clear whether she likes all bikes or only red ones (two *coordinators* in a sentence); "I will travel by plane or ferry and car", that can be interpreted as "I will travel by plane or I will travel by ferry and car" or as "I will travel by plane or ferry and by car" (another example of *coordination*: a coordinator related to a modifier). Somehow all three have to do with associativity.

A requirement contains an *anaphoric* ambiguity when an element of a sentence depends for its reference on another, antecedent, element and it is not clear to which antecedent it refers, as in "The procedure shall convert the 24 bit image to an 8 bit image, then display it in a dynamic window" [12]. Anaphoras are another category of syntactic ambiguity and in most cases can be identified looking for pronouns.

Vagueness is a lexical defect and it is sometimes considered not as an ambiguity but as a separate category, in fact it does not lead to different interpretations, but to a loose interpretation. For completeness and uniformity we keep it in this list and treat it as an ambiguity. An example is "The system must support many users at the same time": what is meant by many? 5? 1000?, without a clarification you will never be able to say if the system will have satisfied the requirement.

Temporal ambiguities are special cases of vagueness, that occur when a temporal quantification is missing. Examples are: "A message shall be sent after activation of the registration function". How long after? "The Project Team must demonstrate mockups of UI changes to project stakeholders <u>early</u> in the development process". How early?

Comparatives and *superlatives* are syntactic ambiguities: a comparator without a term of comparison; a superlative with a vague or no context. Although, also in this case we can identify terms that play the role of indicators. Examples are: "The feature X may be used to provide a <u>better</u> grade of service to certain users". "Any emissions radiating into the driver's cab and other on-board equipment from the exterior aerial shall meet the industry standards to the <u>highest</u> possible degree".

We have seen that *coordinators* can lead to ambiguities in some syntactic constructions, *disjunctions* instead are a separate case because they are inherently ambiguous, as in "The user information contained in the confirmation message shall be the engine number <u>or</u> train number (if registered)".

An *Escape clause* contains a condition or a possibility and makes a requirement weaker, vague and unverifiable. An example is "The system, if possible, updates the data every 2 h".

Similarly, the use of a weak verb (e.g. may or may not) leads to a *weakness* and offers an excuse for not implementing the requirement.

Quantifiers are a notable cause of ambiguity, for different orders of reasons: they can be originated from generalizations of the matter; the analysts may not have defined the universe of quantification; the analysts may not have defined the scope of the quantifier itself. The sentence "<u>All</u> pilots have <u>a</u> red car" contains two ambiguities: universe of discourse, the sentence is true in a Ferrari F1 box, not in general, and scope, the pilots share the same car?

Underspecification is a syntactical ambiguity that occurs when a generic term is used that lacks of a specifier. For instance in "It shall be possible for the driver to store <u>numbers</u> in the radio", it must be made clear what the numbers are. In this case the context of the requirement makes clear that they were train identification numbers.

When using a *passive voice* it is possible to omit the agent, in fact, often we change an active form to passive so that we don't have to decide on the subject of the action (which becomes the agent in the passive form). This generates a (syntactic) ambiguity that must be resolved by deciding who is the subject of the action. As an example "The fire alarm must <u>be issued</u>" will become "The system must issue the fire alarm".

2.1 QuARS

QuARS - Quality Analyzer for Requirement Specifications – is a tool for analyzing NL requirements in a systematic and automatic way by means of NLP techniques with a focus on ambiguity detection [9,10].

QuARS performs an automatic linguistic analysis of a requirements document in plain text format, based on a given quality model. Its output indicates the defective requirements and highlights the words that reveal the defect.

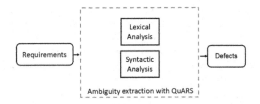

Fig. 1. The QuARS architecture for defect identification.

The defect identification process is divided into two, independent, parts (Fig. 1). The first part, *lexical analysis*, detects candidate defective terms using a set of dictionaries. Lexical analysis permits to capture *disjunction, optionality, subjectivity, vagueness, optionality, weakness, quantifiers, and temporal* defects. The second part is *syntactical analysis*, that captures *implicity and underspecification* defects.

Other features of QuARS are (i) metrics derivation for evaluating the quality of NL requirements; (ii) the capability to modify existing dictionaries, and to add new dictionaries for new indicators; (iii) view derivation, to identify and collect together those requirements belonging to given functional and non functional characteristics, by defining specific requirements.

In Table 1 we present the indicators used by QuARS to detect defects of lexical and syntactic ambiguity in NL sentences. Table 2 shows how they correspond to the ambiguity classes listed above.

Table 1. Defects detected by QuARS

Ambiguity class	Indicators
vagueness	clear, easy, strong, good, bad, adequate...
subjectivity	similar, have in mind, take into account,...
disjunction	or, and/or
optionality	or, and/or, possibly, eventually, case, if possible, if appropriate...
temporal	before, after, as soon as, then...
weakness	can, could, may, ...
implicity	demonstrative adjectives or pronouns
underspecification	wordings missing a qualification (e.g.: interface or information without a qualifier, such as user and price, respectively)

Table 2. Ambiguity classes detected by QuARS

Ambiguity classes	QuARS detection classes
Analytical	n.a.
Attachment	n.a.
Coordination	n.a.
Anaphoric	n.a.
Vagueness	vagueness
Temporal	temporal
Comparatives	n.a.
Superlatives	n.a.
Disjunction	disjunction
Escape Clause	optionality, weakness
Quantifiers	quantifiers
Underspecification	underspecification
Passive voice	n.a.

3   Steam Boiler and QuARS

Based on the insight that thorough comparison between the various design formalisms proposed in the literature as well as some amount of unification would be crucial both to industrial takeover and to scientific progress in the field, an international seminar on "Methods for Semantics and Specification" was held at Schloß Dagstuhl, Wadern, Germany, on June 5–9, 1995. The seminar took the form of a "competition" between different researchers who had been invited as representatives of their particular methods. The competition was on the *Steam Boiler control* specification problem drafted by J.-R. Abrial, E. Börger, and H. Langmaack, which has been posed to the participants as a common case study [1,2].

We consider here the natural language specification of the Steam Boiler case study as if it was the requirements document specifying the system, structuring the text as a set of numbered sentences. The resulting document is made of 39 requirements (see the Appendix). We have applied QuARS to analyse this document looking for potential ambiguities, then the output of the tool has been manually examined to distinguish true ambiguities from false positives (FP). Table 3 shows a summary of the true ambiguity defects.

The *vagueness* analysis revealed 4 defects, 3 of which are true ambiguities and only one FP. The table shows req.29, which is ambiguous because *satisfactory* is not an accurate measure for water level, and req.39 is ambiguous because it is unclear what the *appropriate* actions are. Here the communication problems can be solved by providing ranges so that it is possible to evaluate predicates $satisfactory(water\_level)$ and $appropriate(action)$.

The *quantifiers* analysis revealed 7 defects, one of which is a true ambiguity and 6 are FP. The table shows Req.1, in which it is not clear what "*each* physical unit is connected to a control unit" means.

The *disjunction* analysis revealed 9 defects, all occurrences of *or*: 1 truly ambiguous and 8 FP. Req. 33, containing the ambiguous occurrence, introduces a case of non-determinism.

The *weakness* analysis revealed 8 defects: 2 true ambiguities (both occurrences of *can*) and 6 FP: in Req. 13 it is not clear if the program must send the program-ready signal or can avoid doing so. Req. 17 is ambiguous about the program's obligation to adjust water levels.

The search for *temporal* ambiguities returned 10 defects, out of which 9 true ambiguities and 1 FP. Most ambiguities are due, as in req. 8, to the presence of wording *as soon as*. The analyst must decide whether *as soon as* means the next step or within a some time. Similarly, requirements 11 and 37 must be clarified because of the terms *then* and *after*.

The *optionality* analysis did not reveal any defect, while *implicity* and *underspecification* reveald some defects that were false positives.

4   Solving Communication Mismatch Problems

QuARS helps to find in the natural language specification critical points that can create communication problems as they can lead to multiple interpretations. Depending on the defect class, the range of possible interpretations changes. We address the output of the QuARS analysis of the steam boiler and indicate, for each requirement, the possible interpretations and the corresponding formalization, that depend on the class of defect. This way we provide, through examples, a systematic means to consider, for each ambiguity class as defined in Table 2, the possible interpretations and their formalization, so that they can be compared on a mathematical ground. The formalization makes use of first order logic and temporal logics, which are well suited to express in a rigorous manner the particular aspects to which a single requirement refers.

4.1   Formalising Vagueness and Quantifiers

A vague term may represent an abstraction, as *satisfactory* in the example below, which is made concrete by a set of possible instances.

Req 29 "The rescue mode is the mode in which the program tries to maintain a satisfactory water level despite of the failure of the water level measuring unit"

We need to clarify what *satisfactory* means, providing a measure and a range of values. For example, let us assume it is between N1 and N2, included.

Table 3. Analysis with QuARS of the steam boiler requirements

---

————— QuARS [Lexical] vagueness ANALYSIS —————.
The line number:
29. the rescue mode is the mode in which the program tries to maintain a satisfactory water level despite of the failure of the water level measuring unit. .
is defective because it contains the wording: satisfactory
The line number:
39. once the program has reached the emergency stop mode, the physical environment is then responsible to take appropriate actions, and the program stops.
is defective because it contains the wording: appropriate
—- QuARS [Lexical] vagueness Statistics (on "boiler.txt" file): —————
Number of evaluated sentences: 39 Number of defective sentences: 4

————— QuARS [Lexical] Quantifiers ANALYSIS —————.
The line number:
1. the program communicates with the physical units through messages which are transmitted over a number of dedicated lines connecting each physical unit with the control unit.
In first approximation, the time for transmission can be neglected.
is defective because it contains the wording: each
—- QuARS [Lexical] Quantifiers Statistics (on "boiler.txt" file): —————
Number of evaluated sentences: 39 Number of defective sentences: 7

————— QuARS [Lexical] Disjunction ANALYSIS —————.
The line number:
33. as soon as the water measuring unit is repaired, the program returns into mode degraded or into mode normal.
is defective because it contains the wording: or
—- QuARS [Lexical] Disjunction Statistics (on "boiler.txt" file): —————
Number of evaluated sentences: 39 Number of defective sentences: 9

————— QuARS [Lexical] weakness ANALYSIS —————.
The line number:
13. as soon as a level of water between N1 and N2 has been reached the program can send continuously the signal program-ready to the physical units until it receives the signal physical_units_ready which must necessarily be emitted by the physical units.
is defective because it contains the wording: can
The line number:
17. as soon as the water level is below N1 or above N2 the level can be adjusted by the program by switching the pumps on or off.
is defective because it contains the wording: can
—- QuARS [Lexical] weakness Statistics (on "boiler.txt" file): —————
Number of evaluated sentences: 39 Number of defective sentences: 8

————— QuARS [Lexical] temporal ANALYSIS —————.
The line number:
8. as soon as this message has been received the program checks whether the quantity of steam coming out of the steam-boiler is really zero.
is defective because it contains the wording: as soon as
The line number:
11. if the quantity of water in the steam-boiler is below w then the program activates a pump to fill the steam-boiler.
is defective because it contains the wording: then
The line number:
37. this mode can also be reached after detection of an erroneous transmission between the program and the physical units.
is defective because it contains the wording: after
—- QuARS [Lexical] temporal Statistics (on "boiler.txt" file): —————
Number of evaluated sentences:39 Number of defective sentences:10

If we consider as a satisfying range the closed interval [N1..N2], using first order logic we may write:

$$\forall x.\ (satisfactory\_water\_level(x) \leftrightarrow (x \geq N1 \wedge x \leq N2))$$

and then formalise Req 29 as follows:

$$rescue\_mode \rightarrow \exists x.(water\_level(x) \wedge satisfactory\_water\_level(x))$$

Quantifiers can make a requirement ambiguous for two reasons at least: the universe of discourse and scope.

Req 1 "The program communicates with the physical units through messages which are transmitted over a number of dedicated lines connecting each physical unit with the control unit. in first approximation, the time for transmission can be neglected."

Here the universe for *each* is clear, and it is the set of physical units. The problem is the scope and the two possible interpretations are: "each physical unit is connected to a distinguished control unit" and "all the physical units are connected to a common control unit", formalized, respectively, by:

$$\forall x.\exists y.(connected(x,y) \wedge \forall r, s.((connected(r,s) \wedge x \neq r) \rightarrow y \neq s))$$
$$\exists y.\forall x.connected(x,y)$$

4.2 Formalising Disjuncion, Weakness and Temporal Ambiguities: CTL

We consider a *branching-time* temporal logic that is a subset of CTL - Computation Tree Logic [8][1]:

$$\phi ::= \ |\ true\ |\ p\ |\ \neg\phi\ |\ \phi \vee \phi\ |\ AX\phi\ |\ EX\phi\ |\ AF\phi\ |\ EF\phi\ |\ AG\phi\ |\ EG\phi$$

Each CTL operator is a pair of symbols. The first one is either A ("for All paths"), or E ("there Exists a path"). The second is one of X ("neXt state"), F ("in a Future state", i.e. in a state of the path), G ("Globally in the future", i.e. in all states of the path).

Req 33 "as soon as the water measuring unit is repaired, the program returns into mode degraded or into mode normal."

This requirement is ambiguous because of the presence of *as soon as* that can be interpreted as *next step* or as *eventually* and formalised accordingly as follows:

---

[1] The definition we provide is not minimal, and some operators can be derived by other. We use this definition for simplicity and readability.

$$AG(water\_munit\_repaired \rightarrow AX(mode\_degraded \vee mode\_normal))$$
$$AG(water\_munit\_repaired \rightarrow AF(mode\_degraded \vee mode\_normal))$$

The formalization of the disjunction *or* makes non-determinism explicit: this step makes the analyst aware of the condition, if non-determinism is intended, then the specification itself will contain a disjunction, otherwise the requirement must be disambiguated.

Req 17 "as soon as the water level is below N1 or above N2 the level can be adjusted by the program by switching the pumps on or off."

Besides the temporal ambiguity, that we interpret here as *next step*, there is a weak verb, namely *can*. As in the previous case, if the use of a weak verb is intentional, then the specification will take this into account, using an appropriate modality:

$$AG((water\_below\_N1 \vee water\_above\_N2) \rightarrow EX\,switched\_pumps)$$

otherwise the requirement is disambiguated, for instance, imposing to switch the pumps:

$$AG((water\_below\_N1 \vee water\_above\_N2) \rightarrow AX\,switched\_pumps)$$

Note also that, for more precision, the formula can be split to indicate the exact action taken in the two cases of low and high water, respectively:

$$AG(water\_below\_N1 \rightarrow AX\,switched\_pumps\_on)$$
$$AG(water\_above\_N2 \rightarrow AX\,switched\_pumps\_off)$$

Req 37 "This mode can also be reached after detection of an erroneous transmission between the program and the physical units."

There are three defects in this requirement:

This, that is a false positive since it is easily disambiguated reading requirement 36: it refers to the emergency stop mode;

can as above, the weakness is left in the formalization, using path quantifier "E", or removed using modality "A".

after: it is not clear if it refers to the next state or to a future one, hence either "F" or "X" are to be used.

Metric Temporal Logic. Also, when specifying requirements with temporal expressions, the analyst's first decision is whether to model them in terms of computational steps (LTS) or use a real-time model. For example, *as soon as* can be the next step or within a time limit expressed in seconds or milliseconds. Requirement 33 "as soon as the water measuring unit is repaired, the program returns into mode degraded or into mode normal" for instance could be interpreted as:

"within n millisecond after the water measuring unit is repaired, the program returns into mode degraded or into mode normal."

and specified in a metric temporal logic [11].

5   Conclusions

In this paper we have briefly shown an approach to address the problem, often discussed by Egon Boerger, of grounding the software development process on a formal basis able to capture the intended behaviour by avoiding the typical ambiguity that natural language requirements inherently contain. The presented approach follows a different research path with respect to the one mastered by Egon, in which a ground model by means of ASM is developed first with the aim of solving potential ambiguities. We focus instead on the ambiguity sources in the natural language expression of requirements, exploiting a NLP tool to point out potential ambiguities, then recurring to focused logic formulae expressing the different interpretations that may solve them.

Acknowledgements. The research has been partially supported by the project "IT-MaTTerS" (Methods and Tools for Trustworthy Smart Systems), MIUR PRIN 2017FTXR7S.

A   Steam Boiler Requirements

1. The program communicates with the physical units through messages which are transmitted over a number of dedicated lines connecting each physical unit with the control unit. In first approximation, the time for transmission can be neglected.
2. The program follows a cycle and a priori does not terminate.
3. This cycle takes place each five seconds and consists of the following actions: Reception of messages coming from the physical units; Analysis of informations which have been received; Transmission of messages to the physical units.
4. In first approximation, all messages coming from (or going to) the physical units are supposed to be received (emitted) simultaneously by the program at each cycle.
5. The program operates in different modes, namely: initialization, normal, degraded, rescue, emergency stop.
6. The initialization mode is the mode to start with.
7. The program enters a state in which it waits for the message STEAM-BOILER-WAITING to come from the physical units.
8. As soon as this message has been received the program checks whether the quantity of steam coming out of the steam-boiler is really zero.
9. If the unit for detection of the level of steam is defective—that is, when u is not equal to zero—the program enters the emergency stop mode.
10. If the quantity of water in the steam-boiler is above N2 the program activates the valve of the steam-boiler in order to empty it.
11. If the quantity of water in the steam-boiler is below W then the program activates a pump to fill the steam-boiler.
12. If the program realizes a failure of the water level detection unit it enters the emergency stop mode.

13. As soon as a level of water between N1 and N2 has been reached the program can send continuously the signal PROGRAM-READY to the physical units until it receives the signal PHYSICAL_UNITS_READY which must necessarily be emitted by the physical units.
14. As soon as this signal has been received, the program enters either the mode normal if all the physical units operate correctly or the mode degraded if any physical unit is defective.
15. A transmission failure puts the program into the mode emergency stop.
16. The normal mode is the standard operating mode in which the program tries to maintain the water level in the steam-boiler between N1 and N2 with all physical units operating correctly.
17. As soon as the water level is below N1 or above N2 the level can be adjusted by the program by switching the pumps on or off.
18. The corresponding decision is taken on the basis of the information which has been received from the physical units.
19. As soon as the program recognizes a failure of the water level measuring unit it goes into rescue mode.
20. Failure of any other physical unit puts the program into degraded mode.
21. If the water level is risking to reach one of the limit values Mi or M2 the program enters the mode emergency stop.
22. This risk is evaluated on the basis of a maximal behaviour of the physical units.
23. A transmission failure puts the program into emergency stop mode.
24. The degraded mode is the mode in which the program tries to maintain a satisfactory water level despite of the presence of failure of some physical unit. It is assumed however that the water level measuring unit in the steam-boiler is working correctly. The functionality is the same as in the preceding case.
25. Once all the units which were defective have been repaired, the program comes back to normal mode.
26. As soon as the program sees that the water level measuring unit has a failure, the program goes into mode rescue.
27. If the water level is risking to reach one of the limit values Mi or M2 the program enters the mode emergency stop.
28. A transmission failure puts the program into emergency stop mode.
29. The rescue mode is the mode in which the program tries to maintain a satisfactory water level despite of the failure of the water level measuring unit.
30. The water level is estimated by a computation which is done taking into account the maximum dynamics of the quantity of steam coming out of the steam-boiler.
31. For the sake of simplicity, this calculation can suppose that exactly n liters of water, supplied by the pumps, do account for exactly the same amount of boiler contents (no thermal expansion).
32. This calculation can however be done only if the unit which measures the quantity of steam is itself working and if one can rely upon the information which comes from the units for controlling the pumps.

33. As soon as the water measuring unit is repaired, the program returns into mode degraded or into mode normal.

34. The program goes into emergency stop mode if it realizes that one of the following cases holds: the unit which measures the outcome of steam has a failure, or the units which control the pumps have a failure, or the water level risks to reach one of the two limit values.

35. A transmission failure puts the program into emergency stop mode.

36. The emergency stop mode is the mode into which the program has to go, as we have seen already, when either the vital units have a failure or when the water level risks to reach one of its two limit values.

37. This mode can also be reached after detection of an erroneous transmission between the program and the physical units.

38. This mode can also be set directly from outside.

39. Once the program has reached the Emergency stop mode, the physical environment is then responsible to take appropriate actions, and the program stops.

References

1. Abrial, J.-R.: Steam-boiler control specification problem. In: Abrial, J.-R., Börger, E., Langmaack, H. (eds.) Formal Methods for Industrial Applications. LNCS, vol. 1165, pp. 500–509. Springer, Heidelberg (1996). https://doi.org/10.1007/BFb0027252

2. Abrial, J.-R., Börger, E., Langmaack, H. (eds.): Formal Methods for Industrial Applications. LNCS, vol. 1165. Springer, Heidelberg (1996). https://doi.org/10.1007/BFb0027227

3. Berry, D., Kamsties, E., Krieger, M.: From contract drafting to software specification: Linguistic sources of ambiguity - a handbook version 1.0. (2003)

4. Börger, E.: Why use evolving algebras for hardware and software engineering? In: Bartosek, M., Staudek, J., Wiedermann, J. (eds.) SOFSEM 1995. LNCS, vol. 1012, pp. 236–271. Springer, Heidelberg (1995). https://doi.org/10.1007/3-540-60609-2_12

5. Börger, E.: The ASM ground model method as a foundation of requirements engineering. In: Dershowitz, N. (ed.) Verification: Theory and Practice. LNCS, vol. 2772, pp. 145–160. Springer, Heidelberg (2003). https://doi.org/10.1007/978-3-540-39910-0_6

6. Börger, E.: The abstract state machines method for modular design and analysis of programming languages. J. Log. Comput. **27**(2), 417–439 (2017)

7. Chowdhury, G.G.: Natural language processing. Annu. Rev. Inf. Sci. Technol. **37**(1), 51–89 (2003)

8. Clarke, E.M., Emerson, E.A.: Design and synthesis of synchronization skeletons using branching time temporal logic. In: Kozen, D. (ed.) Logic of Programs 1981. LNCS, vol. 131, pp. 52–71. Springer, Heidelberg (1982). https://doi.org/10.1007/BFb0025774

9. Fabbrini, F., Fusani, M., Gnesi, S., Lami, G.: An automatic quality evaluation for natural language requirements. In: Proceedings of the 7th International Workshop on Requirements Engineering: Foundation for Software Quality (REFSQ), vol. 1, pp. 4–5 (2001)

10. Gnesi, S., Lami, G., Trentanni, G.: An automatic tool for the analysis of natural language requirements. Comput. Syst. Sci. Eng. **20**(1) (2005)
11. Koymans, R.: Specifying real-time properties with metric temporal logic. Real Time Syst. **2**(4), 255–299 (1990)
12. Yang, H., De Roeck, A.N., Gervasi, V., Willis, A., Nuseibeh, B.: Analysing anaphoric ambiguity in natural language requirements. Requir. Eng. **16**(3), 163–189 (2011)

ASM Specification and Refinement
of a Quantum Algorithm

Flavio Ferrarotti[1]($^{(\boxtimes)}$) and Sénen González[2]

[1] Software Competence Center Hagenberg, Hagenberg, Austria
`flavio.ferrarotti@scch.at`
[2] TM Connected, Linz, Austria

Abstract. In this paper we use the Abstract State Machine (ASM) method for high-level system design and analysis created by Egon Börger to formally specify Grover's quantum database search algorithm, stepwise refining it from its highest abstraction level down to its implementation as a quantum circuit. Our aim is to raise the question of whether the ASM method in general and quantum ASMs in particular can improve the current practices of quantum system engineering; providing accurate high-level modelling and linking the descriptions at the successive stages of quantum systems development through a chain of rigorous and coherent system models at stepwise refined levels of abstraction.

1 Introduction

Prof. Egon Börger's method for high-level system design and analysis [4,5], which is built on the solid mathematical base of Gurevich's abstract state machines (ASMs), is without any doubt among the most notorious and lasting contributions to the development of the discipline of rigorous software engineering. The ASM method has proven its strength for the rigorous specification and validation of a wide class of systems, from embedded hardware-software systems to complex multi-agent systems.

A unique feature of the ASMs rigorous systems engineering method (ASM method for short) is the fact that it covers design, specification, verification by reasoning techniques and validation by simulation and testing within a single theoretical and conceptual framework. At a high level of abstraction it provides accurate and mathematically precise modelling which at the same time can be intuitively understood by non-experts. Moreover, starting from high level system specifications, which can be correctly understood as a kind of high level pseudocode, the method enables to link the successive stages of system development in an organic and efficiently maintainable chain of rigorous and coherent system models at stepwise-refined levels of abstraction.

The research reported in this paper was supported by the Austrian Research Promotion Agency (FFG) through the COMET funding for the Software Competence Center Hagenberg.

© Springer Nature Switzerland AG 2021
A. Raschke et al. (Eds.): Börger Festschrift, LNCS 12750, pp. 135–146, 2021.
https://doi.org/10.1007/978-3-030-76020-5_8

Despite its success in many different areas of systems engineering, the ASM method has, up to our knowledge, not yet been considered for high-level design and analysis of quantum systems. This is rather surprising given the fact that it is known since a long time that a very general notion of quantum algorithm can be precisely and faithfully captured by quantum ASMs [8]. Of course, this does not automatically implies that the ASM methods can be successfully used for quantum systems engineering, but it is a strong indication in that sense. We believe that this Festschrift celebrating Prof. Egon Börger 75 birthday is therefore an excellent venue to raise this issue.

Moreover, the development of quantum systems and algorithms are notoriously difficult tasks that involve complex design flows composed of steps such as synthesis, mapping, or optimizations. All these tasks need to be completed before a high level conceptual quantum algorithm can be executed on an actual device. This process results in many different model descriptions at various levels of abstraction, which usually significantly differ from each other. The complexity of the underlying design problem makes it even more important to not only provide efficient solutions for single steps, but also to verify that the originally intended functionality is preserved throughout all levels of abstraction. At the same time there is the challenge of finding abstractions that expose key details while hiding enough complexity [6]. We conjecture that the ASM method can be extended to provide an appropriate and universal mathematical tool to meet, if not all, most of these challenges. In this work we present, in terms of quantum ASMs, a complete and detailed specification and refinement of Grover's database search algorithm [9], which is a very influential and well known quantum algorithm. The aim is to provide concrete evidence that the ASM method can indeed be successfully used for rigorous engineering of quantum systems.

The paper is organized as follows. In Sect. 2 we introduce the necessary background from quantum computing, emphasising in particular the model of quantum circuits adopted for the definition of quantum ASMs in [8]. The actual model of quantum ASM is described in Sect. 3. Our main contribution, i.e., the complete specification and refinement from high-level modelling down to the level of implementation of Grover's algorithm, is presented in Sect. 4. Finally, we provide a brief conclusion in Sect. 5.

2   Quantum Computation

There are several good books such as [11] that provide a solid background on quantum computation. Here we give a brief introduction to the main concepts used in this paper.

In quantum mechanics vectors are written using brackets as follows:

$$|\psi\rangle = \begin{pmatrix} a_1 \\ a_2 \\ \vdots \\ a_n \end{pmatrix} \qquad |\alpha\rangle = \begin{pmatrix} b_1 \\ b_2 \\ \vdots \\ b_n \end{pmatrix} \qquad \begin{aligned} \langle\psi| &= (\bar{a}_1, \bar{a}_2, \ldots, \bar{a}_n) \\ \langle\alpha| &= (\bar{b}_1, \bar{b}_2, \ldots, \bar{b}_n) \end{aligned}$$

where $\bar{a}$ denotes the complex conjugate of a. This notation was the idea of Paul Dirac and is also known as $\langle$bra|ket$\rangle$ notation. Following this notation, the *inner product* of vectors $|\psi\rangle$ and $|a\rangle$ is the scalar $\langle a|\psi\rangle = a_1\bar{b}_1 + a_2\bar{b}_2 + \cdots + a_n\bar{b}_n$.

There are several theoretical models of quantum computing such as quantum circuits, quantum Turing machines, quantum automata and quantum programming. All these theoretical models define an state of a quantum algorithm as a superposition of states of a classical computation model. In mathematical terms a *quantum state* of a quantum algorithm is represented by a unit-length vector $|\psi\rangle$ in an n-dimensional Hilbert space $\mathbb{C}^n$.

For finite-dimensional vector spaces such as $\mathbb{C}^n$, Hilbert spaces are no different than inner product spaces. Formally, a Hilbert space must be an inner product space that is also a complete metric space. That is, a Hilbert space H must be complete with respect to the *norm* $\|v\|\sqrt{\langle v|v\rangle}$ (i.e., the length of the vector in the vector space) induced by the inner product (i.e., every Cauchy sequence in H has a limit in H).

Quantum bits (Qubits for short) are similar to standard bits only in that they are also base-2 numbers, and they take on the value 0 or 1 when measured and thus collapsed to a classical state. Unlike classical bits, in its uncollapsed, quantum state, a qubit is in a superposition of the values 0 and 1. Mathematically, a qubit is represented as a vector in the two dimensional Hilbert space $\mathbb{C}^2$ with orthonormal basis vectors $|0\rangle$ and $|1\rangle$. Hence, the superposition state $|\psi\rangle$ of a qubit is represented as a linear combination of those basis vectors: $|\psi\rangle = a_0|0\rangle + a_1|1\rangle$ with $\|a_0\|^2 + \|a_1\|^2 = 1$, where a_0 and a_1 are, respectively, the complex scalar *amplitudes* of measuring $|0\rangle$ and $|1\rangle$. Complex numbers are required to fully describe the superposition of states and interference or entanglement inherent in quantum systems. Since the squares of the absolute values of the amplitudes of states in a quantum system must add up to 1, we can think of them as probabilities in the sense that $\|a_i\|^2$ represents the chance that when a given quantum state is measured (i.e., when the superposition is collapsed) then the value i is observed.

Similar to classical computers, quantum computers use quantum registers made up of multiple qubits. When collapsed, quantum registers are bit strings whose length determines the amount of information they can store. An n-*qubit quantum register* is a vector $|\psi\rangle$ in the 2^n-dimensional Hilbert space $(\mathbb{C}^2)^{\otimes n} = \mathbb{C}^2 \otimes \cdots \otimes \mathbb{C}^2$ with orthonormal basis $\{|x\rangle : x \in \{0,1\}^n\}$ which we assume to be lexicographically ordered from $|00\cdots0\rangle$ as first basis vector to $|11\cdots1\rangle$ as last basis vector. The state of such a quantum register has then the form

$$|\psi\rangle = \sum_{x\in\{0,1\}} a_x|x\rangle = a_{00\ldots0}|00\cdots0\rangle + \cdots + a_{11\ldots1}|11\cdots1\rangle$$

with $\sum_{x\in\{0,1\}}\|a_x\|^2 = 1$.

The most widely used model of quantum computation is the *quantum circuit model* in which a quantum circuit is a sequence of quantum gates, and a quantum algorithm is a general model that combines classical algorithms with quantum circuits and measurement steps.

A *quantum gate on m qubits* is a unitary transformation U on a 2^m dimensional Hilbert space $(\mathbb{C}^2)^{\otimes m}$. Such transformations can be described by unitary matrices, or linear operators, applied to a quantum register by tensoring the transformation matrix with the matrix representation of the register. All linear operators that correspond to quantum logic gates must be unitary. If a complex matrix U is unitary, then it must hold that $U^{-1} = U^*$, where U^* is the conjugate transpose of U. It follows that $UU^* = U^*U = I$, where I is the identity matrix. Unitary operators preserve the inner product of two vectors and the composition of two unitary operators is also unitary. Notice that a transformation matrix can be seen as a function $U : \{0,1\}^m \times \{0,1\}^m \to \mathbb{C}$, where $U(x,y)$ is the probability amplitude of the transition from base state $|x\rangle$ to base state $|y\rangle$.

No finite collection of gates allows us to express precisely every unitary transformation, not even on $\mathbb{C}^2$, since there are uncountably many such transformations. But there are finite gate sets B which are universal in the sense that, for any unitary transformation U on any $(\mathbb{C}^2)^{\otimes m}$, an arbitrarily close approximation can be synthesized (see [14] among others). The following are well known examples of quantum gates which we will use in this paper.

– *Hadamard gate:* Defined by the following matrix:

$$H = \frac{1}{\sqrt{2}} \begin{pmatrix} 1 & 1 \\ 1 & -1 \end{pmatrix}$$

When applied to a qubit with the value $|0\rangle$ or $|1\rangle$, it induces an equal superposition of the states $|0\rangle$ and $|1\rangle$. More precisely $H|0\rangle = (|0\rangle + |1\rangle)/\sqrt{2}$ and $H|1\rangle = (|0\rangle - |1\rangle)/\sqrt{2}$.

– *Pauli X gate:* Aka. *Not* gate. It is defined by the matrix:

$$Not = \begin{pmatrix} 0 & 1 \\ 1 & 0 \end{pmatrix}$$

This gate simply switches the amplitudes of $|0\rangle$ and $|1\rangle$, i.e., $Not|0\rangle = |1\rangle$ and $Not|1\rangle = |0\rangle$.

– *Toffoli gate:* This gate together with the Hadamard gate can be considered universal in the quantum computing sense [14]. It applies to three distinct qubits. Two work as controls and one works as target. We can define Toffoli on the basis vectors $|c_1\rangle|c_2\rangle|t\rangle$ of $(\mathbb{C}^2)^{\otimes 3}$, where $|c_1\rangle$ and $|c_2\rangle$ are controls and $|t\rangle$ is the target. If $|c_1\rangle$ and $|c_2\rangle$ are both $|1\rangle$ then the new value of $|t\rangle$ is $|t \oplus 1\rangle$ (where $\oplus$ is addition modulo 2), i.e., the value of $|t\rangle$ is "flipped". If $|c_1\rangle$ or $|c_2\rangle$ (or both) is $|0\rangle$, then $|t\rangle$ remains unchanged. The controls qubits are never changed.

3 Quantum ASMs

We assume that the concept of Abstract State Machine (ASM) is well known [5]. We also assume that the reader is familiar with the differences between sequential [10] and parallel ASMs [1,2,7]. In this work we consider quantum ASMs as defined in [8]. As shown there, quantum ASMs faithfully capture a broad class of quantum algorithms as defined by a set of machine independent and language independent postulates. The discussion on whether these postulates truly capture the idea of quantum algorithm in an intuitive and commonly accepted way is of course fascinating, but it is beyond the scope of this paper. For our purpose here, it is enough to note that quantum ASMs provide a good starting point to explore the application of the ASM method to analyse and design quantum algorithms.

The general idea of quantum algorithm captured by quantum ASMs can be informally described as follows [8]:

- A quantum algorithm combines classical algorithms with quantum circuits.
- The collection of quantum gates used by a quantum algorithm is fixed.
- The quantum circuits used by a quantum algorithm may work with any number n of qubits.
- A quantum circuit on some number n of qubits is applied to a computational basis state $|x\rangle = |x_0 \ldots x_{n-1}\rangle$. This results in a state $|\psi\rangle \in (\mathbb{C}^2)^{\otimes n}$ which is usually entangled. The algorithm is at that point in a quantum mode.
- A quantum algorithm can return to a classical mode by applying a measurement step in the computational basis of one or more qubits.

Quantum ASMs are defined in [8] as sequential ASMs extended with two specific rules for unitary transformations and measurement, respectively, and with an expanded background to describe the mathematics of quantum gates and quantum measurement.

The *background $\mathcal{B}$ of states of a quantum ASM* extends the background of states of parallel ASM described in Section 3 in [7] with the basic arithmetic operations on complex numbers, natural numbers, binary strings and functions encoding the transformation matrices of a given set Ω of quantum gates (possibly parametrised if Ω is infinite). Notice that in particular this includes basic operations on multisets of complex values and a summation operator $\sum$ on finite multisets such that if $M \subseteq \mathbb{C}$ is the multiset given by $Mult_M : \mathbb{C} \to \mathbb{N}$, then $\sum M = \sum_{z \in M}(Mult_M(z) \cdot z)$.

A *quantum ASM* is a sequential ASM that operates on states with background $\mathcal{B}$ and is extended with the following two rules:

1. **apply U to qubits $i_1, \ldots, i_m$**   (where $U \in \Omega$)

2. **measure on qubits $i_1, \ldots, i_m$**

The (usually entangled) quantum states $|\psi\rangle$ that will be produced by the unitary transformations resulting from calls to rule (1) as well as the collapsed quantum states resulting from calls to rule (2), are described by a dynamic function $\Psi : 0, 1^* \to \mathbb{C}$ where:

$$|\psi\rangle = \sum_{x \in \{0,1\}^n} \Psi(x)|x\rangle$$

Given a number n of qubits and $x \in 0, 1^n$, the ASM initializes Ψ by setting $\Psi(x) = 1$ and $\Psi(y) = 0$ for $y \neq x$, thus effectively starting with the state of the quantum register collapsed to the vector $|x\rangle$. The semantics of rules 1 and 2 is defined as follows:

Unitary Transformations. Rule

apply U **to qubits** $i_1, \ldots, i_m$ (where $U \in \Omega$)

applies the unitary transformation defined by the quantum gate U to the qubits $i_1, \ldots, i_m$ by updating the corresponding amplitude values in Ψ.

Measurement. Rule

measure on qubits $i_1, \ldots, i_m$

measures the current state with respect to one or more qubits in the computational basis and updates (collapses) the amplitudes of Ψ accordingly. This is done by: writing the current quantum state $|\psi\rangle = \sum_{x \in \{0,1\}^n} \Psi(x)|x\rangle$ as $|\psi\rangle = \sum_{u \in \{0,1\}}^m |\psi_u\rangle$ where $|\psi_u\rangle = \sum_{x:x_{i_1} \cdots x_{i_m} = u_1 \cdots u_m} \Psi(x)|x\rangle$; picking one u according to the probability distribution $p(u) = \sum_{x:x_{i_1} \cdots x_{i_m} = u_1 \cdots u_m} \|\Psi(x)\|^2$; and projecting $|\psi\rangle$ to $(1/\sqrt{p(u)})|\psi_u\rangle$ by updating Ψ accordingly.

Note that to pick u according to the probability distribution we need to assume that the quantum ASM has access to a dynamic function random which provides a random real number in the interval $[0, 1]$. Given the inherent probabilism in quantum mechanical processes, this number needs to be a true random number, *not* a pseudo random number generated by a deterministic process.

Finally, let us point out that the rules for unitary transformations and measurement described above, can both be simulated by standard parallel ASMs as shown in Section 4 in [8]. It follows that every quantum algorithm as defined by the postulates in [8] can be simulated by a corresponding parallel ASM. This means that parallel ASMs are a good starting point to specify quantum algorithms at a high level of abstraction. The ASM specification of Grover's algorithm in the next section supports this claim.

4   ASM Specification of Grover's Algorithm

Grover's Algorithm [9] is a quantum algorithm for a function inversion problem known as the *oracle search problem*. Given access to a *Boolean oracle* $f : \{0,1\}^n \to \{0,1\}$, the problem consists in solving the equation $f(x) = 1$. The quantum algorithm proposed by Grover's can solve this problem (with probability very close to 1) in time $O(\sqrt{N})$, where $N = 2^n$. By comparison the best sequential algorithm for this problem works in linear time $O(N)$.

In the following we assume that $f(x) = 1$ has a unique solution. This is just to simplify the presentation since Grover's algorithm does not actually need it.

Grover's algorithm is usually called a quantum database search algorithm since it can be thought of as performing a search to find the unique element x

that satisfies the condition $f(x) = 1$ among the elements of an unordered set (or database) of size $N = 2^n$.

We start by observing that given a Boolean oracle f, the nondeterministic ASM in Listing 1.1 solves $f(x) = 1$ in just one step.

```
1  choose  x ∈ {0,1}ⁿ  with  f(x) = 1
2     answer := x
```

Listing 1.1. Nondeterministic algorithm for the oracle search problem

Since we assumed that $f(x) = 1$ has a unique solution, the parallel algorithm in Listing 1.2 also solves $f(x) = 1$ in just one step.

```
1  forall  x ∈ {0,1}ⁿ
2    if  f(x) = 1  then
3       answer := x
```

Listing 1.2. Parallel algorithm for oracle search problem

If we want to implement the algorithm specified in Listing 1.1 (or in Listing 1.2), be it in a standard computer or in a quantum computer, and want to be sure that it works exactly as intended, then we need to apply a series of stepwise correct refinements until we reach the required level of abstraction. For standard ASMs this process is well understood and developed [3,12,13]. The same cannot be said for quantum ASMs [8]. To gain intuition on this open problem, we describe next a stepwise refinement of the ASM in Listing 1.2 up to the level of a quantum algorithm which can run in a quantum computer simulator[1]. That is, we apply the ASM method [4,5] to the function inversion problem, refining it up to a step by step quantum ASM specification of Grover's Algorithm.

The goal is to use the inherent parallelism in the quantum effect to construct an efficient quantum algorithm that implements the unbounded parallelism expressed by the standard *forall* rule in Listing 1.2. We start by extending the state with a dynamic function $\Psi : \{0,1\}^* \to \mathbb{C}$ which represents the current quantum state of the quantum register of n qubits necessary for the given search space of size $N = 2^n$. Then we can refine the ASM rule as shown in Listings 1.3–1.7. Note that this new rule formally specifies Grover's algorithm, but at a level of abstraction that is higher than its well known specification in terms of quantum circuits.

```
1  if  state = initial  then
2  par
3      UNIFORMSUPERPOSITION₁
4      i := 1
5      state := iteration
6  endpar
7  if  state = iteration  then
8      par
9          if  i ≤ (π/4)√2ⁿ  then
```

---

[1] see e.g. https://www.ibm.com/quantum-computing/simulator/.

```
10      seq
11          ORACLEREFLECTION₁
12          DIFFUSIONTRANSFORM₁
13      endseq
14      i := i + 1
15    endpar
16    else
17      state := measure
18 if  state = measure  then
19 par
20    MEASUREMENT
21    state := final
22 endpar
```

Listing 1.3. Quantum algorithm for the oracle search problem

The algorithm can be informally described as follows:

1. Set the quantum register into an equal superposition of quantum states (cf. Listing 1.4). This step gives to the 2^n possible bitwise configurations of the n qubits an equal probability of 2^{-n} of being observed when the system is measured.
2. Negate the amplitude of the correct state as determined by the oracle function (cf. Listing 1.5). This step corresponds to a call to a quantum oracle (black-box) that will modify the system (without collapsing it to a classical state) depending on whether we are in the correct configuration, i.e., the state that we are searching for. All other states will remain unchanged. Note that after this step the only state with a negated amplitude will be the target state. However the probability of the system being correct remains the same.
3. Transform the amplitude of each state so that it is as far above average as it was below average before the transformation, and vice versa (cf. Listing 1.6). This is called *diffusion transform* by Grover. This step amplifies the probability of the target state and decreases the probability of all others (the sum of the squares of all the probabilities must add to 1).
4. Repeat steps 2 and 3 until the probability of (choosing) the target state is close to 1 and the probability of the other states is close to 0. This requires $\approx (\phi/4)\sqrt{2^n}$ iterations of steps 2 and 3.
5. Measure the quantum register to determine the result (cf. Listing 1.7). This is done by decomposing the unit interval into 2^n segments $S(x)$ (where $x \in \{0,1\}^n$) of length $\|\Psi(x)\|^2$ and choosing the x for which the random number returned by the function *random* is contained in $S(x)$. Recall that every quantum ASM has access to a dynamic function *random* which provides at each state a truly random number in the interval $[0,1]$.

```
1 UNIFORMSUPERPOSITION₁ =
2    forall  x ∈ {0,1}ⁿ  do
3      Ψ(x) := 1/√2ⁿ
```

Listing 1.4. Set quantum register into an equal superposition of quantum states.

```
1  ORACLEREFLECTION₁ =
2     forall  x ∈ {0,1}ⁿ  do
3        Ψ(x) := (-1)^{f(x)}Ψ(x)
```

Listing 1.5. Negate the amplitude of the correct state.

```
1  DIFFUSIONTRANSFORM₁ =
2     let  average = ∑{{ψ(x) : x ∈ {0,1}ⁿ}}/2ⁿ  in
3        forall  x ∈ {0,1}ⁿ  do
4           ψ(x) := average + (average − Ψ(x))
```

Listing 1.6. Amplify the probability of the target state.

```
1  MEASUREMENT =
2     forall  x ∈ {0,1}ⁿ  do
3        let  p = ‖Ψ(x)‖², l = ∑{{‖Ψ(y)‖² : y ∈ {1,0}ⁿ ∧ y <ₗₑₓ x}}   in
4        if  l ≤ random < l + p  then
5        par
6           Φ(x) := 1
7           answer := x
8        endpar
9        else
10          Ψ(x) := 0
```

Listing 1.7. Measure the quantum register to determine the result.

Notice that the probability $\beta = 1 - \|\Psi(x_0)\|^2$, where x_0 is the target state, of the last measurement step collapsing the quantum register to an incorrect state is *not* 0. This is very unlikely since the algorithm ensures that $\|\Psi(x_0)\|^2$ is very close to 1. In any case, the machine can test the answer and run again the algorithm if it is incorrect. The probability β^k of failing k times consecutively decreases exponentially.

The rule MEASUREMENT in Listing 1.7 does not need further refinement. A quantum computer is supposed to be able to perform such a measurement feasibly in any state $|\psi\rangle = \sum_{\{0,1\}^n} \alpha_x |x\rangle$.

The next step is to refine UNIFORMSUPERPOSITION₁, ORACLEREFLECTION₁ and DIFFUSIONTRANSFORM₁ in Listings 1.4, 1.5 and 1.6, respectively, up to the level of implementation in terms of quantum operations. These are the only remaining sub-rules, apart from MEASUREMENT, which involve unbounded parallelism in the high-level specification of Grover's algorithm in Listing 1.3. We assume that in the initial state the n qubits of the quantum register are initialized to $|0\rangle$, i.e., $\Psi(0^n) = 1$ and $\Psi(x) = 0$ for $x \neq 0^n$.

We refine the rule UNIFORMSUPERPOSITION₁ into UNIFORMSUPERPOSITION₂ as shown in Listing 1.8. Here, H denotes the well known Hadamard quantum operator (aka. "fair coin flip"). Note that many quantum algorithms begin by applying the Hadamard operator to each qubit in a register. In this way the algorithm can start with a clean state in which each of the 2^n possible states of the register have an equal probability of 2^{-n} of being observed when the system is measured.

```
1 UniformSuperposition2 =
2 seq
3    apply H to qubit 1
4    apply H to qubit 2
5    . . .
6    apply H to qubit n
7 endseq
```

Listing 1.8. Set quantum register into an equal superposition of quantum states.

Clearly, execution of UniformSuperposition2 in a quantum computer has asymptotic complexity $\Theta(\log N) = \Theta(\log 2^n) = \Theta(n)$, since it requires a sequence of n applications of the elementary Hadamard gate.

Refinement of OracleReflection1 requires to define a new quantum oracle operator O which depends of the oracle function f. Quantum oracle implementations will often use an extra qubit, since a quantum gate can always be achieved by first transforming the given Boolean oracle function f into a standard Boolean circuit C which computes the reversible function F on the set $\{0,1\}^{n+1}$ defined by $F(x,b) = (x, b \oplus f(x))$, where $\oplus$ denotes addition modulo 2, and then transforming C to the desired quantum oracle O. In this implementation the extra qubit is, however, unnecessary. Thus, we simply write the oracle effect as $O|x\rangle = (-1)^{f(x)}|x\rangle$. The refinement of OracleReflection1 consists in one application of this oracle operator O as shown in Listing 1.9.

```
1 OracleReflection2 =    apply O to qubits 1,...,n
```

Listing 1.9. Negate the amplitude of the correct state.

The rule DiffusionTransform1 actually expresses a quantum operator D that is the reflection of $(\mathbb{C}^2)^{\otimes n}$ in the line spanned by the average vector $|\mu\rangle = \sum_{x \in \{0,1\}^n} |x\rangle/\sqrt{2^n}$, i.e., $D|\mu\rangle = |\mu\rangle$ and $D|v\rangle = -v$ for every vector v orthogonal to $|\mu\rangle$. For our purpose however, it actually suffices to compute $-D$ instead. If the number of iterations made by the algorithm $\lfloor (\pi/4)\sqrt{2} \rfloor$ is odd, then the output vector gets a factor -1, but it makes no difference since collinear vectors represent the same state when they are measured by a quantum machine. It turns out that $-D$ can be written as the rule DiffusionTransform2 in Listing 1.10, thus providing the required refinement of DiffusionTransform1.

```
1 DiffusionTransform2 =
2 seq
3    apply H to qubit 1
4    . . .
5    apply H to qubit n
6    apply Not to qubit 1
7    . . .
8    apply Not to qubit n
```

```
 9    apply c^{n-1}Z to qubits 1,...,n
10    apply Not to qubit 1
11    ...
12    apply Not to qubit n
13    apply H to qubit 1
14    ...
15    apply H to qubit n                              .
16  endseq
```

Listing 1.10. Amplify the probability of the target state.

The gate H in Listing 1.10 is again the well known Hadamard gate. The gate *Not* interchanges two "truth values" $|0\rangle$ and $|1\rangle$ and is defined by the equations: $Not|0\rangle = |1\rangle$ and $Not|1\rangle = |0\rangle$. The remaining gate used in Listing 1.10, i.e., gate $c^{n-1}Z$, multiplies $|1^n\rangle$ by -1 and leaves every other basis vector unchanged. Thus, it leaves every vector orthogonal to $|1^n\rangle$ unchanged. It follows that the operations in lines 6–12 in Listing 1.10 multiply $|0^n\rangle$ by -1 and leave every other vector orthogonal to $|0^n\rangle$ unchanged. Since lines 3–5 as well as lines 13–15 in Listing 1.10 each perform the operation $H^{\otimes n}$ and $(H^{\otimes n})^2$ is the identity operator, it is clear that ORACLEREFLECTION$_2$ multiplies the average vector $|\mu\rangle$ by -1 and leaves every vector orthogonal to $|\mu\rangle$ unchanged. The gate $c^{n-1}Z$ can be expressed by a composition of the standard Hadamard and Toffoli gates. We omit the well known details.

Regarding runtime of the algorithm, note that the runtime of the call to the oracle gate O in Listing 1.9 depends on the specific function f and on the implementation of O. If we view the call to O as one elementary operation, then the total running time of each of the $(\pi/4)\sqrt{2^n}$ calls to the rules ORACLEREFLECTION$_2$ and DIFFUSIONTRANSFORM$_2$ is $\Theta(4n)$ for the two Hadamard transforms and the two $Not^{\otimes n}$ transforms, plus the cost of applying $O(n)$ gates to perform the $c^{n-1}Z$ transform. Therefore, we get that the runtime of the whole algorithm is $O(\sqrt{2^n})$.

5   Conclusion

In this paper we have shown that the ASM method can be suitably used for high level design and analysis of Grover's quantum algorithm, from its high level specification down to its implementation as a quantum circuit. It is clear that the techniques used in this paper for the correct specification and refinement of Grover's algorithm can be replicated for many others quantum algorithms. We think it would be very interesting to carry out a systematic study and classification of the interesting forms of quantum algorithm specification and refinement in terms of quantum ASMs. This should involve the necessary development of a theory of correct quantum ASM refinement that takes into account the fact that quantum algorithm are inherently probabilistic. We conjecture that the development of a full-fledged ASM method for quantum systems, comprising design, formal verification and validation, can be of great benefit for advancing the development of quantum systems.

References

1. Blass, A., Gurevich, Y.: Abstract state machines capture parallel algorithms. ACM Trans. Comput. Log. **4**(4), 578–651 (2003)
2. Blass, A., Gurevich, Y.: Abstract state machines capture parallel algorithms: correction and extension. ACM Trans. Comput. Log. **9**(3), 19:1–19:32 (2008)
3. Börger, E.: The ASM refinement method. Formal Asp. Comput. **15**(2–3), 237–257 (2003)
4. Börger, E., Raschke, A.: Modeling Companion for Software Practitioners. Springer, Heidelberg (2018). https://doi.org/10.1007/978-3-662-56641-1
5. Börger, E., Stärk, R.: Abstract State Machines. Springer, Heidelberg (2003). https://doi.org/10.1007/978-3-642-18216-7
6. Chong, F.T., Franklin, D., Martonosi, M.: Programming languages and compiler design for realistic quantum hardware. Nature **549**(7671), 180–187 (2017). https://doi.org/10.1038/nature23459
7. Ferrarotti, F., Schewe, K., Tec, L., Wang, Q.: A new thesis concerning synchronised parallel computing - simplified parallel ASM thesis. Theor. Comput. Sci. **649**, 25–53 (2016)
8. Grädel, E., Nowack, A.: Quantum computing and abstract state machines. In: Börger, E., Gargantini, A., Riccobene, E. (eds.) ASM 2003. LNCS, vol. 2589, pp. 309–323. Springer, Heidelberg (2003). https://doi.org/10.1007/3-540-36498-6_18
9. Grover, L.K.: A fast quantum mechanical algorithm for database search. In: Miller, G.L. (ed.) Proceedings of the Twenty-Eighth Annual ACM Symposium on the Theory of Computing, Philadelphia, Pennsylvania, USA, 22–24 May 1996, pp. 212–219. ACM (1996)
10. Gurevich, Y.: Sequential abstract-state machines capture sequential algorithms. ACM Trans. Comput. Log. **1**(1), 77–111 (2000)
11. Nielsen, M.A., Chuang, I.L.: Quantum Computation and Quantum Information (10th Anniversary edition). Cambridge University Press, Cambridge (2016)
12. Schellhorn, G.: Verification of ASM refinements using generalized forward simulation. J. UCS **7**(11), 952–979 (2001)
13. Schellhorn, G.: Completeness of fair ASM refinement. Science of Computer Programming **76**(9), 756–773 (2011)
14. Shi, Y.: Both toffoli and controlled-not need little help to do universal quantum computing. Quantum Inf. Comput. **3**(1), 84–92 (2003)

Spot the Difference: A Detailed Comparison Between B and Event-B

Michael Leuschel[(✉)]

Institut für Informatik, Universität Düsseldorf,
Universitätsstr. 1, 40225 Düsseldorf, Germany
michael.leuschel@hhu.de

Abstract. The B landscape can be confusing to formal methods outsiders, especially due to the fact that it is partitioned into classical B for software and Event-B for systems modelling. In this article we shed light on commonalities and differences between these formalisms, based on our experience in building tools that support both of them. In particular, we examine not so well-known pitfalls. For example, despite sharing a common mathematical foundation in predicate logic, set theory and arithmetic, there are formulas that are true in Event-B and false in classical B, and vice-versa.

1 Introduction

B and Event-B are state-based formal methods, where states of a system are modelled as mathematical entities and there are explicit operations which can inspect and modify the state. Other members of this family are abstract state machines (ASMs), TLA$^+$, VDM or Z.

1.1 Classical B

The B-method [3] consists of a formal language along with a methodology for performing refinement and conducting proof. The B-method is rooted in predicate logic, set theory, and arithmetic. B arose out of Z [48], with a focus on tool support and the use of successive refinement to derive provably correct implementations out of high-level specifications. The B-method is arguably one of the industrially more successful formal methods, it being used to develop code for a variety of (mostly railway-based) safety critical systems. The initial industrial use of B was for line 14 in Paris [21], whose product has been adapted by Siemens for many other metro lines worldwide (e.g., [22]). Other successful systems have used the B-method to derive provably correct code, such as Alstom's U400 CBTC (Communication-Based Train Control) system which is used by almost 100 metro lines worldwide.

This initial version of B, as laid out in [3] and supported by ATELIER-B, is now called classical B or also "B for software".

© Springer Nature Switzerland AG 2021
A. Raschke et al. (Eds.): Börger Festschrift, LNCS 12750, pp. 147–172, 2021.
https://doi.org/10.1007/978-3-030-76020-5_9

1.2   Event-B

Out of the experience with classical B, Abrial developed a successor eventually called Event-B (previously the name of B# was also used [4]). Event-B was designed to correct a few issues in classical B, to simplify the language (e.g., making it easier to parse and removing the complex inclusion mechanism) and to make refinement proofs easier to conduct and more scalable to larger models. The main addition though is a more flexible refinement concept targeted at systems modelling. The foundations of Event-B are laid out in the book [5] from 2010, but first ideas were published much earlier, e.g. the notion of events was already presented in 1998 [8] and the idea of extending a kernel language was presented in 2003 [4].

1.3   Tool Support

Classical B was initially supported by the B-Toolkit and by ATELIER-B [16]. Nowadays, only the latter is maintained and used in practice. ATELIER-B provides project management, static checking, proof obligation generation, automatic and interactive proof and code generation for B. A summary of 25 years of development and industrial use of ATELIER-B can be found in [33].

Event-B is supported by the RODIN platform [6], which was initially developed within the Rodin EU project.[1] The RODIN platform provides static checking, proof obligation generation and proof management features. It is maybe less known that ATELIER-B also supports an Event-B dialect. However, formal syntax and semantics have not yet been published. Still ATELIER-B is being used for Event-B modelling in practice, e.g., in [17,18,42,43].

The animator and model checker PROB [36] supports both classical B and Event-B. PROB is also available as a plugin for RODIN, and supports the ATELIER-B version of Event-B.

1.4   Outline

In the remainder of the paper we rely on the book [3] for the semantic foundations of classical B, and on the ATELIER-B handbook [16] for the technical aspects (like concrete syntax).

For Event-B we rely on the book [5] for the mathematical foundations, and [41] for the technical aspects. We thus concentrate on Event-B as supported by RODIN; however, in Sect. 5.1 we discuss the dialect supported by ATELIER-B.

Note that we are not the first to compare Event-B and classical B. Section 2 of [1] contains one page and a half of comparison, while [41] also contains a few isolated comparisons. In this present article, we assemble our view of the differences and insights, acquired during more than 15 year trying to support both formalisms within PROB.

In Sect. 2 we first focus on typing and the fundamental data types. In Sect. 3 we focus on the language for expressions and predicates. Here the languages of

---

[1] See http://rodin.cs.ncl.ac.uk.

classical B and Event-B are very similar, but this is also the area where we find the most subtle and maybe surprising differences. At the level of machines and refinement, there are more marked distinctions between Event-B and classical B. We discuss those in Sect. 4 and also provide a few remarks on other formalisms in Sect. 5.

2   Types

Both classical B and Event-B are based on typed predicate logic: every identifier and expression must have a type that specifies a set of possible values. There are a few base types (such as integers) and type constructors. The set of possible values for two distinct types in disjoint. Typing imposes restrictions on the permissible expressions, but has the benefit of catching some obvious modelling errors [32]. Typing also ensures that classical paradoxes such as the Russell-Zermelo paradox, or concepts such as a set containing itself cannot be expressed in B.

Base Types. Both classical B and Event-B provide the mathematical integers as base type. The only difference lies in the ASCII notation (the Unicode notation is identical):

- Event-B uses `INT` for the mathematical integers, `NAT` for the natural numbers starting at 0 and `NAT1` for the natural numbers starting at 1.
- classical B uses `INTEGER` for the mathematical integers, `NATURAL` for the natural numbers starting at 0 and `NATURAL1` for those starting at 1.

Beware that in classical B, `INT` stands for the *implementable* integers ranging from `MININT` to `MAXINT`. Similarly, `NAT` stands for the implementable natural numbers starting at 0 while `NAT1` is equivalent to `1..MAXINT`. Event-B has no notion of `MININT` and `MAXINT` and no concept of implementable integers.

In addition both classical B and Event-B provide the same base type for booleans (`BOOL`). However, the following base type is only available in classical B and not in Event-B:

- `STRING` the set of strings.

This is typically not a major limitation, one would usually introduce a new deferred set in a context along with the string constants that are required. However, adding a new string constant requires updating the context and its axioms. In classical B one can just use the syntax for string literals. Also, classical B provides no built-in operators apart from equality and disequality. PROB, however, provides a large range of operators (e.g., strings can be concatenated using the ^ operator).

In addition, it is possible to add new *given* sets, which are new base types, disjoint from every other type. Note that Sect. 5.2.6 of [3] (page 281) and Sect. 7.13 of ATELIER-B handbook state that every given set is finite and non-empty. The latter still holds, but the former is not true in Event-B and RODIN, where given sets can be infinite.

Type Constructors. Both classical B and Event-B provide two ways to construct complex types from simpler types A and B:

- the powerset operator $\mathbb{P}(A)$, generating the set of all subsets of A.
- the Cartesian product operator $A \times B$, generating the set of all pairs with first component in A and the second one in B.

A binary relation between A and B has the type $\mathbb{P}(A \times B)$. Similarly, a function from A to B is also represented as a set of pairs and has the same type $\mathbb{P}(A \times B)$.

Functions. The various operators in B for partial or total functions, injections, surjections and bijections thus do not add a new base type. This is different from other formalisms (e.g., TLA$^+$). In the expression $f \in 0..1 \to 0..1$, there is no special treatment required for $\in$: $0..1 \to 0..1$ is simply a set, namely the set of total functions from $0..1$ to itself:[2] $0..1 \to 0..1 = \{\{0 \mapsto 0, 1 \mapsto 0\}, \{0 \mapsto 0, 1 \mapsto 1\}, \{0 \mapsto 1, 1 \mapsto 0\}, \{0 \mapsto 1, 1 \mapsto 1\}\}$.

Records. The following type constructor is missing in Event-B:

- *struct* $(f_1 : T_1, ..., f_k : T_k)$, the set of records with k fields named f_1 to f_k and with types T_1 to T_k.

Some efforts were made to add records to RODIN, e.g., in the form of a record plugin by Colin Snook,[3] which generates construction and accessor functions by axiomatising the required record types in a context. Another proposal can be found here [23].

Sequences. Event-B is also lacking the sequence constructor and the corresponding operations on sequences. Note, however, that in classical B sequences are not a new type: `seq(X)` has the type $\mathbb{P}(\mathbb{Z} \times \tau_X)$, where τ_X is the type of X. Event-B is thus not really lacking the type constructor, but more the associated built-in operations on sequences; we return to this issue later in Sect. 3.2.

Type Inference. RODIN has an improved type-inference over ATELIER-B. Indeed, ATELIER-B requires types of identifiers to be declared using constructions such as $x \in S$ **before** using the identifier. RODIN, on the other hand, uses a more powerful type inference algorithm,[4] meaning that typing can be inferred from usage anywhere in the predicate and via a wider range of operators.

For example, $v = x \cup y \wedge 2 \notin x$ is not accepted by ATELIER-B, but is accepted by RODIN: from $2 \notin x$ one can infer that x is of type $\mathbb{P}(\mathbb{Z})$, from which one can infer in turn that y and v must also have the type $\mathbb{P}(\mathbb{Z})$.

This improved type inference has some ramifications for the Event-B language, enabling simpler expressions; see, e.g., the *id* or *prj1* operators in Sect. 3.2.

---

[2] Alloy's multiplicity annotations cannot be understood so simply in this way; see [30].
[3] https://wiki.event-b.org/index.php/Records_Extension.
[4] Attribute grammars with inherited and synthesised attributes (see [10]).

In one aspect, RODIN is stricter than classical B; it requires that every sub-expression must be given a *ground* type. As such, the predicate $\varnothing = \varnothing$ or the expression $prj1(1 \mapsto \varnothing)$ is rejected by RODIN: it cannot infer a ground type for the respective empty sets. For this, RODIN introduced the `oftype` operator, to enable the user to annotate expressions with a type. One can thus write $\varnothing$ *oftype* $\mathbb{P}(BOOL) = \varnothing$.

Note that PROB provides Hindley-Milner type inference for both classical B and Event-B.

3 Formula Language for Expressions and Predicates

Abrial's book about classical B [3] introduced the ASCII-based abstract machine notation (AMN). Later, Unicode support was added to ATELIER-B and PROB. In this section we concentrate on the core language for formulas, disregarding machine structuring and refinement for the time being.

At this level, both classical B and Event-B distinguish three kinds of formulas:

1. **expressions** which represent a (fixed) value. Expressions have no side-effect.
2. **predicates** which are either true or false, but do no represent a value. B thus syntactically and semantically distinguishes between the boolean value `TRUE` and a true predicate.
3. **substitutions**, also known as statements, which describe how states can be transformed into successor states.

We will examine substitutions in Sect. 4.1. Before looking at the first two classes in more detail, we can make a few general remarks:

– RODIN provides no syntax for comments; comments are stored separately from formulas in the RODIN database. This has the ramification that comments can only be put in certain pre-configure places, and that complex formulas cannot contain comments inside the formula.
– The ASCII version of classical B is notoriously difficult to parse e.g., due to overloading of +, *, comma and semicolon (see Chap. 2 of [39]). These issues with the syntax were corrected in Event-B.
Note that in the for DEFINITIONS in classical B (see Sect. 3.3), the above issues lead to ambiguity: one cannot determine whether a DEFINITION is an expression, predicate or substitution. An extreme case is the definition `d(X)==X` which could actually be either of them, depending on the context of the definition call.

3.1 Predicates

In classical B and Event-B predicates are syntactically different from expressions: the grammar has distinct non-terminal symbols for expressions and predicates (see Appendix B.1 and B.2 in [3] and Sects. 3.2 and 3.3 in [41]). Note, however, that the ATELIER-B parser seems to use an operator parser that mixes together the predicate and expression operators, and only makes the distinction between predicates and expressions *after* the parsing process. This means that certain expressions require parentheses in ATELIER-B, which are not required in RODIN or PROB. For example, 1=2 <=> 2=3 is a valid predicate for RODIN and PROB, but ATELIER-B requires parentheses: (1=2) <=> (2=3) as in ATELIER-B the binary operators = and <=> have the same priority 60 and are left-associative (see Appendix A of [16]).[5]

By default, ATELIER-B requires all identifiers to contain at least two characters. This default setting can be overriden. In Event-B this restriction does not exist. Identifiers can also include special Unicode characters. This is not supported by ATELIER-B (but PROB supports it for both classical B and Event-B).

On the other hand, RODIN performs some additional checks on identifiers, namely that they occur either bound or free but not both. This may give rise to warnings in RODIN, which do not exist in ATELIER-B.

Truth and Falsity as Predicates. The keywords `true` and `false` were added in Event-B. The Unciode equivalents are ⊤ and ⊥ respectively. Note that, both in Event-B and classical B, `TRUE` and `FALSE` are boolean values, which cannot be used as predicates.

ATELIER-B does use the predicates `btrue` and `bfalse`, but they only feature in the proof theory language and only appear inside the proving interface. These keywords cannot be used in B machines, where users then resort to constructs like 1=1 or 1=2. The latest version of PROB has added the predicates `btrue` and `bfalse` also to the core B language, for convenience.

The B-Toolkit [11] had the keyword `true`, which is also used for some examples in Schneider's book on B [45]. This led to confusion, as readers are unable to enter some of the examples into ATELIER-B or PROB.

Priorities and Associativity. In classical B, conjunction and disjunction have the same priority but are left-associative. $a \wedge b \vee c \wedge d$ thus corresponds to $((a \wedge b) \vee c) \wedge d$. This may not correspond to what users expect. E.g, many mathematical text books assume that conjunction binds tighter than disjunction.

In Event-B, conjunction and disjunction also have the same priority, but here they cannot be mixed without parentheses. The formula $a \wedge b \vee c \wedge d$ is thus rejected, which seems sensible to the author and avoids subtle mistakes when writing complex predicates.

Properly parenthesised Event-B predicates with conjunction and disjunction have the same meaning in classical B. This property, however, no longer true for the equivalence operator, where the priority is different. In ATELIER-B, ⇔ binds tighter than conjunction, while in Event-B it is weaker than conjunction.

---

[5] Without parentheses the ATELIER-B parser thus interprets 1=2 <=> 2=3 as the invalid ((1=2) <=> 2)=3. In PROB the grammar specifies that = has two expressions as arguments, and expressions cannot make use of = or <=>. Hence, even without parentheses, 1=2 <=> 2=3 is unambiguously interpreted as (1=2) <=> (2=3).

The following predicate is thus false in ATELIER-B but true in RODIN:

```
2=3 & (1=1) <=> (4=5) & 2=2
```

Note that ATELIER-B does *not* respect Abrial's book [3] here. [3] states on page 26 in Sect. 1.2.5 that $\Leftrightarrow$ binds weaker than conjunction (and implication).

Quantification. Both Event-B and classical B support the same universal and existential quantification. However, the syntax priorities are subtly different. For example, $\exists xx.xx > 2$ or #xx.xx>2 are allowed in RODIN, but generate parse errors in ATELIER-B. In classical B, an existential quantification thus always requires parentheses around the inner predicate: $\exists xx.(xx > 2)$.[6] Similarly, the universal quantification also always requires a parentheses around the inner predicate in classical B. There is thus also no issue with priorities of the quantifiers in classical B.

In Event-B, it is no longer required to put a parentheses around the inner predicates. This also means that the priority has to be specified, which is lower than equivalence and implication (see page 11 of [41]). $\forall x.P \implies Q$ is thus equivalent to $\forall x.(P \implies Q)$ in Event-B.[7]

3.2   Expressions

Expressions in B stand for values and have no side-effects. While predicates contain expressions, expressions can also contain themselves sub-predicates:

- inside the `bool(P)` operator, which converts a predicate to a boolean value.
- inside set comprehensions or lambda abstractions. For example {x | P}.
- inside quantified operators, such as quantified union and intersection.

Minor Syntactic Differences. Let us first look at some syntactic changes made in Event-B, whose aim was to remove some of the potential ambiguities in classical B:

- In classical B the minus operator - is overloaded and can stand for either integer difference or set difference. In Event-B it only stands for integer difference. For set difference one has to use the backslash operator \, which can also be used in classical B.
- In classical B, * is overloaded and can stand for either multiplication or Cartesian product. The solution to this result in a small cascade of differences:
  1. In Event-B, * only stands for arithmetic multiplication.

---

[6] One reason is that classical B allows composed identifiers in the grammar (e.g., xx.xx can refer to variable xx in an included machine xx). Note that, however, $\exists xx.2 < x$ is also not accepted by ATELIER-B.

[7] Note, however, that the statement in Sect. 3.2.3 of [41]: "$\forall x.P \implies Q$ is parsed as $(\forall x.P) \implies Q$ in classical B" is not true: without parentheses $\forall x.P \implies Q$ cannot be parsed at all.

2. The Cartesian product is denoted in ASCII using the ** operator in Event-B. In classical B this operator stands for exponentiation.
3. The integer exponentiation is then denoted using the hat operator ^ in Event-B. In classical B this operator stands for sequence concatenation (which does not exist in Event-B).

Note: in Unicode form some of these differences disappear, e.g., Cartesian product is written as × in both classical B and Event-B.

Expression	ATELIER-B	Event-B	comment
set difference	$A - B$, $A\backslash B$	$A\backslash B$	
integer difference	$x - y$	$x - y$	identical
integer multiplication	$x * y$	$x * y$	identical
Cartesian product	$A * B$	$A * *B$	
integer exponentiation	$x * *y$	$x\ \hat{}\ y$	
sequence concatenation	$x\ \hat{}\ y$	n.a	

Syntax for Division/Multiplication. This point is similar to the treatment of conjunction and disjunction in Sect. 3.1. The syntax of Event-B disallows certain combinations of operators without parentheses, see Table 3.2 on page 19 in [41]. Here is one example, suggested by Leslie Lamport:

- 8 / 4 / 2 is allowed in ATELIER-B and has the value 1, but RODIN rejects the expression and requires parentheses (despite [41] saying on page 19 that the division is left-associative). Apparently this is a bug in RODIN.[8]

Surprisingly, however, the expression 6 / 2 * (1+2) is allowed by RODIN, and has the value 9 in both RODIN and ATELIER-B.[9]

Missing in Event-B. Quite a few classical B operators were removed in Event-B, even though the underlying types still exist.

- Various predicates to iterate over relations are no longer available:
  - the transitive closure operator `closure1`,
  - the reflexive closure operator `closure`, and
  - the iteration operator `iterate`.
- The quantified sum Σ and product Π operators are no longer available in Event-B.
- The `rel` and `fnc` operators to transform functions into relations and vice-versa have been removed in Event-B.

---

[8] Private communication from Laurent Voisin, Paris, 17th September 2019.
[9] See also https://plus.maths.org/content/pemdas-paradox for this particular example.

rel, fnc In classical B we have `fnc({1|->2,1|->3})` = `{1|->{2,3}}` to convert a relation into a function to powersets. The rel operator does the inverse: `{1|->2,1|->3}` = `rel({1|->{2,3}})`. The effect of the `fnc` operator can easily be described constructively by the user for a given relation r:

$$fnc(r) = (\lambda x.x : dom(r)|r[\{x\}])$$

Similarly, the effect of `rel` can be expressed in Event-B as follows:

$$rel(f) = \{x \mapsto y|x : dom(f) \wedge y \in f(x)\}$$

On the downside, however, the user cannot define a polymorphic version in Event-B: for every type one needs to introduce a new version of the operator. Σ, Π, *closure*, ... The quantified sum and product and the closure operators cannot be written so easily. They can be described in an axiomatic fashion, e.g., by proving axioms such as:

$$x \mapsto y \in closure(r) \wedge y \mapsto z \in closure(r) \implies x \mapsto z \in closure(r)$$

Axiomatic definitions make animation with PROB difficult, as the animator has to solve the constraints to infer valid solutions for the given axiomatization of the closure. Currently, these operators are typically introduced using RODIN's theory plug-in [15,26] where custom proof rules and PROB bindings can be stored.

Sequences. Event-B removed support for sequence operators from the core language. In classical B a sequences of length n is a total function from $1..n$ to some range. Here is a brief overview of the missing operators:

- The operators `seq`, `seq1`, `iseq`, `iseq1` to introduce sets of sequences over some domain. E.g., the set of non-empty injective sequences over $0..1$ is `iseq1({0,1})` = `{{1|->0}, {1|->1}, {1|->0,2|->1}, {1|->1,2|->0}}`. The permutation operator `perm` is also no longer available.
  For example, `perm({0,1})` = `{ {0|->0,1|->1}, {0|->1,1|->0} }`.
- Special notation for explicit sequences: `[11,22]` = `{1|->11,2|->22}`
- Operators to deconstruct sequences: first, last, front, tail.
  For example, `tail({1|->11,2|->22})` = `{1|->22}`.
- Operators to combine or modify sequences: ^, `conc`, `rev`.

For a given type of sequence, one can write definitions which mimic these operators. E.g.,

$$seq(T) = \{s|\exists n.n \geq 0 \wedge s \in 1..n \to T\}$$

or

$$first = \lambda x.x \in seq(T)|x(1)$$

Again, these operators are typically introduced using RODIN's theory plug-in [15,26] where custom proof rules and PROB bindings can be stored.

New Operators in Event-B. In Sect. 2 we saw the need for `oftype` operator. There are a few more useful additions to B's expression language in Event-B:

- three more binary operators to construct relations: `<->>` for surjective relations, `<<->` for total relations and `<<->>` for total surjective relations.
- the `finite` operator. `finite(S)` corresponds to `S:FIN(S)` in classical B.
- the partition operator to partition a set. It has flexible arity and provides a handy shortcut for quadratic number of equations. For example, the formula $partition(S, A, B, C)$ corresponds to the predicate

$$S = A \cup B \cup C \ \wedge \ A \cap B = \varnothing \wedge A \cap C = \varnothing \wedge B \cap C = \varnothing$$

Due to the stronger type inference, a more compact operator for identity relations is available. In classical B the identity operator `id` takes an argument, specifying the set over which identity is generated:

$$\texttt{id(BOOL)} = \{\texttt{FALSE} \mapsto \texttt{FALSE}, \texttt{TRUE} \mapsto \texttt{TRUE}\}$$

This operator is still available in Event-B. However, one can also drop the argument and simply write id, which is identity over the base type inferred by (the stronger) type inference.

Pairs. Concerning the syntax, Event-B only allows $a \mapsto b$ or `a |-> b` in ASCII to denote a pair. classical B also allows the comma notation (a, b) for pairs.

Function application always takes one argument in Event-B, the notation `f(x,y)` is not allowed; one has to use $f(x \mapsto y)$ instead. This removes some of the possible confusion in classical B about how many arguments a function takes. See Sect. 3.3.1 of [41] for more discussions and additional motivations.

Note, that PROB always requires parentheses when using comma for pair constructor in classical B. With this rule, the notation $\{1,2\}$ is non-ambiguously a set consisting of *two* numbers, and not a set consisting of *one* pair.

Tuples are nested pairs in both Event-B and classical B. The triple `(a|->b|->c)` thus corresponds to nested pair `(a|->b)|->c`.

Projection Functions for Tuples. To access elements of pairs B provides the projection functions `prj1` and `prj2`. However, the use in classical B is very cumbersome, as one has to provide the domains for the pair's components:

- to access the first position of $(1, 2)$ one has to write:
  `prj1(INTEGER,INTEGER)((1,2))`.
- to access the first position of the triple $(1, 2, 3)$ one has to write:
  `prj1(INTEGER,INTEGER)(prj1(INTEGER*INTEGER,INTEGER)((1,2,3)))`
- to access the first position of $(1, 2, 3, 4)$ the expression gets more convoluted:
  `prj1(INTEGER,INTEGER)(prj1(INTEGER*INTEGER,INTEGER)`
  `(prj1(INTEGER*INTEGER*INTEGER,INTEGER) ((1,2,3,4))))`.

– to access the first position of $(1, 2, 3, 4)$:

```
prj1(INTEGER,INTEGER)(prj1(INTEGER*INTEGER,INTEGER)
(prj1(INTEGER*INTEGER*INTEGER,INTEGER)
(prj1(INTEGER*INTEGER*INTEGER*INTEGER,INTEGER) ((1,2,3,4,5)))))).
```

As one can see, it is very tedious to access components of tuples in classical B, even for tuples just using simple base types like above. In Event-B, the type arguments are luckily dropped, and e.g., to access the first component one can write the last example much more compactly:

$$\texttt{prj1(prj1(prj1(prj1(1} \mapsto \texttt{2} \mapsto \texttt{3} \mapsto \texttt{4} \mapsto \texttt{5))))}$$

But this still remains awkward and is not very readable: records with named fields are much more convenient in this aspect (which are only available in classical B, see Sect. 2). E.g., we simply write r'a to access the first field of a record r = rec(a:1,b:2,c:3,d:4,e:5).

Division, Modulo, Exponentiation and Well-Definedness. Integer division behaves the same in classical B and Event-B. For example, we have (-3 / 2) = -1. In particular, division obeys the rule that

$$a/b \ = \ (-a)/(-b) \ = \ -(-a/b) \ = \ -(a/-b)$$

Thus we have for example: $(-1)/4 = -(1/4) = 0.$[10]

Both Event-B and classical B have well-definedness (WD) conditions for possibly ill-defined expressions [9]. Below we elaborate on some of the differences already discussed in [34].

For the division operator these conditions are identical, i.e., not permitting to divide by 0.

For modulo -3 mod 2 = -1 is well-defined and true in Event-B, but is not well-defined in classical B. But this is not due to a difference in the WD condition, but due to the fact that -3 mod 2 is parsed as -(3 mod 2) in RODIN and (-3) mod 2 in ATELIER-B.

However, for exponentiation Event-B is less permissive than classical B. $(-2)^3$ is allowed in classical B, but not well-defined in Event-B (cf. page 43, Table 5.2 in [41]).

Supposing a and b are well-defined, we have the following well-definedness conditions:[11]

Another subtle difference in treating well-definedness is discussed in [34]. RODIN adds the goals of well-definedness proof obligations as hypotheses to subsequent proof obligations, while ATELIER-B does not. The technique is described in [40] (but is not mentioned in [5,41]) and avoids having to re-prove the same goal multiple times. As a result, it means that discharging WD proof obligations is even more important in RODIN, as otherwise it is very easy to discharge false theorems (see Listing 2.1 in [34]).

---

[10] In other formal languages this may be different; see Sect. 5.2.

[11] The RODIN handbook requires modulo arguments to be non-negative, which is correct; [41] is in error.

Expression	classical B	Event-B	comment
a/b	$b \neq 0$	$b \neq 0$	identical
$a \bmod b$	$a \geq 0, b > 0$	$a \geq 0, b > 0$	identical
a^b	$b \geq 0$	$a \geq 0, b \geq 0$	classical B more permissive

Syntax for Lambda and Quantified Union, Intersection. Maybe surprisingly, the notation for λ and quantified operators is not compatible between Event-B and ATELIER-B: it is generally not possible to write an expression in such a way that it is accepted by both RODIN and ATELIER-B. For example, to define the decrement function over the domain 1..5 and apply it to the value 3 one has to write in ASCII notation:

- `(%x.x:1..5|x-1)(3)` in RODIN
- `%x.(x:1..5|x-1)(3)` in ATELIER-B

More precisely, right-hand side rules of the formal grammar productions in Event-B and classical B are respectively:

- 'λ' ident-pattern '.' predicate '|' expression   in RODIN
- 'λ' ident-pattern '.' '(' predicate'|' expression ')'   in ATELIER-B

As such a parentheses spanning both predicate and expression are not allowed in Event-B but required in classical B.

Another difference concerns lambda abstractions with multiple variables. Here the use of the comma is required for the parameters of the λ in ATELIER-B but not allowed in Event-B; see Sect. 3.3.1, page 13 of [41]. Here is an example:

- `(%x|->y.x:1..5 & y:1..5|x-y)(3|->4)` in RODIN
- `%(x,y).(x:1..5 & y:1..5|x-y)(3,4)` in ATELIER-B
- `%(x,y).(x:1..5 & y:1..5|x-y)(3|->4)` also allowed in ATELIER-B

In addition, Event-B allows arbitrary nested patterns:

$$\lambda x \mapsto (y \mapsto z).x \in 1..5 \wedge y \in 1..5 \wedge z \in 1..4 \mid x * y * z$$

This example cannot be written as a simple lambda expression in classical B, as `%(x,y,z).(P|E)` corresponds to the differently nested $\lambda(x \mapsto y) \mapsto z.P \mid E$ in Event-B.

Finally, Event-B also provides an additional set comprehension construction with an expression term: $\{x.P|E\}$. The equivalent formula is more convoluted to write in classical B and requires the use of an existential quantifier: `{z | #x.(P & z=E)}` in ASCII or $\{z \mid \exists x.(P \wedge z = E)\}$ in Unicode.

3.3 Definitions

ATELIER-B provides a special section where one can add (parameterised) definitions. Definitions are implemented by textual replacement in ATELIER-B. In other words, they are macros with all of their problems (known to seasoned C programmers).

For example, when not using parentheses in the definition bodies, the textual replacement can result in unexpected results due to change of priorities. In the example below, a reader may expect `2*add(0,5)` to be identical to $2*(0+5)$, i.e. 10. However, after textual replacement we obtain the expression `2*0+5` which is equal to 5.

```
DEFINITIONS
  add(xx,yy) == xx+yy
ASSERTIONS
  2*add(0,5) = 10;  // false in Atelier-B
```

Another problem is *variable capture*, when the definition's quantified variables clash with identifiers in the actual arguments of the definition. Take the following example:

```
DEFINITIONS
  egt(xx) == (#yy.(yy:1..99 & xx<yy))  // egt: Exists-Greater-Than
ASSERTIONS
  egt(5); // true
  #yy.(yy:INTEGER & yy=5 & egt(yy)) // false
```

We have that `egt(5)` is true: yy=6 satisfies the predicate. However, the predicate `yy=5 & egt(yy)` is false, as it is textually rewritten to

$$yy=5 \ \& \ (\#yy.(yy:1..99 \ \& \ yy<yy))$$

and there is no solution for `yy<yy`.[12] These problems do not arise very often, but when they do, they are hard to understand and debug. In one instance, I had the surprising result that the value of `{a | a : Sigma}` was different from `Sigma`. I was suspecting a bug in PROB, until I realised that Sigma was introduced by the definition `Sigma == {a,b}`, where a and b were some of the elements of a given set `Symbol`. The set comprehension was thus equivalent to `{a | a : {a,b}}` in turn equivalent to `Symbol`. PROB now warns when such variable captures appear. The price to pay is that PROB does not really treat definitions as macros, every definition body has to be a valid formula; it cannot consist of a partial text of a formula.

Another issue of macros are related to performance: the textual substitution can lead to an explosion of the size of the constructed formula, and can lead in turn to multiple evaluation of the same expression. For example, the following definitions look innocuous:

---

[12] PROB warns when such variable captures appear. The price to pay is that PROB does not really treat definitions as macros, every definition body has to be a valid formula; it cannot consist of a partial text of a formula.

DEFINITIONS
    droplst(ss) == {size(ss)} <<| ss // drop last element

The above works correctly, e.g., to drop the last three elements of a list:

droplst(droplst(droplst([1,2,3,4,5]))) = [1,2]

However, the sequence [1,2,3,4,5] appears $2^3 = 8$ times in the expression after replacement. It is clear that when the argument is a much larger literal value or more definitions are nested this effect can be problematic, in particular for data validation [24]. Also, if the argument to setl is a complicated set comprehension, this expression may be evaluated eight times by a tool like PROB (unless common-subexpression elimination is enabled).

Given the problems with macros, it is understandable that the definitions concept from classical B was not incorporated into Event-B. However, the complete absence of definitions is also problematic in practice. The theory plug-in [15,26] can also be used to introduce definitions. But this is a quite heavyweight solution, so much so that probably nobody uses it for that purpose. Copying and pasting the respective formulas seems to be one solution adopted in practice, but this is not very readable and easy to maintain. As such, hygenic macros [25] (which properly treat quantified variables) would be a nice extension of Event-B. Other formal languages like TLA+ and Alloy provide this feature; see Sects. 3.5 and 5.

3.4   General Missing Features

A few features that are generally useful are missing in both Event-B and classical B when compared to other specification or programming languages. Here is a list of some of them:

IF-THEN-ELSE. While classical B has an IF-THEN-ELSE substitution, it has no such construct for expressions or predicates. For predicates, one can write something like $(A \implies B) \wedge (\neg A \implies C)$, at the cost of duplicating A. For expressions, it is more cumbersome. This construct is particularly useful for defining functions via the lambda construction. It is for example available in TLA+ or Z, and PROB now supports it for classical B as well.

For example, a function to compute the absolute value can be defined as follows:

$$abs = \lambda x.(x \in \mathbb{Z} | IF \ x < 0 \ THEN \ -x \ ELSE \ x \ END)$$

Without this construct more cumbersome encodings are required, e.g.:

$$abs = \lambda x.(x \in \mathbb{N} | x) \cup \lambda x.(x \in \mathbb{Z} \wedge x < 0 | -x)$$

LET. While classical B has a LET substitution, it has no such construct for expressions or predicates. It is also available in TLA+ or Z, and PROB now also supports it for classical B. The construct is particularly useful to structure the definition of more complex set comprehensions or lambda abstractions.

A LET construct *LET x BE x = E IN P END* for a predicate P can be mimicked using existential quantification: $\exists x.x = E \wedge P$.

For expressions, however, it is more cumbersome. Some workaround exists for expressions which happen to be sets: $\bigcup x.(x = E|F)$ is an encoding of *LET x BE x = E IN F END*.

Recursive Functions. While it is in principle possible to define recursive functions in classical B and Event-B (see [37] in the Festschrift for Egon Börger's 60th birthday, presented at Schloß Dagstuhl), the process is still more cumbersome than it should be. PROB provides some support for recursive functions, e.g., one can write a recursive function to compute the sum of a set by defining a constant *sum* by the following equation:[13]

$$sum = (\lambda x.x \subseteq \mathbb{Z}|IF\ x = \varnothing\ THEN\ 0\ ELSE\ min(x) + sum(x \setminus \{min(x)\})\ END)$$

This is, however, there is no support by the B provers yet and PROB does not yet perform any well-foundedness checks.

In classical B, it is possible to use the transitive closure operators to encode a recursive computation (see Sect. 6.2 in [24]). For the example above one would encode a (non-recursive) step function, which receives the set of integers and an accumulator:

$$step = \lambda(s, acc).(s \neq \varnothing\ |\ (s \setminus \{min(s)\}, acc + min(s))$$

We can now use the transitive closure operator (*closure1*) to compute the sum of a set using an initial value of 0 for the accumulator:

$$closure1\,(step)[\{\{1, 2, 3\} \mapsto 0\}] = \{(\varnothing \mapsto 6), (\{2, 3\} \mapsto 1), (\{3\} \mapsto 3)\}$$

The result can be obtained by extracting the value for the base case ($\varnothing$):

$$closure1\,(step)[\{\{1, 2, 3\} \mapsto 0\}](\varnothing) = 6$$

This technique enables proof with ATELIER-B, but is computationally not optimal (the transitive closure derives all intermediate results). It also clearly requires considerable expertise by users.

Here, TLA+ provides a more convenient way of defining new recursive functions.

---

[13] In classical B one can of course just use the Σ operator for this example. Here we just wish to illustrate the various approaches to recursion on a simple example.

Polymorphism. While many of the core B operators are polymorphic a user cannot define polymorphic functions. For example, set union and cardinality are polymorphic and can be applied to sets of different types:

$$card(\{1\} \cup \{2\}) = card(\{TRUE\} \cup \{FALSE\})$$

All variables or constants must have a concrete, fully instantiated type. E.g., a user-defined union function would look like this and work only for the declared type (here sets of integers):

$$myunion = (\lambda x \mapsto y.x \subseteq \mathbb{Z} \wedge y \subseteq \mathbb{Z} | \{z \mid z \in x \vee z \in y\})$$

In classical B, definitions can be polymorphic, as the following example shows:

$$MYUNION(x, y) == \{z \mid z \in x \vee z \in y\}$$

But as we have seen in Sect. 3.3, this approach has some considerable drawbacks.

3.5   Future of B

In future, we would like to extend both classical B and Event-B to address these shortcomings and incorporate the good ideas from other languages:

- making the extensions of PROB as part of the core language for ATELIER-B and RODIN
  - IF-THEN-ELSE for expressions and predicates
  - LET for expressions and predicates
  - better support for strings [24]
- an EXPRESSIONS clause to replace the fragile DEFINITIONS with hygenic macros, derived from the constants and variables of a B machine.
- a way to specify recursive functions, along with a variant for proving well-foundedness. Ideally it should be possible to write polymorphic functions.

The goal would be to obtain a specification language, which is amenable to formal proof and constraint solving, while providing the convenience of a functional programming language.

4   Machines and Refinement

While for predicates and expressions, classical B and Event-B are very close together, there are marked differences at the level of the B machines. Due to space limitations we can only elaborate on some of the differences.

In Event-B, new sets and constants are put into *contexts* while variables and events are put into *machines*. In classical B, there are only machines, which can contain sets, constants, variables and operations.

In Event-B, a machine can include contexts, but cannot include machines. Here, classical B is more flexible, as it allows to include any machine. In contrast to Event-B, a machine can be included multiple times and renamed. E.g., one can include multiple copies of a generic buffer machine and give each copy an individual name. Also, classical B has a relatively complex "ontology" of inclusion keywords: INCLUDES, EXTENDS, USES, SEES and IMPORTS. These various inclusion mechanisms mainly differ by what parts of the included machine can be seen and used. These concepts make sense for enabling compositional proofs of complex software systems, but can be confusing to the B user.

As far as refinement is concerned, a context in Event-B can only refine another context, and a machine can only refine another machine. For contexts this is called *extending* rather than refining.

In summary, the powerful but complex structuring mechanism of classical B has been replaced by just two structuring mechanisms:

- inclusion: a machine can *see* contexts, giving it access to their constants, axioms and theorems.
- refinement: a machine can *refine* another machine and a context can *extend* another context.

There is also a considerable change in terminology:

- operations are now called events, on the account that Event-B is used for systems modelling,
- properties are now called axioms, and
- assertions are now called theorems.

Finally, classical B distinguishes between abstract machines, refinement machines and implementation machines, each with their own particularities. The subset of B allowed in implementation machines is also called B0, for which various code generators are available.

One addition is that in Event-B axioms, invariants, guards and theorems carry labels. These labels are helpful during proof, to identify the source of hypotheses or proof obligations.

Another difference is that theorems can be interleaved with invariants; the order can be relevant for proof (influencing the available hypotheses) and for well-definedness. The same is true for theorems and axioms: there is one section of an Event-B context containing both theorems and axioms and the order is relevant. In classical B, the ASSERTIONS clause is completely separate from the INVARIANTS and PROPERTIES clause, and as such contains both theorems on variables and constants.

4.1 Events, Operations and Substitutions

A B machine has operations, while an Event-B machine has events. Both operations in classical B and events in Event-B consist of substitutions (aka statements). A substitution can modify the state of a B machine, meaning that it can change the values of the variables. In this area Event-B differs quite considerably from classical B. While classical B has a very rich substitution language, including CASE statements or WHILE loops, Event-B follows a minimalistic approach. More precisely, an event in Event-B has just three components:

- an optional list of parameters,
- a guard predicate,
- a list of actions, which are implicitly executed in parallel. The empty action list corresponds to skip, i.e., an event that keeps the state unchanged.

For the actions there are only three substitutions available in Event-B, where x is a variable:

- deterministic assignments of the form x := E, where x is a variable and E an expression.
- non-deterministic assignments of the form x :: S, where x is a variable and S a set expression.
- assignments by predicate of the form $\boldsymbol{x}$:| P, where $\boldsymbol{x}$ is a list of variables and P a predicate.

Witness predicates are associated with some parameters and non-deterministic assignments. We return to this below when talking about refinement.

Some guards can be marked as theorems. This is not possible in classical B, but there is an ASSERT substitution which can serve a similar purpose.

The above form for describing an event is also possible in classical B, but it provides a much richer language. Compared to classical B, we can mention the following most important missing constructs:

- no conditional substitution (IF-THEN-ELSE). In Event-B, this is typically mimicked by multiple events and incorporating the tests into the guard.
- no LET substitution. This drawback can be circumvented by adding a separate parameter to the event. Note that in Event-B, parameters do not correspond to parameters in the generated code, and parameters can be changed in refinements. Hence, this solution is typically quite appropriate.
- no sequential composition. A sequential composition needs to be encoded by separate events. A difference with classical B is that then that the invariant is also checked in the additional intermediate state. Furthermore, interleavings between various events are then possible. Some plugins and extension of RODIN try to solve this limitation by adding back ways to specify a control flow [27] or providing UML state machines [44].
- no while loops. This point is similar to sequential composition. Typically a while loop is modelled in Event-B by introducing additional events. The variant of a while loop can be encoded using the convergence annotations of events.
- no operation call. In Event-B it is not possible to call the event of another machine. Sometimes this limitation can be overcome by refinement: instead

of calling an operation we refine it and add additional behaviour. This solution is, however, not always feasible. Some extensions of RODIN have tried to overcome this [28].

– no preconditions (PRE). Indeed, in Event-B, events have only guards, while classical B allows both guards and preconditions. But as there is no way to call operations of another machine in Event-B, there is typically also no need for preconditions.

4.2   Refinement

A comparison of refinement in various state-based formal methods can be found here: [20] discussing Z and B, and [19] discussing Event-B and ASM. We can only discuss a few important aspects here.

Changing Signatures of Operations. Classical B is designed for software and a high-level abstract machine serves as both specification and provides the signature of the available operations. As such, changing parameters and return values in refinements is not allowed, as it would result in an implementation which is incompatible with the specification.

In Event-B, parameters play another role and there is no concept of an operation call. As such, parameters can be removed, changed and introduced. When parameters are removed, however, a witness predicate has to be provided. This enables to establish a relation between the abstract and concrete parameters while conducting the refinement proofs.

Introducing New Events. In systems modelling it makes sense that during refinement the specifications becomes more concrete and more events become visible. Hence, Event-B allows to introduce new events in refinements. However, to ensure that the abstract model is still a correct abstraction, these new events must refine skip: they are not allowed to change the state of the abstract model, i.e., must be invisible at the abstract level.

In addition, events can be marked as anticipated or convergent. This annotation is often used for new events to ensure termination (to ensure that sooner or later a visible event of the abstraction must occur).

In classical B there is no need to add new events; it would be impossible to call them anyway as they do not figure in the high-level abstract machine.

Splitting and Merging Events. Splitting an operation in software development makes no real sense: we want to derive code and want to know which concrete operation should be used. In systems modelling, however, it is useful to split an event into different variations. Something that is indistinguishable at the high-level, maybe differentiated at the more concrete level. E.g., at an abstract level we may just have an event move_train, but in a more concrete level, with more precise modelling of locations, we may wish to split this event into move_train_forward and move_train_backward).

Refinement Proof Obligations. One big change in Event-B is the refinement proof obligation. In classical B one has to prove for a concrete operation, that a matching abstract event does exist. More precisely, this is done by proving that it is *not* the case that there is a concrete event for which *no* matching abstract event exists. This leads to a proof obligation with two negations [3,45]. While this is a very flexible refinement concept, the resulting proof obligations are not always easy to discharge.

Event-B eliminates the double negation, yielding much simpler proof obligations. The price to pay is that the user has to specify the correct abstract behaviour that matches the concrete behaviour. This is made possible by:

- witness predicates, which specify abstract parameters and values for non-deterministic assignments.
- the simple substitution language, which means that every case of a conditional becomes its own event. As such, one can precisely pinpoint matching abstract events for concrete events. In classical B, the refinement of case statement by another case statement results in combinatorial explosion of cases in the refinement proof.

4.3   Composition and Modularisation

A major missing feature in Event-B is the possibility to include machines and, e.g., call the operations in the included machines. This mechanism is very useful to decompose a system into components. The core language of Event-B only provides refinement as a structuring mechanism for machines. Sometimes this is sufficient to structure a development, but it also implies that various features or subcomponents have to be introduced in a particular order, which the B user has to choose.

For example, to mimic a system which uses a component "traffic light" and "car", one can simply include the corresponding B machines in classical B. In Event-B, one would provide an abstract system model, and refine it while first adding either the "traffic light" or the "car" component. In a second refinement one would add the other component. The order is now fixed, and hard to change. The modelling of the sub-components is weaved into the model of the system and also harder to change and adapt.

There are, however, some extensions of Event-B which address this issue. Most notable are the composition/decomposition approaches:

- shared variable decomposition [2],
- shared event decomposition [46], leading to the [47] decomposition plugin.

These composition concepts are not identical to inclusion in classical B, and are more adapted to systems modelling. It is, however, often possible to do Event-B style compositional modelling in classical B using machine inclusion and operation calls. An example of this style of systems modelling in classical B can be found in [38], with a later translation to a RODIN refinement. The author has also managed to translate a simple example used to illustrate the approach from [2] to inclusion and refinement in classical B.

5   Other Formal Methods

We provide a brief comparison with some other formal methods. Due to space restrictions, we cannot conduct a comprehensive comparison.

5.1   Event-B by Atelier B

As already mentioned, ATELIER-B also supports a dialect of Event-B. Compared to RODIN it provides a much richer language for describing substitutions. Another advantage is the existence of a complete textual representation for machines (even though a formal grammar is unfortunately not available). At some point, the B2RODIN plugin was developed to import such ATELIER-B files.[14]

5.2   TLA+

TLA$^+$ [31] has the same grounding in predicate logic, set theory and arithmetic as B. A major difference, however, is that TLA$^+$ is untyped; see Sect. 2 and the debate in [32]. An undeniable advantage of the untyped nature of TLA$^+$ is that records can be viewed simply as a special function mapping strings to values. It is also possible to use a unique null value to represent partially defined functions.

In contrast to B, the grammar of TLA$^+$ makes no distinction between predicates and expressions. There is thus also no need for conversion predicates like **bool(.)** in B or for the distinction between the booleans and the truth values.

As far as relations are concerned, TLA$^+$ has fewer built-in operators. Also, functions are not sets of pairs and are set apart from relations. As a consequence, one cannot apply set operators to functions. There is, however, a provision for defining recursive functions and new operators (aka polymorphic functions). Combined with the IF-THEN-ELSE and LET constructs for expressions, this provides a convenient way to write functions.

TLA$^+$ uses the term action to denote events. Here TLA$^+$ is much closer to Event-B than classical B. An action in TLA$^+$ is simply described by a predicate, namely the before-after predicate. There is thus no substitution language, and parameters are encoded using existential quantification. There is, however, the Pluscal language which provides programming constructs (like while loops) and which is translated to regular TLA$^+$.

Due to the absence of a substitution language, TLA$^+$ cannot escape the frame problem, and actions need to be annotated with "unchanged" annotations, i.e., specifying which variables are *not* modified by the action. Also, TLA$^+$ provides no refinement methodology and refinement proof obligations. TLA$^+$ is geared towards the verification of temporal logic formulas; something which is not at the heart of classical B or Event-B (but provided by PROB). TLA$^+$ here relies in this aspect on stuttering, which ensures that refinement is possible. See [7] for a discussion and the influence of TLA$^+$ on Event-B.

As a minor note, TLA$^+$ uses floored division, which is different from B. For example $(-1)/4 = -1 \neq 0$.

---

[14] https://www.methode-b.com/en/download-b-tools/rodin/b2rodin/.

5.3 Alloy

Alloy [29] is also rooted in logic and set theory. The semantic rules of Alloy can be expressed in B [30]. Alloy puts even more emphasis on relations than B; the fact that everything is a relation provides an elegant core language. Also, tuples are "flat" in Alloy and not nested pairs as in B. This makes various operations more elegant. In particular the relational join operator can be used to cover various B operators (such as both domain and range).

On the downside, the multiplicity annotations in Alloy are not fully denotational; their meaning depends on the context [30]. Here I find B's approach to specifying partial and total functions more elegant. Also, Alloy is restricted to first-order sets and relations; i.e., one cannot use the Alloy analyser on sets of sets. The restriction, ensures that a translation to propositional logic for SAT solving is feasible.

5.4 ASM, Z

Z [48] is the predecessor of B, but has more data types available. For instance, multisets (bags) or freetypes with constructors are available. The latter correspond to inductive datatypes, as typically available in functional or logic programming languages. In Rodin freetypes can be introduced in the theory plug-in [15,26]. Z also provides for the IF-THEN-ELSE or LET constructs for expressions. A particularity of Z is its schema calculus, which provides a completely different way to organise or structure a specification than B machines.

Abstract state machines (ASMs) [13] have considerable commonalities with B. An interesting difference is that ASMs allow parallel assignments to the same variables, and provides various semantics to reconcile updates. More details about the differences can be found in [35].

6 Discussion and Conclusion

We conclude with a few more discussions about the tools available. The present paper was influenced by the experience of making PROB accept both classical B and Event-B. As we have seen, predicates and expressions of classical B and Event-B are very close, so much so that PROB uses the same interpreter and constraint solver for predicates, expressions of classical B and Event-B (and actually also TLA$^+$ and Z). PROB has a flag (`animation_minor_mode`) which influences some the behaviour of the interpreter, like division and well-definedness conditions for modulo or exponentiation. At the machine level there is a much larger difference; here PROB uses two completely different interpreters for classical B and Event-B.

On the practical side, classical B has multiple code generators, which have been used in a considerable number of industrial projects [14,33]. While Event-B also has a few code generators, none of them have been used in industrial projects yet. Also, in Event-B the control structure must be extracted; in classical B it is explicit expressed using the richer substitution language.

Another practical difference is the lack of a textual representation for machines and contexts in RODIN. In fact, RODIN was built on the idea of iterative development, where users update the RODIN model database, with automatic incremental building and proving occurring in the background. With hindsight one can say that this dream has only been partially fulfilled, and that the lack of a textual representation was more of a hindrance than an enabler. Indeed, there are at least five editors in RODIN, all with their own machine representation. The lack of a textual representation made it difficult to version, refactor and share RODIN models. The Camille editor [12] was developed to address these issues; but its maintenance has proven difficult; in particular synchronising the textual representation with the RODIN database within the evolving Eclipse Modelling Framework (EMF) is a challenge. Camille also does not provide editing for the theories within the theory plugin. A successor, called CamilleX together with a redeveloped theory plugin, will hopefully overcome these challenges.

In our exploration of Event-B and classical B we have stumbled upon various subtle and surprising differences. We have identified shortcomings of one or the other formalisms and have also provided some recommendations for an improved mathematical notation. We hope that this article can help new researchers better understand the B landscape.

Acknowledgements. Egon Börger visited my group at the University of Düsseldorf in summer of 2015. Egon was funded by a renewed Forschungspreis grant of the Humboldt Foundation. This visit was very fruitful and helped me gain a better understanding of ASMs and enabled us to write the ABZ 2016 paper on a compact encoding of ASMs in Event-B. Egon also played an important role in establishing the ABZ conference series, which was instrumental in establishing bridges between the various state-based formalisms and led to considerable cross-fertilization. I hope that this article provides an additional bridge and helps researches travel more easily between the various state-based formalisms.

I also wish to thank Jean-Raymond Abrial, Lilian Burdy, Michael Butler, Stefan Hallerstede, Luis-Fernando Mejia, Sebastian Stock, Laurent Voisin, and Fabian Vu for useful feedback, ATELIER-B and RODIN implementation details and pointers. Finally, an anonymous referee provided a lot of detailed feedback, for which I am grateful.

References

1. Abrial, J.-R.: On B and Event-B: principles, success and challenges. In: Butler, M., Raschke, A., Hoang, T.S., Reichl, K. (eds.) ABZ 2018. LNCS, vol. 10817, pp. 31–35. Springer, Cham (2018). https://doi.org/10.1007/978-3-319-91271-4_3
2. Abrial, J., Hallerstede, S.: Refinement, decomposition, and instantiation of discrete models: application to Event-B. Fundam. Inform. **77**(1–2), 1–28 (2007)
3. Abrial, J.-R.: The B-Book. Cambridge University Press, Cambridge (1996)
4. Abrial, J.-R.: B$^{\#}$: toward a Synthesis between Z and B. In: Bert, D., Bowen, J.P., King, S., Waldén, M. (eds.) ZB 2003. LNCS, vol. 2651, pp. 168–177. Springer, Heidelberg (2003). https://doi.org/10.1007/3-540-44880-2_12
5. Abrial, J.-R.: Modeling in Event-B: System and Software Engineering. Cambridge University Press, Cambridge (2010)

6. Abrial, J.-R., Butler, M., Hallerstede, S., Voisin, L.: An open extensible tool environment for event-B. In: Liu, Z., He, J. (eds.) ICFEM 2006. LNCS, vol. 4260, pp. 588–605. Springer, Heidelberg (2006). https://doi.org/10.1007/11901433_32

7. Abrial, J.-R., Cansell, D., Méry, D.: Refinement and reachability in event-B. In: Treharne, H., King, S., Henson, M., Schneider, S. (eds.) ZB 2005. LNCS, vol. 3455, pp. 222–241. Springer, Heidelberg (2005). https://doi.org/10.1007/11415787_14

8. Abrial, J.-R., Mussat, L.: Introducing dynamic constraints in B. In: Bert, D. (ed.) B 1998. LNCS, vol. 1393, pp. 83–128. Springer, Heidelberg (1998). https://doi.org/10.1007/BFb0053357

9. Abrial, J.-R., Mussat, L.: On using conditional definitions in formal theories. In: Bert, D., Bowen, J.P., Henson, M.C., Robinson, K. (eds.) ZB 2002. LNCS, vol. 2272, pp. 242–269. Springer, Heidelberg (2002). https://doi.org/10.1007/3-540-45648-1_13

10. Aho, A.V., Lam, M.S., Sethi, R., Ullman, J.D.: Compilers: Principles, Techniques, and Tools, 2nd edn. Addison-Wesley, Boston (2007)

11. B-Core (UK) Ltd, Oxon, UK. B-Toolkit, On-line manual (1999). http://sens.cse.msu.edu/Software/B-Toolkit/BKIT/BHELP/BToolkit.html

12. Bendisposto, J., Fritz, F., Jastram, M., Leuschel, M., Weigelt, I.: Developing Camille, a text editor for Rodin. Softw. Prac. Exp. $41(2)$, 189–198 (2011)

13. Börger, E.: Abstract State Machines. Springer, Heidelberg (2003). https://doi.org/10.1007/978-3-642-18216-7

14. Butler, M., et al.: The first twenty-five years of industrial use of the B-method. In: ter Beek, M.H., Ničković, D. (eds.) FMICS 2020. LNCS, vol. 12327, pp. 189–209. Springer, Cham (2020). https://doi.org/10.1007/978-3-030-58298-2_8

15. Butler, M., Maamria, I.: Practical theory extension in event-B. In: Liu, Z., Woodcock, J., Zhu, H. (eds.) Theories of Programming and Formal Methods. LNCS, vol. 8051, pp. 67–81. Springer, Heidelberg (2013). https://doi.org/10.1007/978-3-642-39698-4_5

16. ClearSy. Atelier B, User and Reference Manuals. Aix-en-Provence, France (2009). http://www.atelierb.eu/

17. Comptier, M., Déharbe, D., Perez, J.M., Mussat, L., Thibaut, P., Sabatier, D.: Safety analysis of a CBTC system: a rigorous approach with Event-B. In: Fantechi, A., Lecomte, T., Romanovsky, A.B. (eds.) Proceedings RSSRail 2017, LNCS, vol. 10598, pp. 148–159. Springer, Cham (2017). https://doi.org/10.1007/978-3-319-68499-4_10

18. Comptier, M., Leuschel, M., Mejia, L.-F., Perez, J.M., Mutz, M.: Property-based modelling and validation of a CBTC zone controller in Event-B. In: Collart-Dutilleul, S., Lecomte, T., Romanovsky, A. (eds.) RSSRail 2019. LNCS, vol. 11495, pp. 202–212. Springer, Cham (2019). https://doi.org/10.1007/978-3-030-18744-6_13

19. Derrick, J., Boiten, E.: State-based languages: event-B and ASM. Refinement, pp. 149–176. Springer, Cham (2018). https://doi.org/10.1007/978-3-319-92711-4_8

20. Derrick, J., Boiten, E.: State-based languages: Z and B. Refinement, pp. 121–147. Springer, Cham (2018). https://doi.org/10.1007/978-3-319-92711-4_7

21. Dollé, D., Essamé, D., Falampin, J.: B dans le transport ferroviaire. L'expérience de Siemens Transportation Systems. Technique et Science Informatiques $22(1)$, 11–32 (2003)

22. Essamé, D., Dollé, D.: B in large-scale projects: the canarsie line CBTC experience. In: Julliand, J., Kouchnarenko, O. (eds.) B 2007. LNCS, vol. 4355, pp. 252–254. Springer, Heidelberg (2006). https://doi.org/10.1007/11955757_21

23. Evans, N., Butler, M.: A proposal for records in event-B. In: Misra, J., Nipkow, T., Sekerinski, E. (eds.) FM 2006. LNCS, vol. 4085, pp. 221–235. Springer, Heidelberg (2006). https://doi.org/10.1007/11813040_16

24. Hansen, D., Schneider, D., Leuschel, M.: Using B and ProB for data validation projects. In: Butler, M., Schewe, K.-D., Mashkoor, A., Biro, M. (eds.) ABZ 2016. LNCS, vol. 9675, pp. 167–182. Springer, Cham (2016). https://doi.org/10.1007/978-3-319-33600-8_10

25. Herman, D., Wand, M.: A theory of hygienic macros. In: Drossopoulou, S. (ed.) ESOP 2008. LNCS, vol. 4960, pp. 48–62. Springer, Heidelberg (2008). https://doi.org/10.1007/978-3-540-78739-6_4

26. Hoang, T.S., Voisin, L., Salehi, A., Butler, M.J., Wilkinson, T., Beauger, N.: Theory plug-in for Rodin 3.x. CoRR, abs/1701.08625 (2017)

27. Iliasov, A.: Use case scenarios as verification conditions: event-B/flow approach. In: Troubitsyna, E.A. (ed.) SERENE 2011. LNCS, vol. 6968, pp. 9–23. Springer, Heidelberg (2011). https://doi.org/10.1007/978-3-642-24124-6_2

28. Iliasov, A., et al.: Supporting reuse in event B development: modularisation approach. In: Frappier, M., Glässer, U., Khurshid, S., Laleau, R., Reeves, S. (eds.) ABZ 2010. LNCS, vol. 5977, pp. 174–188. Springer, Heidelberg (2010). https://doi.org/10.1007/978-3-642-11811-1_14

29. Jackson, D.: Alloy: a lightweight object modelling notation. ACM Trans. Softw. Eng. Methodol. 11, 256–290 (2002)

30. Krings, S., Leuschel, M., Schmidt, J., Schneider, D., Frappier, M.: Translating alloy and extensions to classical B. Sci. Comput. Program. 188, 102378 (2020)

31. Lamport, L.: Specifying Systems, the TLA+ Language and Tools for Hardware and Software Engineers. Addison-Wesley, Boston (2002)

32. Lamport, L., Paulson, L.C.: Should your specification language be typed. ACM Trans. Program. Lang. Syst. 21(3), 502–526 (1999)

33. Lecomte, T., Deharbe, D., Prun, E., Mottin, E.: Applying a formal method in industry: a 25-year trajectory. In: Cavalheiro, S., Fiadeiro, J. (eds.) SBMF 2017. LNCS, vol. 10623, pp. 70–87. Springer, Cham (2017). https://doi.org/10.1007/978-3-319-70848-5_6

34. Leuschel, M.: Fast and effective well-definedness checking. In: Dongol, B., Troubitsyna, E. (eds.) IFM 2020. LNCS, vol. 12546, pp. 63–81. Springer, Cham (2020). https://doi.org/10.1007/978-3-030-63461-2_4

35. Leuschel, M., Börger, E.: A compact encoding of sequential ASMs in event-B. In: Butler, M., Schewe, K.-D., Mashkoor, A., Biro, M. (eds.) ABZ 2016. LNCS, vol. 9675, pp. 119–134. Springer, Cham (2016). https://doi.org/10.1007/978-3-319-33600-8_7

36. Leuschel, M., Butler, M.J.: ProB: an automated analysis toolset for the B method. STTT 10(2), 185–203 (2008). https://doi.org/10.1007/s10009-007-0063-9

37. Leuschel, M., Cansell, D., Butler, M.: Validating and animating higher-order recursive functions in B. In: Abrial, J.-R., Glässer, U. (eds.) Rigorous Methods for Software Construction and Analysis. LNCS, vol. 5115, pp. 78–92. Springer, Heidelberg (2009). https://doi.org/10.1007/978-3-642-11447-2_6

38. Leuschel, M., Mutz, M., Werth, M.: Modelling and validating an automotive system in classical B and event-B. In: Raschke, A., Méry, D., Houdek, F. (eds.) ABZ 2020. LNCS, vol. 12071, pp. 335–350. Springer, Cham (2020). https://doi.org/10.1007/978-3-030-48077-6_27

39. Mariano, G.: Évaluation de Logiciels Critiques Développés par la Méthode B: Une Approche Quantitative. Ph.D. thesis, Université de Valenciennes et Du Hainaut-Cambrésis, December 1997

40. Mehta, F.: A practical approach to partiality – a proof based approach. In: Liu, S., Maibaum, T., Araki, K. (eds.) ICFEM 2008. LNCS, vol. 5256, pp. 238–257. Springer, Heidelberg (2008). https://doi.org/10.1007/978-3-540-88194-0_16

41. Métayer, C., Voisin, L.: The event-B mathematical language (2009). http://wiki.event-b.org/index.php/Event-B_Mathematical_Language

42. Sabatier, D.: Using formal proof and B method at system level for industrial projects. In: Lecomte, T., Pinger, R., Romanovsky, A. (eds.) RSSRail 2016. LNCS, vol. 9707, pp. 20–31. Springer, Cham (2016). https://doi.org/10.1007/978-3-319-33951-1_2

43. Sabatier, D., Burdy, L., Requet, A., Guéry, J.: Formal proofs for the NYCT line 7 (flushing) modernization project. In: Derrick, J., et al. (eds.) ABZ 2012. LNCS, vol. 7316, pp. 369–372. Springer, Heidelberg (2012). https://doi.org/10.1007/978-3-642-30885-7_34

44. Said, M.Y., Butler, M., Snook, C.: A method of refinement in UML-B. Softw. Syst. Model. **14**(4), 1557–1580 (2013). https://doi.org/10.1007/s10270-013-0391-z

45. Schneider, S.: The B-Method: An introduction. Palgrave Macmillan, London (2001)

46. Silva, R., Butler, M.: Shared event composition/decomposition in event-B. In: Aichernig, B.K., de Boer, F.S., Bonsangue, M.M. (eds.) FMCO 2010. LNCS, vol. 6957, pp. 122–141. Springer, Heidelberg (2011). https://doi.org/10.1007/978-3-642-25271-6_7

47. Silva, R., Pascal, C., Hoang, T.S., Butler, M.J.: Decomposition tool for Event-B. Softw. Pract. Exper. **41**(2), 199–208 (2011)

48. Spivey, J.M.: The Z Notation: A Reference Manual. Prentice-Hall, Hoboken (1992)

Some Thoughts on Computational Models: From Massive Human Computing to Abstract State Machines, and Beyond

J. A. Makowsky$^{(\boxtimes)}$

Department of Computer Science, Technion-IIT, Haifa, Israel
`janos@cs.technion.ac.il`

Abstract. I sketch what I think led to the emergence of Abstract State Machines. A central role in this is played by the work of Ashok Chandra and David Harel on computable queries in databases. I also define Chandra-Harel Algebras, and analyse Blum-Shub-Smale computability over these algebras.

Keywords: Computational models · Register machines · Program schemas · Abstract State Machines · Blum-Shub-Smale over relational structures

1 Happy Birthday, Egon Börger!

Dear Egon!
Our scientific paths crossed several times starting in the late seventies of the last century. I am happy, they did. I remember our encounters as fruitful and illuminating, and I thank you, Egon, for what my memory retained from these encounters.

We were both working in Germany in this period, you in Münster and Dortmund, and I in Berlin with frequent visits to Israel. Both of us came from Mathematical Logic. You worked between 1969–1989 in Logic and Complexity Theory. You were one of the pioneers in applying logical methods in computer science. I worked in 1971–1984 in Model Theory, both classical and abstract. But already in 1978 I started looking for applications of Model Theory in computer science. My path led me through database theory and specification of abstract data types to the foundations of logic programming and PROLOG. Your path also passed through logic programming and PROLOG, via the analysis of the classical decision problems in logic.

Together with E. Grädel and Y. Gurevich you are one of the authors of the fundamental monograph on the Classical Decision Problem [16], reprinted again in 2001 as [15]. You were the enthusiastic initiator of the conference series CSL (Computer Science Logic) in 1987 which led to the creation of its umbrella

© Springer Nature Switzerland AG 2021
A. Raschke et al. (Eds.): Börger Festschrift, LNCS 12750, pp. 173–186, 2021.
https://doi.org/10.1007/978-3-030-76020-5_10

organization EACSL in 1992 in Dagstuhl, with again you leading the initiative, supported by Klaus. Ambos-Spies, Yuri Gurevich, Moshe Vardi, myself, and many other attendees of the Dagstuhl seminar, see [13].

Since 1990 your research has mainly focused on Software Technology. You were co-pioneering the development and the industrial applications of the Abstract State Machines Method. In your CV posted at your homepage you write that "these applications were aimed at controllable construction and maintenance of hardware/software systems with a focus on rigorously relating requirements capture by high-level models to detailed design and their analysis (using both mathematical verification and experimental validation)". This is when our paths diverged. From 1995 on I became more involved with applications of Finite Model Theory to combinatorial counting, which led me back to more mathematical questions. Unfortunately, since then we have not met for almost 30 years.

I am happy to be able to dedicate the reflections below to the celebration of your 75th birthday. I hope your enthusiasm for research remains unbroken, even in spite of the COVID pandemic. Stay healthy and productive until 120.

2   The Logical Origins of the AMS Method

In 2002, Egon Börger published an excellent historical sketch and annotated bibliography on the origins and the development of the AMS method, [14]. He traces the origins of the ASM concept to various papers by Y. Gurevich, starting in 1984. This is correct when looking at the published papers. However, starting in 1978, during a one-year stay in Jerusalem, I started exploring model theoretic methods in computer science, working on the suggestion of E. Shamir with C. Beeri on the foundation of database dependencies. This led to [22]. Some of my unpublished work with C. Beeri later found its way into M. Vardi's PhD Thesis. From Fall 1980 on both Yuri and I participated in the Logic Year organized by the Institute of Advanced Studies in Jerusalem. The research program mostly centered around S. Shelah's research topics in Model Theory. Y. Gurevich, J. Baldwin, J. Stavi, and I also worked on our contributions to the encyclopedic volume on Model Theoretic Logics [4], Y. Gurevich on Monadic Second Order Theories [33], J. Baldwin on Definable Second Order Quantifiers [3], J. Stavi on Second Order Logic (a chapter which remained unfinished and was not included in the volume) and I on three chapters: one on Compact Logics, one with D. Mundici on Abstract Equivalence Relation, and one, originally planned with S. Shelah, on the Foundations of Abstract Elementary classes (aka Abstract Embedding Relations) [50–52]. It was during the Logic Year 1980 that Yuri and I discovered our parallel interest in applications of logic to theoretical computer science. Between 1978 and 1982 I also had intensive discussions with J. Stavi touching on abstract model theory, second order logic and the foundations of computer science. This was an exciting period, and its impact on Yuri's and my own future work cannot be underestimated.

In 1982 I was invited speaker at the Logic Colloquium in Florence, where I spoke about *Model theoretic issues in theoretical computer science*. A written version was made public in [48] and published as [49]. Already in 1980, during the Logic Year in Jerusalem, Y. Gurevich got interested in my work in computer science and kept "interrogating" me on my views.

In fall 1984 Egon Börger organized a series of courses in Udine under the title *Trends in Theoretical Computer Science* which led to the book [12]. The speakers were K. Ambos-Spies, K. Apt, E. Börger, P. Flajolet, Y. Gurevich, M. Karpinski, P. Martin-Loef, E Shamir, E. Specker and M. Vardi. All the papers in the book are landmark papers[1]. However, for mysterious reasons, the book is not in the databases of google.scholar, and also otherwise difficult to retrieve. Y. Gurevich's contribution is widely quoted as a preprint or offprint. E. Börger produced an outstanding book, which most definitely deserves to be reprinted even today.

In this paper I would like to sketch what I think led ultimately to Gurevich's definition of Abstract State Machines. In retrospect, it all amounted to finding models of computation of the right level of abstraction. A common theme in this quest appears to be looking at two models of computation:

- Computations as performed by register machines, where the contents of the registers may be a bit, a natural number, a real number, a relation, or even first order or higher order structure.
- Computations, as described by Ianov schemes, are more commonly known as program schemes.

The view I present here is not meant to be historical, but conceptual. It is also not always technically precise, and it does not always address the original motivation the respective authors had in mind. What I try to sketch is a line of development which ultimately led to the formulation of Abstract State Machines.

3   MHC-Machines

When I was an undergraduate from 1967 to 1971 there was no Computer Science Department at the Institute of Technology in Zurich, ETHZ. However, computing machinery had a long tradition in Zurich. On the initiative of E. Stiefel, ETH acquired the ZUSE-4 in 1944. A detailed history can be found in the encyclopedic book *Milestones in Analog and Digital Computing* by Herbert Bruderer [17] and in [37]. In 1968 I took a compulsory course, Numeric Analysis I, given by Peter Läuchli but based on a course designed by Heinz Rutishauser. The nascent computing group in the Mathematics Department of ETHZ consisted of E. Stiefel, H. Rutishauser, P. Läuchli and N. Wirth.

---

[1] The book is divided into four parts. Part I. Logic and complexity, with [2,9,35,60], Part II. Database Theory, with [61]. Part III. Analysis of algorithms with [10], and Concurrency and distributed algorithms, with [8,11]. M. Karpinski and P. Martin-Loef did not contribute to the book.

E. Stiefel   H. Rutishauser   P. Läuchli   N. Wirth

H. Rutishauser is credited for inventing compilers and being the driving force behind the definition of ALGOL 60. N. Wirth, who joined the group in 1968, later won the Turing Award for creating the programming languages ALGOL W, Euler, Pascal, Modula, Modula-2, Oberon, Oberon-2, Oberon-07, Oberon System.

There was no talk about the foundations of Computability in this course. It was mostly about basic numeric analysis, but its novelty was the use of a punch card driven computer for our homework. We were given a manual of ALGOL 60 and a four-hour tutorial session on how to punch cards, use the manual and submit the homework. I learned PASCAL in 1973 from the mimeographed yet unpublished version of [44].

The logical aspects of computing were taught at this time by E. Specker and H. Läuchli, Peter's younger brother. Their joint seminar was also attended by the retired Paul Bernays, the gray eminence of Logic in Zurich. In his active time in Zurch he had three students with major impact on theoretical computer science: Corrado Böhm, J. Richard Büchi and Erwin Engeler. During his time in Göttingen before the rise of the Nazis P. Bernays supervised M. Schönfinkel, G. Gentzen and S. MacLane, and he hosted R. Péter for several months working with her on the foundations of Computability, published after WWII in 1951 as [54], see also [55]. The 1951 book was praised by M. Davies as the first comprehensive treatment of recursive function theory.

P. Bernays' influence on modern computer science is still vastly underestimated. I learned about various notions of Computability from the Zurich logicians, but only in the context of the undecidability of the Decision Problem, or Recursion Theory.

In the 1969 course on numeric analysis we were given a model of computation inspired by the practice of human computing during World War II, a model I would like to call MHC: Massive Human Computing. This was a model of highly parallel computing. Each human (during WWII mostly women) acted as an unbounded register matching moving floating point numbers between registers and performing arithmetic operations with the help of a mechanical device like a CURTA or an ARITHMOMETER. There was a central authority, the MASTER, who assigned computing plans to the human computers. Once the plan was executed, the human returned the results to the MASTER, and was given a new computing plan, possibly containing registers with previously computed numbers.

Curta
National Museum of Computing

Felix Arithmometer
Computer History Museum

When I worked for a subcontractor of the European Space Agency developing a simulation program for geostationary satellites, MHC was my computation model. In fact, I developed computation plans in the form of flow charts which were afterwards translated by a programmer into FORTRAN and punched into cards. They were then run on a Control Data machine after being carefully checked both for mathematical errors and compilability. Executing the programs then was prohibitively expensive.

4   Program Schemes and Unbounded Register Machines

Between 1950 and 1960 many researchers attempted to make the notion of computing (in contrast to computability) precise. Iuri I Ianov initiated a model of computation (the Ianov schemes), which was based on flowcharts and led later, under the influence of M. Paterson's thesis, to the widely studied *abstract program schemes*. Ianov's work is [39–41]. It was followed by [19,42,53,56]. Early concerns in the development of program schemes were how to prove properties of programs readable by humans, rather than analyzing Turing machines. For my narrative here I would like to stress the conceptual development leading from Ianov schemes, via Paterson's work, to the work by A. Chandra.

Another line of developments emanating from similar concerns culminated in the definition of *unbounded register machines*. A good and widely accepted formal treatment is given in [58]. There is also a good discussion of early attempts to define register machine on the Wikipedia page *Register Machines* [63]. We are not interested in the exact historic developments here.

The original register machines differ in details of how to model control. For us their main common feature is what can be put into the registers, and what operations can be performed. Natural choices are $\mathbb{N}$, the non-negative integers together with the arithmetic operations addition, subtraction, multiplication and equality and order for comparisons. A register machine model is considered complete if it computes (enumerates) exactly the recursive functions (sets) of natural numbers. If the registers are allowed to contain finite objects which are codable and decodable by natural numbers, one usually defines completeness via this coding.

Iuri I. Ianov   John C. Shepherdson

5   Computability over Abstract Structures

Inspired by J. Shepherdson's work, [58], E. Engeler in 1967, and independently, H. Friedman in 1969, [29,31], formulated a notion of computability over a fixed relational structure. J. Shepherdson, [57], discusses these two papers, and also relates them to program schemes as defined in [19,46].

H. Friedman   E. Engeler

The fixed relational structure was intended to be the real numbers, or some other algebraic structure used in traditional mathematics. Originally, these papers did not generate vast interest. In the first ten years they were hardly quoted outside the circles interested in generalized recursion theory. It was the period of emerging complexity theory based on the Turing model of computation, the emergence of the **P** vs **NP** question, and also the period when Generalized Recursion Theory developed in different directions.

Only in 1989 did L. Blum, M. Shub and S. Smale reinvent register machines for the real numbers, and more generally, for arbitrary rings and fields [6]. Their work was based on earlier work by S. Smale [59]. Its main novelty was the discovery of an analog of the **P** vs **NP** question over arbitrary fixed structures $\mathcal{A}$. We come back to their work at the end of this paper and will refer to this computational model as BSS-computing over $\mathcal{A}$. Similarly, S. Abiteboul and V. Vianu looked at, what they call *generic* computations for relational databases, [1].

6   Computable Queries in Databases

Around 1980 there were four papers which triggered anew Yuri's and my interest in questions of computability:

A. Chandra's and D. Harel's seminal work on computable queries in databases [18, 20, 21], and N. Immerman's and M. Vardi's work on polynomial time computable queries [43, 62].

A. Chandra and D. Harel, 1987 in Hawaii

A. Chandra and D. Harel combined two fundamental ideas. They used some kind of register machines where the registers could hold finite relations over a fixed (infinite) universe, and they required that the register machine should act invariantly on isomorphic relations. In their first paper they considered a simple version of their query language, which is untyped. [20] was a breakthrough paper if only for being a proof of concept. In [18] a typed version is presented which also allows a closer look at complexity issues.

N. Immerman   M. Vardi

The papers by N. Immerman and M. Vardi were first presented at the same conference in 1982. They combined the approach by A. Chandra and D. Harel with earlier work by R. Fagin, N. Jones and A. Selman [30, 45], who characterized classes of finite relational structures recognizable in **NP** as those classes definable in existential second order logic. N. Immerman and M Vardi gave similar characterization for classes of ordered finite relational structures recognizable in in **P** using formalisms inspired by work of A. Chandra and D. Harel. This line of research has very old roots in the Spectrum Problem formulated by H. Scholz in 1953. For a state of the art account of the developments arising from Scholz's problem, see [28]. We note that E. Börger also worked on Spectrum Problem in 1983 [7].

Yuri's first paper leading to the evolution of ASM was published in 1983 as [32]. This was followed by [35] in the volume edited by E. Börger. My first paper in this direction was [25], finally published as [27]. In 1987 I helped R. Herken to choose contributors for his book project *The Universal Turing Machine – A Half Century Survey* [38], reprinted several times. Relevant to our narrative were two papers: Y. Gurevich's [34] and my own paper co-authored with E. Dahlhaus [26].

7   Logic vs Engineering

Personally, I am convinced that the general atmosphere of the Logic Year 1980 in Jerusalem and these developments emanating from evolving database theory played an important role in the maturing of the concepts leading to the Abstract State Machines. Yuri Gurevich was hired in 1982 by the CSEE Department of the University of Michigan, where he was exposed to the challenge of teaching and learning from practically minded students and colleagues. According to him, it was this challenge which influenced the further evolution of the ASM concepts. Y. Gurevich views AMS as an engineering discipline. I have described the logical background which influenced the formulation of AMS. In Y. Gurevich's account of the origins of AMS [36], he also writes:

> We hope that the story of the ASM project will support the maxim that there is nothing more practical than good theory.

8   P vs NP for Query Languages

In this last section I want to discuss what happens when we merge computability of query languages with the approach of Blum-Shub-Smale in the unit cost model, see [5].

M. Shub, L. Blum, F. Cucker and S. Smale

A similar approach was also pursued by S. Abiteboul and V. Vianu [1]. They write:

> The machines described here model computations where a structures is accessed through an abstract interface. They were used to describe generic complexity classes.

They introduce complexity classes GEN − P and GEN − NP and show them to be different, and conclude that:

> the results point to a trade-off between complexity [in the Turing model] and computations with an abstract interface.

This also applies to our case below. It may be not of any practical significance, but it also sheds light on the above-mentioned trade-off. It also leads to interesting questions in the realm of *l'art pour l'art*, i.e., testing our available mathematical and logical tools. The exact relationship between [1] and BSS-inspired approach will be discussed in [47].

We first introduce a structure inspired by the work of A. Chandra and D. Harel. It is an algebra of finite relations over a fixed countable domain D, which we call a CH-algebra over D, denoted by $CH(D)$.

We then use this algebra to give an alternative definition of computable queries, which disregards the size of the relations. BSS-computability with $CH(D)$ allows us to formulate deterministic and non-deterministic polynomial time computability over $CH(D)$, denoted by $\mathbf{P}_{CH(D)}$ and $\mathbf{NP}_{CH(D)}$. Finally, we sketch a proof which shows that $\mathbf{P}_{CH(D)} \neq \mathbf{NP}_{CH(D)}$. Details will be presented in [47].

8.1   CH-Algebras

Let D be an infinite set. We define a structure $CH(D)$ as follows: A finite relation over D is a subset of A^n with A a finite subset of D. The universe $U(D)$ of $CH(D)$ consists of all finite relations over D.

More formally,

$$U(D)_n = \{R \subset A^n : A \subset D, A \text{ finite}\}$$

and

$$U(D) = \bigcup_n U(F)_n$$

For a relation r we define the active domain $AD(r)$ of r to be the subset

$$AD(r) = \{d \in D : \text{ there is tuple } x \in r \text{ which contains } d\}$$

We now equip $U(D)$ with the following (partial) operations.

Constants:
(i) There is a constant $[]_0$ for the relation of arity zero.

Unary functions: There are three unary operations:
(i) $\uparrow$: For a relation r, $r \uparrow$ is obtained from r adding a column to the right, i.e., by forming $r \times AD(r)$.
(ii) $\downarrow$: For a relation r of arity ≥ 1, $\downarrow r$ is obtained from r deleting the left most column i.e., projecting the left most column away.
(iii) $\sim$: For a relation r of arity $k \geq 2$, $r^\sim$ is obtained from r by interchanging the two right most columns.

Binary functions: There two binary operations:
(i) $\cup$: For two relations r and s of the same arity $r \cup s$ is their union.
(ii) $\setminus$: For two relations r and s, r *of the same arity* $\setminus s$ is their set difference.

To make the operations total, we can set $[]_0$ whenever the operations are not defined.

The set $U(D)$ equipped with these constants and operations is called the CH-Algebra over D, denoted by $CH(D)$:

$$CH(D) = (U(D), \uparrow, \downarrow, \tilde{\ }, \cup, \backslash, []_0)$$

The computable queries in the sense of Chandra-Harel are functions which map a finite sequence of relations onto a relation and which can be represented by the programming language they define. The BSS computable functions over a CH-Algebra coincide with the computable queries.

8.2   BSS-Computations in Unit Cost

Let $\mathcal{A}$ be a first order structure. In the unit cost model of BSS-computations, an input of size n consists of a sequence of n elements of $\mathcal{A}$. By abuse of notation, we denote the set of finite sequences of elements of $\mathcal{A}$ by $\mathcal{A}^*$. Decision problems are given as functions from $\mathcal{A}^*$ to $\mathcal{A}$ with values in a two-element set, so one has to assume that in $\mathcal{A}$ there are at least two constant terms. Non-determinism is based on guessing sequences of elements in $\mathcal{A}^*$ of size polynomial in the size of the input. A function $F : \mathcal{A}^* \to \mathcal{A}$ is computable if there is a time bound $t : \mathbb{N} \to \mathbb{N}$ such that there is a BSS-machine which computes F on input of size n in time $t(n)$.

We denote the class of problems over $\mathcal{A}$ recognizable in deterministic BSS-polynomial time by $\mathbf{P}_{\mathcal{A}}$, and by $\mathbf{NP}_{\mathcal{A}}$, and the class of problems over $\mathcal{A}$ recognizable in non-deterministic BSS-polynomial. In this model it is possible that a decision problem is in $\mathbf{NP}_{\mathcal{A}}$ but not decidable (computable). One example for such an $\mathcal{A}$ is the ring of integers $\mathbb{Z}$, see [5].

8.3 $\mathbf{P}_{CH(D)}$ vs $\mathbf{NP}_{CH(D)}$

In the unit cost model of BSS-computing over $CH(D)$ each relation has the same cost, namely 1. As a consequence of this, a relational structure with m relations can be represented by a $m + 1$ tuple. In other words, every finite graph has cost 2. We look at the problem 3COL that asks whether an input graph (V, E) is 3-colorable.

Proposition 1. 3COL *is in* $\mathbf{NP}_{CH(D)}$.

Proof. Guess three set $C_1, C_2, C_3 \subset V$ and verify that $C_i^2 \cap E = \emptyset$.

Proposition 2. 3COL *is not in* $\mathbf{P}_{CH(D)}$.

Proof. In fact it is not even computable. Assume for contradiction that there is a deterministic BSS-machine M which checks whether a graph $(V, E) \in$ 3COL. M runs in constant time, as all inputs are of size 2. So M performs a finite number of test. A test checks whether two relations r and s defined by terms t and u in $CH(D)$ are equal. In this case the terms use as constants only the input graph (V, E), the vertex set and the edge relation. If a relation is definable by a term in $CH(D)$ then it is also definable by a first order formula ϕ_t of some quantifier rank q_t.

Let q be bigger that all the quantifier ranks used in the tests performed by M. Let (V_1, E_1) be in 3COL and let (V_2, E_2) not be in 3COL, but such that they cannot be distinguished by first order formulas of quantifier rank $\leq q$. Two such graphs exist because it is well known that 3COL is not first order definable. In this case M accepts or rejects both graphs, a contradiction.

We conclude:

Theorem 1. $\mathbf{P}_{\mathrm{CH}(D)} \neq \mathbf{NP}_{\mathrm{CH}(D)}$.

Acknowledgements. I would like to thank Y. Gurevich, M. Prunescu and K.-D. Schewe for some critical remarks, and L. Beklemishev and D. Harel for the permission to include pictures they provided. All pictures our either in the public domain or courtesy their respective copyright owners.

The pictures of E. Stiefel, H. Rutishauser, P. Läuchli, N. Wirth and E. Engeler are taken from their respective pages of ETHZ. The pictures of the Curta and the Arithmometer belong to the National Museum of Computing and the Computer History Museum respectively. The picture of Yu. Ianov was provided by Alexey Yashunsky from the Keldysh institute were Ianov worked all his life. The pictures of J. Shepherdson is taken from his obituary [23]. The pictures of H. Friedman, N. Immerman and M. Vardi are from their Wikipedia entries. Finally, the pictures of L. Blum, F. Cucker, M. Shub and S. Smale was provided by them and also appears in the collected papers of S. Smale [24].

References

1. Abiteboul, S., Vianu, V.: Generic computation and its complexity. In Proceedings of the Twenty-Third Annual ACM Symposium on Theory of Computing, pp. 209–219 (1991)
2. Ambos-Spies, K.: Polynomial time degrees of NP-sets. In: Börger, E.B (ed.) Trends in Theoretical Computer Science, pp. 95–142. Computer Science Press, Rockville (1988)
3. Baldwin, J.T.: Definable second-order quantifiers. In: Barwise, J., Feferman, S. (eds.) Model-Theoretic Logics, Perspectives in Mathematical Logic, Chapter 19. Springer (1985)
4. Barwise, J., Feferman, S. (eds.): Model-Theoretic Logics. Perspectives in Mathematical Logic. Springer, New York (1985)
5. Blum, L., Cucker, F., Shub, M., Smale, S.: Complexity and Real Computation. Complexity and Real Computation. Springer, New York (1998). https://doi.org/10.1007/978-1-4612-0701-6
6. Blum, L., Shub, M., Smale, S.: On a theory of computation and complexity over the real numbers. Bull. Am. Math. Soc. **21**, 1–46 (1989)
7. Börger, E.: Spektral problem and completeness of logical decision problems. In: Börger, E., Hasenjaeger, G., Rödding, D. (eds.) LaM 1983. LNCS, vol. 171, pp. 333–356. Springer, Heidelberg (1984). https://doi.org/10.1007/3-540-13331-3_50
8. Börger, E.: Distributed network algorithms. In: Börger, E. (ed.) Trends in Theoretical Computer Science, pp. 347–378. Computer Science Press (1988)
9. Börger, E.: Logic as machine: complexity relations between programs and formulas. In: Börger, E. (ed.) Trends in Theoretical Computer Science, pp. 59–94. Computer Science Press (1988)

10. Börger, E.: Mathematical methods in the analysis of algorithms and data structures. In: Börger, E. (ed.) Trends in Theoretical Computer Science, pp. 225–304. Computer Science Press (1988)
11. Börger, E.: Proving correctness of concurrent programs: a quick introduction. In: Börger, E.B. (ed.) Trends in Theoretical Computer Science, pp. 305–346. Computer Science Press (1988)
12. Börger, E.: (ed.): Trends in Theoretical Computer Science. Computer Science Press (1988). Papers presented in Udine at a Conference of the Same Title, 24 September–5 October (1984)
13. Börger, E.: Ten years of CSL conferences (1987–1997). Bull. Eur. Assoc. Theor. Comput. Sci. **63**, 62–64 (1997)
14. Börger, E.: The origins and the development of the ASM method for high level system design and analysis. J. Univ. Comput. Sci. **8**(1), 2–74 (2002)
15. Börger, E., Grädel, E., Gurevich, Y.: The Classical Decision Problem. Springer, Heidelberg (2001)
16. Borger, E., Gurevich, Y., Graedel, E.: The Classical Decision Problem. Springer, Heidelberg (1997)
17. Bruderer, H.: Milestones in Analog and Digital Computing. Springer, Cham (2020). https://doi.org/10.1007/978-3-030-40974-6
18. Chandra, A., Harel, D.: Structure and complexity of relational queries. J. Comput. Syst. Sci. **25**(1), 99–128 (1982)
19. Chandra, A.K.: Generalized program schemas. SIAM J. Comput. **5**(3), 402–413 (1976)
20. Chandra, A.K., Harel, D.: Computable queries for relational data bases (preliminary report). In: Proceedings of the Eleventh Annual ACM Symposium on Theory of Computing, pp. 309–318 (1979)
21. Chandra, A.K., Harel, D.: Computable queries for relational data bases. J. Comput. Syst. Sci. **21**, 156–178 (1980)
22. Chandra, A.K., Lewis, H.R., Makowsky. J.A.: Embedded implicational dependencies and their inference problem. In: Proceedings of the Thirteenth Annual ACM Symposium on Theory of computing, pp. 342–354 (1981)
23. Crossley, J.N.: John Cedric Shepherdson (1926–2015). https://www.thebritishacademy.ac.uk/documents/1539/08-Shepherdson.pdf
24. Cucker, F., Wong, R.: The Collected Papers of Stephen Smale. World Scientific (2000)
25. Dahlhaus, E., Makowsky. J.A.: Computable directory queries: the choice of programming primitives for SETL-like programming languages (1985)
26. Dahlhaus, E., Makowsky, J.A.: Gandy's principles for mechanisms as a model of parallel computation. In: Herken, R. (ed.), The Universal Turing Machine: A Half-Century Survey, pp. 83–288. Kammerer & Unverzagt, Berlin and Oxford University Press (1988)
27. Dahlhaus, E., Makowsky, J.A.: Query languages for hierarchic databases. Inf. Comput. **101**(1), 1–32 (1992)
28. Durand, A., Jones, D.A., Makowsky, J.A., More, M.: Fifty years of the spectrum problem: survey and new results. Bull. Symbol. Logic **18**(4), 505–553 (2012)
29. Engeler, E.: Algorithmic properties of structures. Math. Syst. Theor. **1**(2), 183–195 (1967)
30. Fagin, R., Karp, R.M.: Complexity of computation. Am. Math. Soc. **7**, 43–74 (1974)
31. Friedman, H.: Algorithmic procedures, generalized turing algorithms, and elementary recursion theory. In: Studies in Logic and the Foundations of Mathematics, vol. 61, pp. 361–389. Elsevier (1971)

32. Gurevich, Y.: Algebras of feasible functions. In: 24th Annual Symposium on Foundations of Computer Science (SFCS 1983), pp. 210–214. IEEE (1983)
33. Gurevich, Y.: Monadic second-order theories. In: Barwise, J., Feferman, S. (eds.) Model-Theoretic Logics, Perspectives in Mathematical Logic, Chapter 19. Springer (1985)
34. Gurevich, Y.: Algorithms in the world of bounded resources. In: Herken, R. (ed.) The Universal Turing Machine: A Half-Century Survey, pp. 407–416. Kammerer & Unverzagt and Oxford University Press, Oxford and New York (1988)
35. Gurevich, Y.: Logic and the challenge of computer science. In: Börger, E. (ed.) Trends in Theoretical Computer Science, Principles of Computer Science Series, Chapter 1. Computer Science Press (1988)
36. Gurevich, Y.: Abstract state machines: an overview of the project. In: Seipel, D., Turull-Torres, J.M. (eds.) FoIKS 2004. LNCS, vol. 2942, pp. 6–13. Springer, Heidelberg (2004). https://doi.org/10.1007/978-3-540-24627-5_2
37. Gutknecht, M.H.: Numerical analysis in Zurich: 50 years ago. Swiss Mathematical Society 1910–2010, pp. 279–290 (2010)
38. Herken, R. (ed.): The Universal Turing Machine: A Half-Century Survey. Kammerer % Unverzagt, Berln and Oxford University Press (1988). Reprinted by Oxford University Press, 1988, 1992, reprinted by Springer Verlag, Vienna (1995)
39. Ianov, T.I.: On matrix program schemes. Commun. ACM 1(12), 3–6 (1958)
40. Ianov, J.I.: On the equivalence and transformation of program schemes. Commun. ACM 1(10), 8–12 (1958)
41. Ianov, J.I.: The logical schemes of algorithms. Probl. Cybernet. (USSR) 1, 82–140 (1960)
42. Igarashi, S.: On the logical schemes of algorithms. Inf. Process. Jpn. 3, 12–18 (1963)
43. Immerman, N.: Relational queries computable in polynomial time (extended abstract). In: Proceedings of the Fourteenth Annual ACM Symposium on Theory of Computing (1982)
44. Jensen, K., Wirth, N.: Pascal user manual and report (1974)
45. Jones, N.D., Selman, A.I.: Turing machines and the spectra of first-order formulas. J. Symbol. Logic 39(1), 139–150 (1974)
46. Luckham, D.C., Park, D.M., Paterson, M.S.: On formalised computer programs. J. Comput. Syst. Sci. 4(3), 220–249 (1970)
47. Makowsky, J.A.: BSS computing over Chandra-Harel algebras (in preparation)
48. Makowsky, J.A.: Model Theoretic Issues in Theoretical Computer Science, Part I: Relational Data Bases and Abstract Data Types. Technical Report, Computer Science Department, Technion (1983)
49. Makowsky, J.A.: Model theoretic issues in theoretical computer science, Part I: relational data bases and abstract data types. In: Studies in Logic and the Foundations of Mathematics, vol. 112, pp. 303–343. Elsevier (1984)
50. Makowsky, J.A.: Abstract embedding relations. In: Barwise, J., Feferman, S. (eds.) Model-Theoretic Logics, Perspectives in Mathematical Logic, Chapter 20. Springer (1985)
51. Makowsky, J.A.: Compactness, embeddings and definability. In: Barwise, J., Feferman, S. (eds.) Model-Theoretic Logics, Perspectives in Mathematical Logic, Chapter 18. Springer (1985)
52. Makowsky, J.A., Mundici. D.: Abstract equivalence relations. In: Barwise, J., Feferman, S. (eds.) Model-Theoretic Logics, Perspectives in Mathematical Logic, Chapter 19. Springer (1985)
53. Paterson, M.S.: Equivalence problems in a model of computation. PhD thesis, Trinity College, University of Cambridge (1967)

54. Péter, R.: Rekursive funktionen, Budapest, vol. 57 (1951)
55. Péter, R.: Rekursive Funktionen in der Computer Theorie, Budapest (1976)
56. Rutledge, J.D.: On Ianov's program schemata. J. ACM (JACM) **11**(1), 1–9 (1964)
57. Shepherdson, J.C.: Computation over abstract structures: serial and parallel procedures and Friedman's effective definitional schemes. In: Studies in Logic and the Foundations of Mathematics, vol. 80, pp.445–513. Elsevier (1975)
58. Shepherdson, J.C., Sturgis, H.E.: Computability of recursive functions. J. ACM (JACM) **10**(2), 217–255 (1963)
59. Smale, S.: The fundamental theorem of algebra and complexity theory. Bull. Am. Math. Soc. **4**(1), 1–36 (1981)
60. Specker, E.: Application of logic and combinatorics to enumeration problems. In: Börger, E. (ed.) Trends in Theoretical Computer Science, pp.141–169. Computer Science Press, 1988. Reprinted. In: Ernst Specker, Selecta, Birkhäuser, pp. 324–350 (1990)
61. Vardi, M.: Fundamentals of dependency theory. In: Trends in Theoretical Computer Science, pp. 171–224. Computer Science Press (1988)
62. Vardi, M.Y.: The complexity of relational query languages. In: Proceedings of the Fourteenth Annual ACM Symposium on Theory of Computing, pp. 137–146 (1982)
63. Wikipedia: Register machine. https://en.wikipedia.org/wiki/Register_machine. Accessed 13 Mar 2021

Analysis of Mobile Networks' Protocols Based on Abstract State Machine

Emanuele Covino and Giovanni Pani[(✉)]

Dipartimento di Informatica, Universitá di Bari, Bari, Italy
{emanuele.covino,giovanni.pani}@uniba.it

Abstract. We define MOTION (MOdeling and simulaTIng mObile ad-hoc Networks), a Java application based on the framework ASMETA (ASM mETAmodeling), that uses the ASM (Abstract State Machine) formalism to model and simulate mobile networks. In particular, the AODV (Ad-hoc On-demand Distance Vector) protocol is used to show the behaviour of the application.

Keywords: Abstract State Machines · Mobile ad-hoc networks

1 Introduction

Mobile Ad-hoc NETwork (MANET) is a technology used to establish and to perform wireless communication among both stationary and mobile devices in absence of physical infrastructure [1]. While stationary devices cannot change their physical location, mobile devices are free to move randomly: they can enter or leave the network and change their relative positions. Thus, the network lacks a predictable topology. Each device is able to broadcast messages inside its radio range only; outside this area, communication is possible only by means of cooperation between intermediate devices. They can act as initiator, intermediate and destination of a communication. This research area is receiving attention in the last few years, in the context of smart mobile computing, cloud computing and Cyber Physical Systems ([21] and [13]).

One of the most popular routing protocols for MANETs is the Ad-hoc On-demand Distance Vector (AODV, [22]), and several variants have been introduced in order to reduce communication failures due to topology changes. For example, Reverse-AODV (R-AODV, [18] and [8]) overcomes this problem by building all possible routes between initiator and destination: in case of failure of the primary route (typically the shortest one), communication is still provided by the alternative routes. More recently, variants have been proposed to cope with congestion issues ([17] and [10]) and to improve the security on communications, using cryptography to secure data packets during their transmission (Secure-AODV, [29]), and adopting the so-called *trust methods*, in which nodes are part of the communication if and only if they are considered trustworthy (Trusted-AODV, [19] and [10]).

© Springer Nature Switzerland AG 2021
A. Raschke et al. (Eds.): Börger Festschrift, LNCS 12750, pp. 187–198, 2021.
https://doi.org/10.1007/978-3-030-76020-5_11

The technology of Mobile Ad-hoc NETwork (MANET) raises several problems about the analysis of performance, synchronization and concurrency of the network. Moreover, the request of computing services characterized by high quality levels, broad and continuous availability, and inter-operability over heterogeneous platforms, increases the complexity of the systems' architecture. Therefore, it is important to be able to verify qualities like responsiveness, robustness, correctness and performance, starting from the early stages of the development. To do this, many studies are executed with the support of simulators [3,25,27]. They are suitable to evaluate performance and to compare different solutions, implementing the network at a low abstraction level, and considering only a limited range of scenarios. The simulators, by their intrinsic nature, cannot provide specification at higher level, and they cannot support proofs of correctness, of synchronization and of deadlock properties. They measure performances, but they cannot model MANETs with a higher abstraction level of specification.

To do this, formal methods that model the process are needed. For instance, the process-calculus [24], CMN (Calculus of Mobile Ad Hoc Networks, [20]), and AWN (Algebra for Wireless Networks, [12]) capture essential characteristic of nodes, such as mobility or packets broadcasting. Petri nets have been employed to study the modeling and verification of routing protocols [28] and the evaluation of protocols performance [11]. With respect to process calculi, state-based models provide a suitable way of representing algorithms, and they are typically equipped with tools, such as CPN Tools [16], that allow to simulate the algorithms, directly. However, we believe that proposed state-based models lack expressiveness: basically, they provide only a single level of abstraction, and cannot support refinements to executable code. Instead, this characteristic is intrinsic in the ASM model. Even if formal methods are satisfactory for reasoning about correctness properties, they rarely are useful for studying performance properties [9]. Generally speaking, correctness properties are formally proved, while the performance properties are investigated through simulations of the system.

Our aim is to use the ASM formalism to study formal properties, and to use MOTION as a tool for evaluating performance properties. The ASM approach provides a way to describe algorithms in a simple abstract pseudo-code, which can be translated into a high-level programming language source code, as in [7] and in [15].

In Sect. 2, we recall concepts and definitions related to the Abstract State Machine's model. In Sect. 3, we describe three mobile network's protocols: AODV (Ad-hoc On-demand Distance Vector), N-AODV (NACK-based AODV), and BN-AODV (Black hole-free N-AODV). In Sect. 4, we introduce the definition and specific behaviour of MOTION, with respect to the ASM's model of the previous network protocols. Conclusions and future work can be found in Sect. 5.

2   Abstract State Machines

An Abstract State Machine (ASM, [7]) M is a tuple (Σ, S, R, P_M). Σ is a *signature*, that is a finite collection of names of total functions; each function has arity n, and the special value *undef* belongs to the range (*undef* represents an undetermined object, the default value). Relations are expressed as particular functions that always evaluate to *true*, *false* or *undef*.

S is a finite set of *abstract states*. The concept of abstract state extends the usual notion of state occurring in finite state machines: it is an algebra over the signature Σ, i.e. a non-empty set of objects together with interpretations of the functions in Σ. Pairs of function names together with values for their arguments are called *locations*: they are the abstraction of the notion of memory unit. Since a state can be viewed as a function that maps locations to their values, the current configuration of locations, together with their values, determines the current state of the ASM.

R is a finite set of *rule declarations* built starting from the *transition rules* skip, update ($f(t_1, t_2, \ldots, t_n) := t$), conditional (**if** ϕ **then** P **else** Q), let (**let** $x = t$ **in** P), choose (**choose** x **with** ϕ **do** P), sequence (P **seq** Q), call ($r(t_1, \ldots, t_n)$), block (P **par** Q) (see [7] for their operational semantics). The rules transform the states of the machine, and they reflect the notion of transition occurring in traditional transition systems. A distinguished rule P_M, called the *main rule* of the machine, represents the starting point of the computation.

A *move* of a ASM, in a given state, consists of the execution of all the rules whose conditions are true in that state. Since different updates could affect the same location, it is necessary to impose a consistency requirement: a set of updates is said to be *consistent* if it contains no pairs of updates referring to the same location. Therefore, if the updates are consistent, the result of a move is the transition of the machine from the current state to another; otherwise, the computation doesn't produce a next state. A *run* is a (possibly infinite) sequence of moves: they are iterated until no more rules are applicable.

The aforementioned notions refer to the *basic* ASMs. However, there exist some generalisations, e.g. Parallel ASMs and Distributed ASMs (DASMs) [15]. Parallel ASMs are basic ASMs enriched with the rule forall x with ϕ do P, to express the simultaneous execution of the same ASM P over x satisfying the condition ϕ. A Distributed ASM is intended as a finite number of independent agents, each one executing its own underlying ASM: it is capable of capturing the formalization of multiple agents acting in a distributed environment. A run, which is defined for sequential systems as a sequence of computation steps of a single agent, is defined as a partial order of moves of finitely many agents, such that the three conditions of co-finiteness, sequentiality of single agents, and coherence are satisfied. Roughly speaking, a global state corresponds to the union of the signatures of each ASM together with interpretations of their functions.

3   MANET and Routing Protocols

Mobile Ad-hoc NETworks are networks of autonomous mobile nodes whose topology is not predefined. Each node has a transmission radio range within which it can transmit data to other nodes, directly. Because of the potential movements of the nodes, the routes connecting them can change rapidly.

Several routing protocols have been proposed; among them, the *Ad-hoc On-demand Distance Vector* (AODV) is one of the most popular. Indeed, a large number of simulation studies are dealing with it, representing a reliable baseline for comparison to the results of simulations executed with MOTION. Moreover, we add two variants of AODV: *NACK-based Ad-hoc On-demand Distance Vector* (N-AODV, [4]), that improves the awareness that each host has about the network topology, and *Blackhole-free N-AODV* (BN-AODV, [5]), that detects the presence of malicious nodes leading to a blackhole attack.

3.1   Ad-hoc On-Demand Distance Vector (AODV)

This routing protocol has been defined in [22]: it is a reactive protocol that combines two mechanisms, namely the *route discovery* and the *route maintenance*, in order to store some knowledge about the routes into *routing tables*. The routing table associated with each node is a list of all the discovered (and still valid) routes towards other nodes in the network, together with other information. In particular, for the purposes of the present paper, an entry of the routing table of the node i concerning a node j includes: the *address* of j; the last known *sequence number* of j; the *hop count* field, expressing the distance between i and j; and the *next hop* field, identifying the next node in the route to reach j.

The sequence number is an increasing number maintained by each node, that expresses the freshness of the information about the respective node. When an *initiator* wants to start a communication session towards the ·*destination*, it checks if a route is currently stored in its routing table. If so, the protocol ends and the communication starts. Otherwise, the initiator broadcasts a control packet called *route request* (RREQ) to all its neighbors.

An RREQ packet includes the initiator address and broadcast id, the destination address, the sequence number of the destination (i.e., the latest available information about the destination), and the hop count, initially set to 0, and increased by each intermediate node. The pair <*initiator address*; *broadcast id*> identifies the packet, uniquely; this implies that duplications of RREQs already handled by nodes can be ignored.

When an intermediate node n receives an RREQ, it creates the routing table entry for the initiator, or updates it in the fields related to the sequence number and to the next hop. Then, the process is iterated: n checks if it knows a route to the destination with corresponding sequence number greater than the one contained into the RREQ (this means that its knowledge about the route is more recent). If so, n unicasts a second control packet (the *route reply*, RREP) back to the initiator. Otherwise, n updates the hop count field and broadcasts once more the RREQ to all its neighbors.

The process successfully ends when a route to the destination is found. While the RREP travels towards the initiator, routes are updated inside the routing tables of the traversed nodes, creating an entry for the destination, when needed. Once the initiator receives back the RREP, the communication can start. If the nodes' movements break a link (i.e., a logical link stored in a routing table is no more available), a route maintenance is executed in order to notify the error and to invalidate the corresponding routes: to this end the control packet *route error* (RERR) is used.

3.2   NACK-Based AODV (N-AODV)

One of the main disadvantages of the AODV protocol is the poor knowledge that each node has about the network topology. In fact, each node n is aware of the existence of a node m only when n receives an RREQ, either originated by, or directed to m. In order to improve the network topology awareness, the NACK-based AODV routing protocol has been proposed and modeled by means of a Distributed ASM in [4].

This protocol is a variant of AODV: it adds a *Not ACKnowledgment* (NACK) control packet in the route discovery phase. Whenever an RREQ originated by n and directed to m is received by the node p that doesn't know anything about m, p unicasts the NACK to n. The purpose of this control packet is to state the ignorance of p about m. In this way, n (as well as all the nodes in the path to it) receives fresh information about the existence and the relative position of p. Therefore, on receiving the NACK, all the nodes in the path to p add an entry in their respective routing tables, or update the pre-existing entry. N-AODV has been experimentally validated through simulations, showing its efficiency and effectiveness: the nodes in the network actually improve their knowledge about the other nodes and, in the long run, the number of RREQ decreases, with respect to the AODV protocol.

3.3   Black Hole-Free N-AODV (BN-AODV)

All routing protocols assume the trustworthiness of each node; this implies that MANETS are very prone to the *black hole attack* [26]. In AODV and N-AODV a black hole node produces fakes RREPs, in which the sequence number is as great as possible, so that the initiator establishes the communication with the malicious node, and the latter can misuse or discard the received information. The black hole can be supported by one or more *colluders*, that confirm the trustworthiness of the fake RREP. The Black hole-free N-AODV protocol [5] allows the honest nodes to intercept the black holes and the colluders, thanks to two control packets: each intermediate node n receiving an RREP must verify the trustworthiness of the nodes in the path followed by the RREP; to do this, n produces a *challenge packet* (CHL) for the destination node, and only the latter can produce the correct *response packet* (RES). If n receives RES, it sends the RREP, otherwise the next node towards the destination is a possible black hole.

4   MOTION

4.1   Development and Behavior

As stated before, MOTION (MOdeling and simulaTIng mObile ad-hoc Networks) is a Java application that allows to specify the simulation parameters, to execute the network described, and to collect the output data of the simulation.

To define MOTION, we have used the ASM-based method consisting in development phases, from requirements' specification to implementation. Some environments support this method, and among them the ASMETA (ASM mETA-modeling) framework [2,14]. This framework is characterized by logical components that capture the requirements by constructing the so-called *ground models*, i.e. representations at high level of abstraction that can be graphically depicted. Starting from ground models, hierarchies of intermediate models can be built by stepwise refinements, leading to executable code: each refinement describes the same system at a finer granularity. The framework supports both verification, through formal proof, and validation, through simulation.

MOTION is developed within the ASMETA framework thanks to the abstract syntax defined in the AsmM metamodel; the behavior of the MANET is modelled using the AsmetaL language, and then the network is executed by the AsmetaS simulator. Since AsmetaS simulates instances of the model expressed by means of the AsmetaL, the information concerning each instance (number of agents and their features, for instance) must be recorded into the AsmetaL file.

The executions of MOTION and ASMETA are interleaved: MOTION provides the user interface and captures the data inserted by the user, representing the parameters of the simulation. MOTION then includes these data into the AsmetaL file, and it runs AsmetaS. AsmetaS executes an ASM move, simulating the behavior of the network protocol over the current data, and it records the values of the locations in a log file, for each state. At the end of each move the control goes back to MOTION: it gets the information about the results of the ASM move, such as the relative position of the hosts, the sent/received packets, and the values of waiting time, and it records them into the AsmetaL file. Then, MOTION invokes AsmetaS for the next move. Even if this interleaved executions requires a good amount of interaction work, this is done in order to collect the information about the evolution of the network step by step, and to use it for the analysis of the performances and behaviour of the network itself.

At the end of the simulation, MOTION reads the final log file, parses it, and stores the collected results in a csv file. Web pages, with the complete package, can be found at https://sourceforge.net/projects/motion-project/.

4.2 Defining the Mobility Model

A realistic simulation of a MANET should take into account all its features. We have decided that the movement issues, as well as the amplitude of the radio range, are defined within the mobility model. We assume that the whole network topology is expressed by the connections among devices, implicitly, and for each of them we consider only its current neighborhood. More precisely, in MOTION the network topology is expressed by an *adjacency matrix C*, such that $c_{ij} = 1$ if i and j are neighbors, 0 otherwise, for each pair of devices i and j. This implies that we can use concepts and properties of graph theory; for instance, the reachability between two agents a_i and a_j is expressed by the predicate isLinked(a_i, a_j), which evaluates to *true* if there exists a coherent path from a_i to a_j, to *false* otherwise.

Within MOTION, the mobility model is implemented into a Java class that, before executing any ASM move, updates the adjacency matrix. To this end, each c_{ij} is randomly set to 0 or 1, according to a mobility parameter defined by the user (see Sect. 4.4). The new values of the matrix are then set within the AsmetaL file, so that the ASM move can be executed, accordingly.

4.3 The Abstract State Machine-Based Models

The AODV routing protocol has been formally modelled through ASMs in [6] (Chap. 6). It is described as a set of nodes, each one representing a device. A modified version is used in MOTION, that takes into consideration the parameter *Timeout* (that is, the waiting time for the route-reply packet). The high-level definition of MOTION for AODV is:

MAIN RULE AODV =
    forall a ∈ Nodes **do** AODVSPEC(a)

where

AODVSPEC(a) =
  forall dest ∈ Nodes **with** dest ≠ a **do**
    if *WaitingForRouteTo*(a, dest) **then**
      if *Timeout*(a, dest) > 0 **then**
        Timeout(a, dest) := *Timeout*(a, dest)-1
      else
        WaitingForRouteTo(a, dest) := false
    if *WishToInitiate*(a) **then** PREPARECOMM
    if not *Empty*(Message) **then** ROUTER

If the device needs to start a communication (i.e. the predicate *WishToIni-tiate* evaluates to true), then PREPARECOMM is called. The predicate *WaitingForRouteTo* expresses that the discovery process previously started is still running; in this case, if the waiting time for RREP is not expired (i.e., *Timeout*() > 0), the time-counter is decreased. Finally, if the device has received a message (either RREQ, RREP or RERR), ROUTER is called, with

ROUTER = ProcessRouteReq
          ProcessRouteRep
          ProcessRouteErr

where each process expresses the behavior of the device, depending on the type of the message received.

The main difference between the previous model and the ASM model for N-AODV concerns ROUTER, that includes the call to PROCESS-NACK, in order to unicast the NACK packet, if needed.

The BN-AODV model is more structured, because it has to describe the behavior of three different kinds of agents: honest hosts, black holes, and colluders. So, the main rule has the form:

MAIN RULE BN-AODV=
    forall a ∈ Blackhole **do** BLACKHOLESPEC(a)
    forall a ∈ Colluder **do** COLLUDERSPEC(a)
    forall a ∈ Honest **do** HONESTSPEC(a)

where HONESTSPEC describes the behavior of the honest nodes, and it's analogous to AODVSPEC. BLACKHOLESPEC and COLLUDERSPEC are the specifications for the non-honest nodes and the colluders, respectively. Moreover, ROUTER for the honest nodes must verify the trustworthiness of the received RREPs.

Thanks to the formalization of the protocols, some correctness properties have been proved in the past, such as the starvation freeness for the AODV protocol, the properness of the packet (either NACK or RREP) received back by the initiator of any communication, when it is not isolated for N-AODV, and the capability to intercept blackhole attacks for BN-AODV.

4.4   Specific Behavior of the Tool

A simulation in MOTION is performed in a number of sessions established by the user (10 sessions, Fig. 1), each of which has a duration (50 moves, Fig. 1); during each session, the MANET includes a number of devices defined by the user, that depends on the specific evolution of the network (due to movements, some of them can be disconnected). Moreover, during each session, each device is the initiator for a number of attempts for establishing a communication, each of them towards a destination different from the initiator itself: the user expresses the probability that each device acts as an initiator by setting the parameter *Initiator Probability* (10%, Fig. 1). Thanks to the intrinsic parallelism in the execution of the ASM's rules, more attempts can be simultaneously executed. A communication attempt is considered successful if the initiator receives an RREP packet within the waiting time expressed by the parameter *RREP Timeout*; otherwise, the attempt is considered failed.

In MOTION, the devices mobility is defined by the user by means of two parameters, namely *Initial connectivity* and *Mobility level*. The former defines the initial topology of the MANET: it expresses the probability that each device is directly linked to any other. During the simulation, the devices mobility is expressed by the random redefinition of the values of the *adjacency matrix C*. More precisely, for each pair of devices $<a_i, a_j>$, and for each move of the ASM, the values of C are changed with a probability expressed by *Mobility level*.

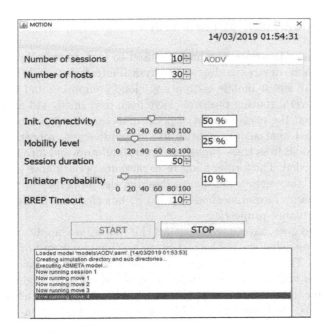

Fig. 1. MOTION user interface for AODV protocol

When the BN-AODV routing protocol is simulated, the MOTION user interface includes the definition of the number of black holes and colluders, and two parameters establishing the increment of the fake sequence number produced by the black holes. Figure 1 shows the current state of the simulation in the panel under the two buttons START and STOP.

From the ASM perspective, there are two different machines, both called by the ASMETA's main rule. The first one is OBSERVERPROGRAM: it is not part of the MANET, but it is used in order to manage the execution. It initializes the locations and data structures for all devices, manages the mobility (setting the initial topology and resetting the adjacency matrix at each move), and updates the counter for the time expiration. The second machine, called by the main rule, is the model of the devices' behavior. Currently, MOTION allows the users to study AODV, N-AODV, and BN-AODV; for all of them, the MANET is modeled by means of a Distributed ASM. In both AODV and N-AODV all the nodes behave in the same way, described by the respective DASM, so the machine specifying the protocol is called; at each move the machine randomly decides if the current agent will initiate new communication attempts by invoking PREPARECOMM, then it acts as a router by processing the proper control packets (with ROUTER).

5   Conclusions and Future Work

Mobile Ad-hoc NETwork is a technology used to perform wireless communications among mobile devices in absence of physical infrastructure. It is widely used in the context of smart mobile computing, cloud computing and Cyber Physical Systems. Several routing protocols have been developed, and problems have been raised about the measurement of performances of these networks, and also about the formal analysis of qualities like responsiveness, robustness, correctness. In order to address these problems, both simulators and formal description methods are needed. The former allow us to measure performance through direct simulation, but they aren't suitable to describe the properties of the networks. On the other hand, formal methods can do it, but they can hardly be used for studying performance properties.

In this paper, we have introduced MOTION, a Java application in which MANET's are modeled as an Abstract State Machine by means of the AsmetaL representation. This representation can be used to prove formal properties of the network, as well as can be simulated by the simulation engine AsmentaS. MOTION can collect the results of this simulation, that can be used for performances' analysis. We have validated MOTION on the Ad-hoc On-Demand Vector protocol and on two of its variants (concerning the host's network topology awareness and the ability to intercept blackhole attacks). Note that MOTION itself has been developed within the ASMETA framework, thanks to the abstract syntax defined in the AsmM metamodel.

A sensible improvement of MOTION could be the definition of a new interface, in which the dynamic evolution of the network, during the computations, is shown (as in [23]). Moreover, a complexity analysis of the network's protocols and the related algorithms could be performed, when the network is represented by means of ASM's. Finally, a change of the structure that represents the connectivity among the nodes (from adjacency matrix to adjacency list, for instance), could lead to a dramatic improvement of the resource-consumption during the simulation of the behaviour of the network.

References

1. Agrawal, D.P., Zeng, Q.-A.: Introduction to Wireless and Mobile Systems. Cengage learning - Fourth Edition, Boston (2016)
2. Arcaini, P., Gargantini, A., Riccobene, E., Scandurra, P.: A model-driven process for engineering a toolset for a formal method. Softw. Practice Experience **41**(2), 155–166 (2011)
3. Basagni, S., Mastrogiovanni, M., Panconesi, A., Petrioli, C.: Localized protocols for ad hoc clustering and backbone formation: a performance comparison. IEEE Trans. Parallel Distrib. Syst. **17**(4), 292–306 (2006). https://doi.org/10.1109/TPDS.2006. 52
4. Bianchi, A., Pizzutilo, S. Vessio, G.: Preliminary description of nack-based ad-hoc on-demand distance vector routing protocol for MANETS. In: 2014 9th International Conference on Software Engineering and Applications (ICSOFT-EA), pp. 500–505. IEEE (2014)

5. Bianchi, A., Pizzutilo, S., Vessio, G.: Intercepting blackhole attacks in MANETs: an ASM-based model. In: Cerone, A., Roveri, M. (eds.) SEFM 2017. LNCS, vol. 10729, pp. 137–152. Springer, Cham (2018). https://doi.org/10.1007/978-3-319-74781-1_10

6. Börger, E., Raschk, A.: Modeling Companion for Software Practitioners. Springer, Heidelberg (2018). https://doi.org/10.1007/978-3-662-56641-1

7. Börger, E., Stärk, R.: Abstract State Machines: A Method for High-Level System Design and Analysis. Springer, Heidelberg (2003). https://doi.org/10.1007/978-3-642-18216-7

8. Bononi, L., D'Angelo, G., Donatiello, L.: HLA-based adaptive distributed simulation of wireless mobile systems. In: Proceedings of the Seventeenth Workshop on Parallel and Distributed Simulation, p. 40. IEEE Computer Society (2003)

9. Calinescu, R., Ghezzi, C., Kwiatkowska, M., Mirandola, R.: Self-adaptive software needs quantitative verification at runtime. Commun. ACM **55**(9), 69–77 (2012)

10. Das, N., Bisoy, S.K., Tanty, S.: Performance analysis of tcp variants using routing protocols of MANET in grid topology. In: Mallick, P.K., Balas, V.E., Bhoi, A.K., Zobaa, A.F. (eds.) Cognitive Informatics and Soft Computing. AISC, vol. 768, pp. 239–245. Springer, Singapore (2019). https://doi.org/10.1007/978-981-13-0617-4_23

11. Erbas, F., Kyamakya, K., Jobmann, K.: Modelling and performance analysis of a novel position-based reliable unicast and multicast routing method using coloured Petri nets. In: 2003 IEEE 58th Vehicular Technology Conference. VTC 2003-Fall, vol. 5, pp. 3099–3104. IEEE (2003)

12. Fehnker, A., van Glabbeek, R., Höfner, P., McIver, A., Portmann, M., Tan, W.L.: A process algebra for wireless mesh networks. In: Seidl, H. (ed.) ESOP 2012. LNCS, vol. 7211, pp. 295–315. Springer, Heidelberg (2012). https://doi.org/10.1007/978-3-642-28869-2_15

13. Garcia-Santiago, A., Castaneda-Camacho, J., Guerrero-Castellanos, J.F., Mino-Aguilar, G., Ponce-Hinestroza, V.Y.: Simulation platform for a VANET using the truetime toolbox: Further result toward cyber-physical vehicle systems. In: IEEE 88th Vehicular Technology Conference (VTC-Fall), pp. 1–5. IEEE (2018)

14. Gargantini, A., Riccobene, E., Scandurra, P.: A metamodel-based language and a simulation engine for abstract state machines. J. UCS **14**(12), 1949–1983 (2008). https://doi.org/10.3217/jucs-014-12-1949

15. Glässer, U., Gurevich, Y., Veanes, M.: Abstract communication model for distributed systems. IEEE Trans. Software Eng. **30**(7), 458–472 (2004). https://doi.org/10.1109/TSE.2004.25

16. Jensen, K., Kristensen, L.M., Wells, L.: Coloured Petri nets and CPN tools for modelling and validation of concurrent systems. Int. J. Softw. Tools Technol. Transfer **9**(3–4), 213–254 (2007)

17. Kaur, N., Singhai, R.: Analysis of traffic impact on proposed congestion control scheme in AODV. In: Wireless Personal Communications, pp. 1–24 (2019)

18. Kim, C., Talipov, E., Ahn, B.: A reverse AODV routing protocol in ad hoc mobile networks. In: Zhou, X., et al. (eds.) EUC 2006. LNCS, vol. 4097, pp. 522–531. Springer, Heidelberg (2006). https://doi.org/10.1007/11807964_53

19. Li, X., Lyu, M.R., Liu, J.: A trust model based routing protocol for secure ad hoc networks. In: 2004 IEEE Aerospace Conference Proceedings, vol. 2, pp. 1286–1295. IEEE (2004)

20. Merro, M.: An observational theory for mobile ad hoc networks. Inf. Comput. **207**(2), 194–208 (2009)

21. Pandian, A.P., Chen, J.I.-Z., Baig, Z.A.: Sustainable mobile networks and its applications. Mobile Networks Appl. **24**(2), 295–297 (2019)
22. Perkins, C.E., Belding-Royer, E.M., Das, S.R.: Ad hoc on-demand distance vector (AODV) routing. In: RFC, vol. 3561, pp. 1–37 (2003). https://doi.org/10.17487/RFC3561
23. Saquib, N., Sakib, M.S.R., Al-Sakib M., Pathan, K.: ViSim: a user-friendly graphical simulation tool for performance analysis of MANET routing protocols. In: Mathematical and Computer Modelling, vol. 53, pp. 2204–2218 (2013)
24. Singh, A., Ramakrishnan, C., Smolka, S.A.: A process calculus for mobile ad-hoc networks. Sci. Comput. Program. **75**(6), 440–469 (2010)
25. Tran, D.A., Raghavendra, H.: Congestion adaptive routing in mobile ad-hoc networks. IEEE Trans. Parallel Distrib. Syst. **17**(11), 1294–1305 (2006). https://doi.org/10.1109/TPDS.2006.151
26. Tseng, F.-H., Chou, L.-D., Chao, H.-C.: A survey of black hole attacks in wireless mobile ad hoc networks. Hum. Centric Comput. Inf. Sci. **1**, 4 (2011)
27. Wu, J., Dai, F.: Mobility-sensitive topology control in mobile ad hoc networks. IEEE Trans. Parallel Distrib. Syst. **17**(6), 522–535 (2006). https://doi.org/10.1109/TPDS.2006.73
28. Xiong, C., Murata, T., Leigh, J.: An approach for verifying routing protocols in mobile ad hoc networks using Petri nets. In: Proceedings of the IEEE 6th Circuits and Systems Symposium on Emerging Technologies: Frontiers of Mobile and Wireless Communication, vol. 2, pp. 537–540. IEEE (2004)
29. Zapata, M.G.: Secure ad hoc on-demand distance vector routing. ACM SIGMOBILE Mobile Comput. Commun. Rev. **6**(3), 106–107 (2002)

What is the Natural Abstraction Level of an Algorithm?

Andreas Prinz$^{(\boxtimes)}$

Department of ICT, University of Agder, Grimstad, Norway
`Andreas.Prinz@UIA.no`

Abstract. Abstract State Machines work with algorithms on the natural abstraction level. In this paper, we discuss the notion of the natural abstraction level of an algorithm and how ASM manage to capture this abstraction level. We will look into three areas of algorithms: the algorithm execution, the algorithm description, and the algorithm semantics. We conclude that ASM capture the natural abstraction level of the algorithm execution, but not necessarily of the algorithm description. ASM do also capture the natural abstraction level of execution semantics.

Keywords: Abstract state machine · Abstraction · Execution · Description · Semantics

1 Introduction

Abstract state machines (ASM) [9], originally called Evolving algebras, [22] enable a high-level and abstract description of computations. ASM can be considered formalized pseudo-code, such that ASM programs are readable even without much introduction. The original purpose of ASM was to improve on the low-level abstraction provided by Turing machines [43], in order to be able to reason better about computability. This original purpose was achieved with the sequential ASM thesis [23]. It was later extended with an ASM thesis for parallel [3,5,40] and distributed computations [13].

From there, ASM were developed into different directions. Egon Börger understood very early that ASM are not only a mathematical tool for computability, but also a tool for system design and analysis. For the practical applicability, several more features were needed for ASM beyond [23], for example time [34] and distributed computations [38].

Another major ingredient for system design is a method to design systems, in this case the ground model approach [10]. This approach enables step-wise systems design, keeping correctness all the way to the final system.

The theoretical track of ASM has achieved a lot of success, and even though there are still details to be sorted out [36,38], this work is well under way. The major difference between the Turing machines approach and ASM is that ASM promise to work on the natural abstraction level for the computation.

© Springer Nature Switzerland AG 2021
A. Raschke et al. (Eds.): Börger Festschrift, LNCS 12750, pp. 199–214, 2021.
https://doi.org/10.1007/978-3-030-76020-5_12

It is of essence for every engineer to work on the right level of abstraction, as problem descriptions are simpler, and solutions are understandable at the right level of abstraction. Sometimes, solutions introduce high level of complexity by being at the wrong level of abstraction [15].

This paper tries to review the concept of abstraction level and identifies the meaning of natural abstraction in three dimensions: algorithm execution, algorithm description, and algorithm (language) semantics. We focus on sequential algorithms, although the conclusions also apply to other kinds of algorithms.

This paper starts with a discussion of the concept of algorithm in Sect. 2 before introducing abstract state machines in Sect. 3. Then we look into abstraction levels in executions in Sect. 4, in descriptions in Sect. 5, and in language semantics in Sect. 6. We conclude in Sect. 7.

2   What is an Algorithm?

Before looking into abstraction levels, we need to agree what an algorithm is. Harold Stone proposes the following definition "...any sequence of instructions that can be obeyed by a robot, is called an algorithm" (p. 4) [42]. Boolos et al. offers a similar definition in [7]: "... explicit instructions such that they could be followed by a computing machine".

This definition includes computer programs, bureaucratic procedures and cook-book recipes. Often, the condition that the algorithm stops eventually is included. In our context, also infinite loops are permitted because we also want to include server programs. Besides, termination is undecidable. Please note that the notion of algorithm relies on a basic set of elementary operations or functions.

Turing machines formalize this informal definition. Gurevich writes in [23]: "... Turing's informal argument in favor of his thesis justifies a stronger thesis: every algorithm can be simulated by a Turing machine ... according to Savage [1987], an algorithm is a computational process defined by a Turing machines".

From the considerations so far, we conclude that an algorithm has a *description* ("a sequence of instructions") and an *execution* ("a computational process"). It is the *semantics* of the description that leads to the execution.

As an example, let's look at the Euclidean algorithm which computes the greatest common divisor gcd from two natural numbers n_1 and n_2. It can be expressed in ASM as follows.

```
IF n₁ > n₂ THEN
    DO IN-PARALLEL
        n₁ := n₂
        n₂ := n₁
    ENDDO
ELSEIF n₁ = 0 THEN
    gcd := n₂
ELSE
    n₂ := n₂ - n₁
ENDIF
```

2.1 Algorithm Execution

For the execution of the Euclidean algorithm we have to know that the parallel execution of assignments is the standard mode in ASM [9]. The ASM code for the Euclidean algorithm describes one step of the algorithm, and it is repeated until there are no more changes.

A sample execution of the ASM algorithm for the numbers $n_1 = 1071$ and $n_2 = 462$ leads to the following sequence of pairs (n_1, n_2): $(1071, 462)$, $(462, 1071)$, $(462, 609)$, $(462, 147)$, $(147, 462)$, $(147, 315)$, $(147, 168)$, $(147, 21)$, $(21, 147)$, $(21, 126)$, $(21, 105)$, $(21, 84)$, $(21, 63)$, $(21, 42)$, $(21, 21)$, $(21, 0)$, $(0, 21)$. The result is then $gcd = 21$. Please note that gcd would be present in all states.

We will look at the execution of algorithms in Sect. 4.

2.2 Algorithm Description

Algorithms can be expressed in many kinds of notation, including natural languages, pseudo code, flowcharts, programming languages or control tables.

When we use Java to express the same algorithm, it looks something like that. Please note the extra temporary variable t for swapping $n1$ and $n2$. In addition, there is an enclosing **while** loop which is not needed in ASM. Java does not provide natural numbers as types; we silently assume that the parameters are non-negative.

```
public static int gcd(int n1, int n2) {
    while (n1 > 0) {
        if (n1 > n2) {
            int t = n1;
            n1 = n2;
            n2 = t;
        } else {
            n2 = n2 - n1;
        }
    }
    return n2;
}
```

Now we look at a version of the algorithm in Lisp. The solution is recursive as is customary in Lisp. Again, we assume that the typing of the parameters is correct.

```
(defun gcd (n1 n2)
    (if (= n1 0)
        n2
    (if (> n1 n2)
        (gcd n2 n1)
        (gcd n1 (- n2 n1))
    )
)
```

Finally, we can also use Prolog for the algorithm as follows. Here, we need to use an extra parameter for the result. This solution is again recursive and the typing is assumed to be correct.

gcd(0, N2, Result):- !, N2=Result.
gcd(N1, N2, Result):- N1 > N2, gcd(N2, N1, Result).
gcd(N1, N2, Result):- N1 =< N2, N2New is N2-N1, gcd(N1, N2New, Result).

We will look at the description (languages) of algorithms in Sect. 5.

2.3 Algorithm Semantics

All possible executions of an algorithm are the semantics of the algorithm. This means that the semantics of the algorithm connects the description of the algorithm with the execution of the algorithm. More precisely, the description is written in a language, and the semantics of the language provides the execution(s), see also [30] and [19].

This way, the semantics of the algorithm description is implied by the semantics of the language which is used for the description. There is not a semantics for each and every description, but a general semantics for the language of the descriptions. Therefore, algorithm semantics is in fact language semantics.

We will look at semantics in Sect. 6.

3 Abstract State Machines

The central concepts in abstract state machines (ASMs) are (abstract) states with locations and transition rules with updates. Their definitions can be found in many sources, including, but not limited to [2,4,6,8,9,12].

An abstract state machine (ASM) program defines abstract states and a transition rule that specifies how the ASM transitions through its states. ASM states are defined using a *signature* of names (function symbols), where each name has an arity. This allows to construct expressions using the names in the usual way. ASM names can be typed in the usual way.

The states are then interpretations of the names over a base set of values. Each name with arity zero is interpreted as a single element of the base set, while each name with arity n is interpreted as an n-ary function. Expressions are interpreted recursively.

ASM names are classified into *static* names whose interpretation does not change (e.g. True), and *dynamic* names which are subject to updates. Each ASM signature includes the predefined static names *True*, *False* and *Undefined*, interpreted as three distinct values. All ASM functions are total, and the special value *Undefined* is used to model partial functions.

An ASM transition rule (program) looks like pseudo-code and can be read without further explanation. The rules include assignments, if, forall, and some other statements. We refer to [9] for a formal definition.

The basic unit of change is an assignment, written as $loc := e$. Executing this assignment means to change the interpretation of the location loc to the value of the expression e in the given state.

Locations $(loc = f(e_1, \ldots, e_n))$ are constructed of an n-ary name (f) and n expressions e_i. More precisely, a unary function symbol u is a location, and any function symbol f with a number of locations l_i as arguments of f is a *location* as well. In each state, each location has a value.

An *update* is given by two locations, one on the right-hand side and one on the left-hand-side. The value of the left-hand side location is replaced by the value of the right-hand side location, such that $lhs = rhs$ will be true in the new state, unless the value of rhs is also changed in the state change. Formally, the assignment creates an update, which is a pair of a location and a value. All the applicable updates are collected into an update set, thereby implementing the parallel execution mode. Applying the update set to the current state (executing it) leads to the changes in the next state.

An ASM run starts with an *initial state*, being a state as defined above. For each state, the transition rule produces an update set which leads to the next state, thereby creating a sequence of states. Each state change (or step or move) *updates* the interpretation of its symbols given by a set of updates according to the assignments used.

4   Executing Algorithms

For the execution of algorithms, we need to look at the runtime, which is basically the same as operational semantics [18,26,35]. Runtime has two aspects, namely runtime structure including a set of initial states and runtime changes (steps) [37,39]. These same aspects are also identified in the sequential time postulate in [23], which postulates the existence of a set of states including initial states, and a one-step transformation function between states. We look into states and steps in the sequel.

4.1   Runtime Structure (States)

There is agreement between the theoretical [23][1] and the practical [39] understanding of runtime states as follows.

- States have a structure (States are first-order structures).
- The possible runtime states are fixed (All states have the same vocabulary and no transformations change the base set of states).
- There are several ways to implement states (They are closed under isomorphisms).

---

[1] This is given by the abstract state postulate.

The difference between theoretical and practical runtime states is that states are object structures in [39], while they are value structures in [23]. This difference is not serious, as objects can be considered as object IDs, and properties of objects are then functions over objects IDs. As usual, methods of objects just get an implicit parameter which is their enclosing object.

There are two perspectives to runtime structure, namely low-level (defined by the machine), and high-level (defined by the language). Low-level structure is given by the general von-Neumann architecture [33] which involves a CPU, a memory unit and input and output devices. High-level structure depends on the language used. As an example, for Java the runtime structure includes a set of objects, a program counter, threads, a stack, and exception objects [29]. There is also a part of the high-level runtime structure that depends on the algorithm itself, for example objects of Java classes. For Prolog, the runtime structure includes a (local) stack with environments and choice points, a heap (global stack) with terms, and a trail with variable bindings as described in the Warren Abstract Machine [44].

Of course, a computation cannot be run on an abstract or a virtual machine, some real (physical) machine has to be there to do the work. For example, the Java virtual machine (JVM) is typically implemented on top of a general-purpose machine, which again is based on machine code, which again is based on circuits, which again is based on electronics, which again is based on electrons, which again is based on quarks. A similar argument can be made for ASM, where the semantics of ASM has to be implemented on a standard computer. In ASM, a state change is done as one step, whereas in an implementation on a real computer, it would amount to a series of steps.

Which of these levels is the natural level for the algorithm? We can safely assume that the natural level is the highest of them, in the JVM example it would be the level of JVM operations. This means that the language of formulating the algorithm is essential, as the runtime structure can be different for different languages. Writing the same algorithm in Prolog versus in Java would imply serious changes in the runtime, i.e. the execution of the algorithm is different.

Although it is easy to forget, the runtime structure also needs a specification of the initial runtime state.

We see that ASMs provide the flexibility to use or define structures that fit the user's natural understanding of the algorithm. ASM makes explicit the implicit runtime elements of typical programming languages, e.g. the program counter. This is possible because ASM does not have any fixed runtime elements implied by the language.

4.2   Runtime Changes

Based on the runtime structure, runtime changes define what happens at runtime (dynamics), i.e. what is a computation step and what changes are done. As the runtime structure is given, only the changes are relevant. This relates to a finite set of changes on locations in the runtime structure, as already defined in [23] by the bounded exploration postulate. Bounded exploration has not been important for practical considerations of runtime structure, as boundedness is implied by the underlying machine. In practical terms, ways to express the changes have been more important.

Minsky [31] has demonstrated that Turing completeness requires only four instruction types: conditional GOTO, unconditional GOTO, assignment, and HALT. There is one implied instruction which is sequential composition. Nowadays, GOTO is considered bad and related to "spaghetti code", so ASM introduce the same level of Turing completeness using structured programming with update (assignment), parallel composition, and if-then-else. For ease of writing, also a let and a forall are provided. In ASM, sequential composition is not available because there is no predefined program counter. HALT is implicit as the execution stops when there are no more changes, i.e. the update set is empty[2].

The ASM algorithm could also be written using different syntax, for example traditional programming language syntax. Using Java syntax, we can express the ASM Euclidean algorithm (syntactically) as follows. Warning: This is *not* Java, just Java syntax for ASM.

```
if (n1 > n2) {
    n1 = n2;
    n2 = n1;
} else if (n1 == 0) {
    gcd = n2;
} else {
    n2 = n2 - n1;
}
```

Remember that the execution mode is parallel here. We have changed the names such that they fit the Java conventions.

This formulation reveals that the syntax is not too important for ASM and it has not been focused upon much. Instead, constructs in ASM are often considered abstract syntax that can be written in different ways, as is customary in mathematics. In the abstract syntax of ASM, we need locations, updates, choices and parallel blocks.

In each state, the complete ASM program is executed. This deviates slightly from the idea of regular programming languages, where only the current statement is executed, identified by the program counter. For example, during the execution of the Java code as given in Sect. 1, the program counter keeps track of the current code position during execution. In addition, there is an extra temporary variable t for the swap between $n1$ and $n2$.

The execution in Lisp includes a number of function activation records to keep track of all the recursive calls. We have a similar situation with Prolog execution, which adds a number of variable unifications into the runtime structure.

---

[2] In some sense, this turns HALT from a syntactic element into a semantic element. Minsky would have been able to avoid the HALT if there was a rule that the execution stops when moving (GOTO) out of the program.

In-place state transformations can also be expressed by transformation languages like QVT [16] or ATL [25]. We do not go into more detail of those languages, as they do not add more possibilities than the languages we already discussed.

Operational semantics languages also provide possibilities to express runtime changes. The Euclidean algorithms can be expressed using SOS [35] as follows.

$$\frac{\langle n_1 > n_2, s \rangle \Rightarrow true}{(s) \longrightarrow (s \uplus \{n_1 \mapsto n_2, n_2 \mapsto n_1\})}$$

$$\frac{\langle n_1, s \rangle \Rightarrow 0}{(s) \longrightarrow (s \uplus \{gcd \mapsto n_2\})}$$

$$\frac{\langle n_1 > n_2, s \rangle \Rightarrow false, \langle n_1 > 0, s \rangle \Rightarrow true}{(s) \longrightarrow (s \uplus \{n_2 \mapsto n_2 - n_1\})}$$

In this situation, ASM are an effective formalization of pseudo-code, as they are dedicated to describing one transition only[3].

What is needed for the runtime changes is navigation of the runtime structure for reading and writing of locations. In SOS [35] and also in ASM, the current program is outside the runtime state. That is possible in ASM, as the program is constant and it is always applied as one. SOS wants to keep the current execution position, but does not use a program counter. Instead, the program is a parameter of the SOS rules. Changing the PC amounts to changing the program for the next runtime state.

We see that ASM allow an explicit description of the runtime changes based on the explicit description of the runtime structure. This works for all kinds of runtime changes, be it program executions or movements of knitting needles. If the algorithm includes a sequence of actions, advancing through the sequence step by step can be considered the natural level of abstraction, such that an implicit program counter would be needed. This is standard in imperative programming languages, and can also be provided by extended ASM variants. We discuss this aspect of the user perspective in the next section.

5   Describing Algorithms

When describing an algorithm, the focus is not on the execution of the algorithm, but the understanding on the part of the user. Algorithms are ubiquitous, and we find them nearly in all aspects of life. What is the natural level of abstraction in this case? In a first attempt, we distinguish three abstraction levels of algorithm descriptions, see also [41].

High-level descriptions are given by prose and ignore implementation details like a UML use case description.

---

[3] There are also advanced ASM concepts to handle structured executions, often called Turbo-ASM [9].

An implementation description is still prose and details how the high-level
description is turned into something executable, for example as UML classes
and activity diagrams.

A formal description gives all the detail and makes the algorithm executable,
for example in Java.

It is possible to have an executable understanding of all these levels, but they
differ in the level of abstraction and detail. ASMs contribute to lifting the level
of formality and executability even to the high-level descriptions. The same is
achieved with model-driven development, see Sect. 6.

One might argue that ASM fit the bill again, as they are proven to provide
the natural level of abstraction for algorithms as discussed in Sect. 4. However,
this is only true from the point of view of the machine. In many cases, this
is the same point of view as for the designer and the user. As an example,
computer scientists often think of algorithms in terms of the machine, and there
the argument with ASM applies.

However, many other users do not look at algorithms from the point of view
of the machine. Examples are algorithms that describe recipes, or knitting algorithms, or algorithms to calculate loan security, or how to assemble a piece of
furniture, see Fig. 1. Typically, there are experts that know how to cook or to
knit, and they will describe their algorithms in a way related to their expertise.
This is usually connected to the area of domain-specific languages (DSL) [20].

Fig. 1. Different sample algorithm descriptions

There is extensive research in the area of DSL, and the general result is that
a good DSL captures the concepts of the domain in question, rather than the
concepts of the underlying machine. Instead, there is a transformation process
from the DSL to some lower-level language in terms of model-drivel development
(MDD) [1], which is standardized in the model-driven architecture (MDA) of
OMG [27]. ASM can be related here as a transformation target language.

ASM cannot and will not be the universal description language, because it
is impossible to have just one language for all purposes. The language with the
natural level of abstraction has to be found and developed in the domain where it

is used, together with its users. The examples in Fig. 1 are algorithm descriptions in DSLs that are not readily captured by ASM syntax.

Of course, ASMs were never intended to replace languages, especially not their concrete syntax. However, ASMs can be used to take the abstract syntax of a language and define its semantics. This has been done with UML state diagrams [11], the programming language Java [14], the specification language SDL [21], and could also be done with the music sheet and a knitting pattern. We discuss this aspect in the next section.

Even for regular algorithms, ASM are missing several essential language features of modern languages, for example classes, exceptions, namespaces, generics, inheritance, and iterators. These features might not be needed for simple algorithms, but they are essential for system-level complex algorithms.

We see that ASMs do not provide the description of algorithms in a concrete syntax on the natural level of abstraction in the same way as DSLs. Of course, this is not the intention of ASMs. Using ASM, we can define the behavior, as discussed in Sect. 4. We look at how ASMs can define languages (DSLs) in the next section.

6   Language Semantics

In connection with DSLs, there is a need to describe languages formally. How else would a DSL come to life if not using a description. Typically, meta-languages are used to describe languages, see for example the well-known OMG stack of modelling languages in Fig. 2.

Level	Example	Description
M3	MOF	Defines a language for specifying metamodels
M2	UML	Defines a language for specifying models
M1	model of a bank	Defines a language that describes a semantic domain
M0	a runtime state of the bank model	Contains runtime instances of the model elements defined in the model

Fig. 2. OMG stack

In the OMG stack, specifications (descriptions of algorithms - Sect. 5) are placed on level M1, while the language they are written in is on level M2. An algorithm written in ASM would be on M1, while ASM itself is on M2. The execution of the algorithm (Sect. 4) is on the base level M0. The level M3 is dedicated to meta-languages, i.e. languages that are used to describe languages. Often, meta-languages are already languages on their own, such they could be placed both on M2 and on M3. The definition of the meta-languages themselves is done using the same meta-languages, where the language definition languages used are found on M3, while the languages defined are on M2 (bootstrapping).

Is ASM a good language to describe languages? To answer this question, we have to consider what it takes to define a language, i.e. which meta-languages we need. As it turns out, there are several elements that need to be described for a language, namely abstract syntax (also called structure), concrete syntax, constraints being a part of structure, and semantics (translational or operational) [24,32], see Fig. 3. We will consider these aspects one by one.

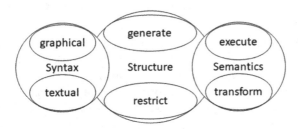

Fig. 3. Language aspects

6.1 Structure (Abstract Syntax)

The abstract syntax of a language contains the concepts of the language and their relationships with each other. Class diagrams are the method of choice to describe abstract syntax, as shown in MOF [17]. Even though ASM also allow describing structure, class diagrams are not supported in ASM. However, it is possible to use MOF diagrams to show ASM structure definitions. This way, MOF gets an ASM semantics. It should be noted that abstract syntax typically entails an abstract syntax tree, and tree structures can be expressed using ASM.

This way, ASM has support for the abstract syntax of structure definitions, but not the concrete syntax as given by MOF. Moreover, classes with inheritance are not supported by ASM, which is mainly a typing issue.

A second part of Structure is related to constraints, often expressed as OCL formulas. Logical formulas are well within the capabilities of ASM, so this part would be possible to express. The main part of the logical formulas is a way to navigate the syntax tree, and this is commonly done using expressions. More advanced DSLs for name resolution [28] are beyond the capacities of ASM. Still the semantics of all these languages can be formalized using ASM.

6.2 (Concrete) Syntax

Concrete syntax has two main forms, namely textual syntax and graphical syntax. Textual syntax is commonly given by grammars, which ASMs do not provide. Again, using grammars for analysis will finally lead to syntax trees, which can be expressed by ASM. Still, the notation of choice in this case would be grammars. Graphical syntax could be given by graph grammars, which again cannot be written in ASM. A similar argument as before applies also here.

As for concrete syntax, ASMs do not provide the concrete syntax on the natural abstraction level. However, the semantics of grammars can be described using ASM.

6.3 Semantics

We consider two essential kinds of semantics, translational semantics and execution semantics (operational semantics), see [32] for a more detailed discussion of other kinds of semantics.

Translational Semantics refers to semantics that is given as a translation into a languages which has a given semantics already. Semantically, a translation is a simple function and it could be given by various forms of function definition. It has become customary to define transformations between abstract syntax, such that the connection between the language constructs becomes visible. In principle, ASM can define functions, but in order to define structural transformations, more dedicated languages should be used [16,25].

Dedicated transformation languages allow the specification of input and output patterns for the transformation. In addition, templates can be used to specify the result of the transformation. The semantics of transformation languages is often a function or a series of functions.

As with the previous language definition elements, ASM are able to capture the aspects semantically, but do not provide the syntax on the natural abstraction level.

Execution Semantics describes how a program is executed at runtime. It includes the runtime structure and the runtime changes as discussed in Sect. 4. ASM are very well suited to describe runtime with both runtime state and runtime changes. This is already discussed in Sect. 4. This is also the way that language semantics is given using ASM, see for example [14] and [21].

There are only few dedicated languages for the definition of execution semantics, and ASM provides all features that are needed. For application of ASM in an object-oriented language definition context, where both the language structure and the runtime environment are object-oriented, the availability of classes and inheritance in ASM would be an advantage.

SOS [35] is a DSL for the description of execution semantics. The example of SOS in Sect. 4 shows that its expressive power is comparable to ASM.

6.4 Summary

ASM shines for the formulation of execution semantics on the natural level of abstraction, which relates very well to its power in describing algorithm executions. This implies that the semantics of all meta-languages can be formalized using ASM. On the syntax side, DSLs are on a more natural abstraction level. The same applies to transformations.

7   Conclusion

We have considered the abstraction level of Abstract State Machines and whether the ASM capture the natural abstraction level of an algorithm. We have looked into three aspects of natural abstraction level, namely abstraction of executions, abstraction of descriptions, and abstraction of language semantics.

As it turns out, ASM are on the correct level of abstraction for algorithm execution, which is already established in [23]. The consideration of runtime environments brings the same result from a different perspective.

For the description of algorithms, ASMs cannot provide the correct abstraction level, as this depends on the application domain of the algorithm. Domain-specific languages are the way to provide such good descriptions, and no single language can provide the correct abstraction level.

This leads to the discussion how languages can be formalized, and whether ASM are on a natural abstraction level as a meta-language. Language design has several areas, and ASM are not on the right abstraction level for abstract syntax and concrete syntax. ASM can be used for some aspects of constraints and of transformation semantics. However, the strength of ASM is that it is on the natural abstraction level for operational semantics, which essentially is the same as repeating that ASM are on the natural abstraction level for algorithm execution.

When we connect these results to the OMG modelling levels as presented in Fig. 2, then ASM is strong on level M0 (executions), and not strong on level M1 (descriptions). On level M2, the strength of ASM is again on the execution semantics side, i.e. the connection of the description with the executions. We can interpret this such that ASM is a semantic language with little concern for syntax. It provides support to explicitly capture executions on the correct level of abstraction, and it avoids predefined execution patterns like a program counter.

This way, ASMs give just the right level of freedom for describing all execution situations on the natural level of abstraction.

References

1. Bennedsen, J., Caspersen, M.E.: Model-driven programming. In: Bennedsen, J., Caspersen, M.E., Kölling, M. (eds.) Reflections on the Teaching of Programming. LNCS, vol. 4821, pp. 116–129. Springer, Heidelberg (2008). https://doi.org/10.1007/978-3-540-77934-6_10. http://link.springer.com/book/10.1007%2 F978-3-540-77934-60
2. Blass, A., Gurevich, Y.: Ordinary interactive small-step algorithms I. ACM Trans. Comput. Log. **7**(2), 363–419 (2006)
3. Blass, A., Gurevich, Y.: Abstract state machines capture parallel algorithms: correction and extension. ACM Trans. Comput. Log. **9**(3) (2008). https://doi.org/10.1145/1352582.1352587
4. Blass, A., Gurevich, Y., Rosenzweig, D., Rossman, B.: Interactive small-step algorithms II: abstract state machines and the characterization theorem. Log. Methods Comput. Sci. **3**(4) (2007). https://doi.org/10.2168/LMCS-3(4:4)2007

5. Blass, A., Gurevich, Y.: Abstract state machines capture parallel algorithms. ACM Trans. Comput. Log. (TOCL) **4**(4), 578–651 (2003). https://doi.org/10.1145/937555.937561

6. Blass, A., Gurevich, Y.: Persistent queries in the behavioral theory of algorithms. ACM Trans. Comput. Log. (TOCL) **12**(2), 16:1–16:43 (2011). https://doi.org/10.1145/1877714.1877722

7. Boolos, G.S., Burgess, J.P., Jeffrey, R.C.: Computability and Logic, 5th edn. Cambridge University Press, Cambridge (2007). https://doi.org/10.1017/CBO9780511804076

8. Börger, E., Cisternino, A. (eds.): Advances in Software Engineering. LNCS, vol. 5316. Springer, Heidelberg (2008). https://doi.org/10.1007/978-3-540-89762-0

9. Börger, E., Stärk, R.: Abstract State Machines - A Method for High-Level System Design and Analysis. Springer, Heidelberg (2003). https://doi.org/10.1007/978-3-642-18216-7

10. Börger, E.: Construction and analysis of ground models and their refinements as a foundation for validating computer-based systems. Appl. Formal Methods **19**(2), 225–241 (2007). https://doi.org/10.1007/s00165-006-0019-y

11. Börger, E., Cavarra, A., Riccobene, E.: On formalizing UML state machines using ASMs. Inf. Softw. Technol. **46**(5), 287–292 (2004). https://doi.org/10.1016/j.infsof.2003.09.009

12. Börger, E., Raschke, A.: Modeling Companion for Software Practitioners. Springer, Heidelberg (2018). https://doi.org/10.1007/978-3-662-56641-1

13. Börger, E., Schewe, K.D.: Concurrent abstract state machines. Acta Informatica **53**(5), 469–492 (2016). https://doi.org/10.1007/s00236-015-0249-7

14. Börger, E., Schulte, W.: A programmer friendly modular definition of the semantics of Java. In: Alves-Foss, J. (ed.) Formal Syntax and Semantics of Java. LNCS, vol. 1523, pp. 353–404. Springer, Heidelberg (1999). https://doi.org/10.1007/3-540-48737-9_10

15. Brooks Jr, F.P.: No silver bullet essence and accidents of software engineering. Computer **20**(4), 10–19 (1987). https://doi.org/10.1109/MC.1987.1663532. http://ieeexplore.ieee.org/xpl/articleDetails.jsp?arnumber=1663532

16. Editor OMG: Meta object facility (MOF) 2.0 query/view/transformation specification, version 1.1. Object Management Group (2011). http://www.omg.org/spec/QVT/1.1/

17. Editor OMG: Meta object facility (MOF). Object Management Group (2016). https://www.omg.org/spec/MOF

18. Felleisen, M., Hieb, R.: The revised report on the syntactic theories of sequential control and state. Theor. Comput. Sci. **103**(2), 235–271 (1992)

19. Fischer, J., Møller-Pedersen, B., Prinz, A.: Real models are really on M0- or how to make programmers use modeling. In: Proceedings of the 8th International Conference on Model-Driven Engineering and Software Development-Volume 1: MODELSWARD, pp. 307–318. INSTICC, SciTePress (2020). https://doi.org/10.5220/0008928403070318

20. Fowler, M.: Domain Specific Languages, 1st edn. Addison-Wesley Professional, Boston (2010)

21. Glässer, U., Gotzhein, R., Prinz, A.: The formal semantics of SDL-2000: status and perspectives. Comput. Netw. **42**(3), 343–358 (2003). https://doi.org/10.1016/S1389-1286(03)00247-0

22. Gurevich, Y.: Evolving algebras 1993: lipari guide. In: Specification and Validation Methods, pp. 231–243. Oxford University Press (1995)

23. Gurevich, Y.: Sequential abstract-state machines capture sequential algorithms. ACM Trans. Comput. Log. (TOCL) **1**(1), 77–111 (2000). https://doi.org/10.1145/343369.343384
24. Harel, D., Rumpe, B.: Meaningful modeling: what's the semantics of "semantics"? Computer **37**(10), 64–72 (2004). https://doi.org/10.1109/MC.2004.172
25. Jouault, F., Allilaire, F., Bézivin, J., Kurtev, I.: ATL: a model transformation tool. Sci. Comput. Program. **72**(1–2), 31–39 (2008). https://doi.org/10.1016/j.scico.2007.08.002
26. Kahn, G.: Natural semantics. In: Brandenburg, F.J., Vidal-Naquet, G., Wirsing, M. (eds.) STACS 1987. LNCS, vol. 247, pp. 22–39. Springer, Heidelberg (1987). https://doi.org/10.1007/BFb0039592
27. Kleppe, A., Warmer, J.: MDA Explained. Addison-Wesley, Boston (2003)
28. Konat, G., Kats, L., Wachsmuth, G., Visser, E.: Declarative name binding and scope rules. In: Czarnecki, K., Hedin, G. (eds.) SLE 2012. LNCS, vol. 7745, pp. 311–331. Springer, Heidelberg (2013). https://doi.org/10.1007/978-3-642-36089-3_18
29. Lindholm, T., Yellin, F., Bracha, G., Buckley, A.: The Java Virtual Machine Specification, Java SE 8 Edition, 1st edn. Addison-Wesley Professional, Boston (2014)
30. Madsen, O.L., Møller-Pedersen, B.: This is not a model. In: Margaria, T., Steffen, B. (eds.) ISoLA 2018. LNCS, vol. 11244, pp. 206–224. Springer, Cham (2018). https://doi.org/10.1007/978-3-030-03418-4_13
31. Minsky, M.L.: Computation: Finite and Infinite Machines. Prentice-Hall Inc., Englewood Cliffs (1967)
32. Mu, L., Gjøsæter, T., Prinz, A., Tveit, M.S.: Specification of modelling languages in a flexible meta-model architecture. In: Proceedings of the Fourth European Conference on Software Architecture: Companion Volume, ECSA 2010, pp. 302–308. Association for Computing Machinery, New York (2010). https://doi.org/10.1145/1842752.1842807
33. von Neumann, J.: First draft of a report on the EDVAC. In: Randell, B. (ed.) The Origins of Digital Computers. MCS, pp. 383–392. Springer, Heidelberg (1982). https://doi.org/10.1007/978-3-642-61812-3_30
34. Ouimet, M., Lundqvist, K.: The timed abstract state machine language: abstract state machines for real-time system engineering. J. UCS **14**, 2007–2033 (2008)
35. Plotkin, G.D.: A structural approach to operational semantics. J. Log. Algebraic Program. **60–61**, 17–139 (2004). https://doi.org/10.1016/j.jlap.2004.05.001. Structural Operational Semantics
36. Prinz, A.: Distributed computing on distributed memory. In: Khendek, F., Gotzhein, R. (eds.) SAM 2018. LNCS, vol. 11150, pp. 67–84. Springer, Cham (2018). https://doi.org/10.1007/978-3-030-01042-3_5
37. Prinz, A., Møller-Pedersen, B., Fischer, J.: Object-oriented operational semantics. In: Grabowski, J., Herbold, S. (eds.) SAM 2016. LNCS, vol. 9959, pp. 132–147. Springer, Cham (2016). https://doi.org/10.1007/978-3-319-46613-2_9
38. Prinz, A., Sherratt, E.: Distributed ASM - pitfalls and solutions. In: Aït-Ameur, Y., Schewe, K.D. (eds.) ABZ 2014. LNCS, vol. 8477, pp. 210–215. Springer, Heidelberg (2014). https://doi.org/10.1007/978-3-662-43652-3_18
39. Scheidgen, M., Fischer, J.: Human comprehensible and machine processable specifications of operational semantics. In: Akehurst, D.H., Vogel, R., Paige, R.F. (eds.) ECMDA-FA 2007. LNCS, vol. 4530, pp. 157–171. Springer, Heidelberg (2007). https://doi.org/10.1007/978-3-540-72901-3_12

40. Schewe, K.-D., Wang, Q.: A simplified parallel ASM thesis. In: Derrick, J., et al. (eds.) ABZ 2012. LNCS, vol. 7316, pp. 341–344. Springer, Heidelberg (2012). https://doi.org/10.1007/978-3-642-30885-7_27
41. Sipser, M.: Introduction to the Theory of Computation, 3rd edn. Course Technology, Boston (2013)
42. Stone, H.S.: Introduction to Computer Organization and Data Structures. McGraw-Hill, New York (1972)
43. Turing, A.M.: On computable numbers, with an application to the Entscheidungsproblem. Proc. Lond. Math. Soc. **s2-42**(1), 230–265 (1937). https://doi.org/10.1112/plms/s2-42.1.230. https://londmathsoc.onlinelibrary.wiley.com/doi/abs/10.1112/plms/s2-42.1.230
44. Warren, D.S.: WAM for everyone: a virtual machine for logic programming, pp. 237–277. Association for Computing Machinery and Morgan & Claypool (2018). https://doi.org/10.1145/3191315.3191320

The ASMETA Approach to Safety Assurance of Software Systems

Paolo Arcaini[1](✉)[ID], Andrea Bombarda[2][ID], Silvia Bonfanti[2][ID],
Angelo Gargantini[2][ID], Elvinia Riccobene[3][ID], and Patrizia Scandurra[2][ID]

[1] National Institute of Informatics, Tokyo, Japan
arcaini@nii.ac.jp
[2] University of Bergamo, Bergamo, Italy
{andrea.bombarda,silvia.bonfanti,angelo.gargantini,
patrizia.scandurra}@unibg.it
[3] Università degli Studi di Milano, Milan, Italy
elvinia.riccobene@unimi.it

Abstract. Safety-critical systems require development methods and processes that lead to provably correct systems in order to prevent catastrophic consequences due to system failure or unsafe operation. The use of models and formal analysis techniques is highly demanded both at design-time, to guarantee safety and other desired qualities already at the early stages of the system development, and at runtime, to address requirements assurance during the system operational stage.

In this paper, we present the modeling features and analysis techniques supported by ASMETA (ASM mETAmodeling), a set of tools for the Abstract State Machines formal method. We show how the modeling and analysis approaches in ASMETA can be used during the design, development, and operation phases of the assurance process for safety-critical systems, and we illustrate the advantages of integrated use of tools as that provided by ASMETA.

1 Introduction

Failures of safety-critical systems could have potentially large and catastrophic consequences, such as human hazards or even loss of human life, damage to the environment, or economic disasters. There are many well-known examples of critical failures in application areas such as medical devices, aircraft flight control, weapons, and nuclear systems [40,41].

To assure safe operation and prevent catastrophic consequences of system failure, safety-critical systems need development methods and processes that lead to provably correct systems. Rigorous development processes require the use of formal methods, which can guarantee, thanks to their mathematical foundation, model preciseness, and properties assurance.

P. Arcaini is supported by ERATO HASUO Metamathematics for Systems Design Project (No. JPMJER1603), JST. Funding Reference number: 10.13039/501100009024 ERATO.

A. Raschke et al. (Eds.): Börger Festschrift, LNCS 12750, pp. 215–238, 2021.
https://doi.org/10.1007/978-3-030-76020-5_13

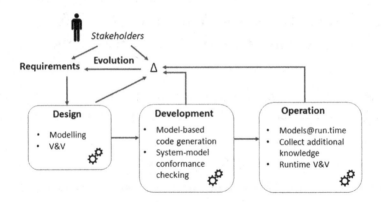

Fig. 1. Assurance process during system's life cycle

However, modern safety-critical software systems usually include physical systems and humans in the loop, as for example Cyber-Physical Systems (CPSs), and, therefore, "system safety" is not only "software safety" but may depend on the use of the software within its untrusted and unreliable environment. Reproducing and validating real usage scenarios of such systems at design- or at development- time is not always possible. Their behavior under certain circumstances cannot be completely validated without deploying them in a real environment, where all relevant uncertainties and unknowns caused by the close interactions of the system with their users and the environment can be detected and resolved [28,41]. Therefore, an important aspect of the software engineering process for safety-critical systems is providing evidence that the requirements are satisfied by the system during the *entire* system's life cycle, from inception to and throughout operation [49]. As envisioned by the Models@run.time research community, the use of models and formal analysis techniques is fundamental at design-time to guarantee reliability and desired qualities already at the early stages of the system development, but also at runtime to address requirements assurance during the system operational stage.

Providing assurances that a system complies with its requirements demands for an analysis process spanning the whole life cycle of the system. Figure 1 outlines such a process, showing the three main phases of *Design*, *Development*, and *Operation* of a system life cycle. During the system development phase, models created, validated, and verified during the design phase are eventually used to derive correct-by-construction code/artifacts of the system and/or to check that the developed system conforms to its model(s). During the operation phase, models introduced at design-time are executed in tandem with the system to perform analysis at runtime. In this assurance process, stakeholders and the system jointly derive and integrate new evidence and arguments for analysis (Δ); system requirements and models are eventually adapted according to the collected knowledge. Hence, requirements and models evolve accordingly throughout the system life cycle.

This assurance process requires the availability of formal approaches having specific characteristics in order to cover all the three phases: models should possibly be executable for high-level design validation and endowed with properties verification mechanisms; operational approaches are more adequate than denotational ones to support (automatic) code generation from models and model-based testing; state-based methods are suitable for co-simulation between model and code and for checking state conformance between model state and code state at runtime. In principle, different methods and tools can be used in the three phases; however, the integrated use of different tools around the same formal method is much more convenient than having different tools working on input models with their own languages.

This article presents, in a unified manner, the distinctive modeling features and analysis techniques supported by ASMETA (ASM mETAmodeling) [13,17], a modeling and analysis framework based on the formal method Abstract State Machines (ASMs) [26,27], and how they can be used in the three phases of the assurance process (see Fig. 1). ASMETA adopts a set of modeling languages and tools for not only specifying the executable behavior of a system but also for checking properties of interest, specifying and executing validation scenarios, generating prototype code, etc. Moreover, runtime validation and verification techniques have been recently developed as part of ASMETA to allow runtime assurance and enforcement of system safety assertions.

The remainder of this article is organized as follows. Section 2 explains the origin of the ASMETA project, recalls some basic concepts of the ASM method, and overviews the ASMETA tools in the light of the assurance process. The subsequent sections describe analysis techniques and associated tooling strategies supported by ASMETA for the safety assurance process: Sect. 3 for the design phase, Sect. 4 for the development phase, and Sect. 5 for the operation phase. Section 6 concludes the paper and outlines future research directions.

2   The ASMETA Approach

This section recalls the origin of the ASMETA project [17] and the basic concepts of the ASM method it is based on; we also overview the set of tools in the light of the assurance process.

2.1   Project Description

The ASMETA project started roughly in 2004 with the goal of overcoming the lack of tools supporting the ASMs. The formal approach had already shown to be widely used for the specification and verification of a number of software systems and in different application domains (see the *survey of the ASM research* in [27]); however, the lack of tools supporting the ASM method was perceived as a limitation, and there was skepticism regarding its use in practice.

The main goal when we started the ASMETA project, encouraged by the Egon Börger suggestion, was to develop a textual notation for encoding ASM models. We exploited the (at that time) novel *Model-driven Engineering* (MDE) approach [45] to develop an abstract syntax of a modeling language for ASMs [36] in terms of a metamodel, and to derive from it a user-facing textual notation to edit ASM models. Then, from the ASM metamodel – called *Abstract State Machine Metamodel* (AsmM) – and by exploiting the runtime support for models and model transformation facility of the open-source Eclipse-based MDE IDE EMF, ASMETA has been progressively developed till now as an Eclipse-based set of tools for ASM model editing, visualization, simulation, validation, property verification, and model-based testing [13].

In order to support a variety of analysis activities on ASM models, ASMETA integrates different external tools, such as the NuSMV model checker for performing property verification and SMT solvers to support correct model refinement verification and runtime verification. To this purpose, ASMETA mainly supports a black-box model composition strategy based on *semantic mapping* [35,39], i.e., model transformations realize semantic mappings from ASM models (edited using the textual user-facing language AsmetaL) to the input formalism of the target analysis tool depending on the purpose of the analysis, and then lift the analysis results back to the ASM level.

ASMETA is widely used for research purposes (also by groups different from the development teams [1,14,19,48]) and as teaching support in formal methods courses at the universities of Milan and Bergamo in Italy.

Case Studies. ASMETA has been applied to different case studies in several application domains; moreover, a wide repository of examples, many of which are benchmarks presented by Egon Börger in his dissemination work on the ASM method, are available on line[1]. Specifically, ASMETA has been applied in the context of medical devices (PillBox [20], hemodialysis device [3], amblyopia diagnosis [2], PHD Protocol [21]), software control systems (Landing Gear System [12], Automotive Software-Intensive Systems [5], Hybrid European Rail Traffic Management System [37]), cloud- [14] and service-based systems [42,43], Self-adaptive systems [15,16].

2.2   Abstract State Machines: Background Concepts

The computational model at the base of the ASMETA framework is that of the Abstract State Machines (ASMs) formal method. It was originally introduced by Yuri Gurevich as *Evolving Algebras* [38], but it was Egon Börger who renamed the approach as ASMs – viewed as an extension of Finite State Machines (FSMs) –, and disseminated it as a method for the high-level design and analysis of computing systems [26,27].

ASM *states* replace unstructured FSM control states by algebraic structures, i.e., domains of objects with functions and predicates defined on them. An ASM *location*, defined as the pair (*function-name, list-of-parameter-values*), represents

---

[1] Repository https://github.com/asmeta/asmeta/tree/master/asm_examples.

the ASM concept of basic object container, and the couple (*location, value*) is a memory unit; an ASM state can be thus viewed as a set of abstract memories.

State transitions are performed by firing *transition rules*, which express the modification of functions interpretation from one state to the next one and, therefore, they change location values. Location *updates* are given as assignments of the form $loc := v$, where loc is a location and v its new value. They are the basic units of rules construction. By a limited but powerful set of *rule constructors*, location updates can be combined to express other forms of machine actions as: guarded actions (`if-then`, `switch-case`), simultaneous parallel actions (`par` and `forall`), sequential actions (`seq`), non-deterministic actions (`choose`).

Functions that are not updated by rule transitions are *static*. Those updated are *dynamic*, and distinguished in *monitored* (read by the machine and modified by the environment), *controlled* (read and written by the machine), *shared* (read and written by the machine and its environment).

An ASM *computation* (or *run*) is defined as a finite or infinite sequence $S_0, S_1,$ $\dots, S_n, \dots$ of states of the machine, where S_0 is an initial state and each S_{n+1} is obtained from S_n by firing the set of all transition rules invoked by a unique *main rule*, which is the starting point of the computation.

It is also possible to specify state *invariants* as first-order formulas that must be true in each computational state. A set of safety assertions can be specified as model invariants, and a model state is *safe* if state invariants are satisfied.

ASMs allow modeling different computational paradigms, from a *single* agent to distributed *multiple* agents. A *multi-agent ASM* is a family of pairs $(a, ASM(a))$, where each a of a predefined set *Agent* executes its own machine $ASM(a)$ (specifying the agent's behavior), and contributes to determine the next state by interacting synchronously or asynchronously with the other agents.

ASMs offer several advantages w.r.t. other automaton-based formalisms: (1) due to their *pseudo-code format*, they can be easily understood by practitioners and can be used for high-level programming; (2) they offer a precise system specification at any desired *level of abstraction*; (3) they are *executable models*, so they can be co-executed with system low-level implementations [43]; (4) *model refinement* is an embedded concept in the ASM formal approach; it allows for facing the complexity of system specification by starting with a high-level description of the system and then proceeding step-by-step by adding further details till a desired level of specification has been reached; each refined model must be proved to be a correct refinement of the previous one, and checking of such relation can be performed automatically [11]; (5) the concept of ASM *modularization*, i.e., an ASM without the main firing rule, facilitates model scalability and separation of concerns, so tackling the complexity of big systems specification; (6) they support synch/async multi-agent compositions, which allows for *modeling distributed and decentralized software systems* [16].

2.3   Tool-Support for Safety Assurance

Figure 2 gives an overview of the ASMETA tools by showing their use to support the different activities of the safety assurance process depicted in Fig. 1.

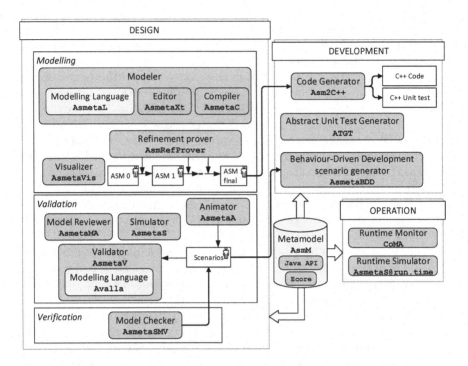

Fig. 2. ASMETA tool-set

At *design-time*, ASMETA provides a number of tools for model editing and visualization (the modeling language `AsmetaL` and its editor and compiler, plus the model visualizer `AsmetaVis` for graphical visualization of ASM models), model validation (e.g., interactive or random simulation by the simulator `AsmetaS`, animation by the animator `AsmetaA`, scenario construction and validation by the validator `AsmetaV`), and verification (e.g., static analysis by the model reviewer `AsmetaMA`, proof of temporal properties by the model checker `AsmetaSMV`, proof of correct model refinement by `AsmRefProver`).

At *development-time*, ASMETA supports automatic code and test case generation from models (the code generator `Asm2C++`, the unit test generator `ATGT`, and the acceptance test generator `AsmetaBDD` for complex system scenarios).

Finally, at *operation-time*, ASMETA supports runtime simulation (the simulator `AsmetaS@run.time`) and runtime monitoring (the tool `CoMA`).

The analysis techniques and associated tooling strategies supported by ASMETA are described in more detail in the next sections and they are applied to the *one-way traffic light* case study introduced in [27].

3   ASMETA@design-time

In order to assure the safety of software systems, system design is the first activity supported by ASMETA. During this phase, users can model the desired system

```
asm oneWayTrafficLight
import StandardLibrary
signature:
    enum domain LightUnit = {LIGHTUNIT1 | LIGHTUNIT2}
    enum domain PhaseDomain = { STOP1STOP2 | GO2STOP1 |
    STOP2STOP1 | GO1STOP2 }
    enum domain Time = {FIFTY | ONEHUNDREDTWENTY | LESS}
    dynamic controlled phase: PhaseDomain
    dynamic controlled stopLight: LightUnit −> Boolean
    dynamic controlled goLight: LightUnit −> Boolean
    dynamic monitored passed: Time
definitions:
    rule r_stop1stop2_to_go2stop1 =
        if phase=STOP1STOP2 then
            if passed = FIFTY then
                par
                    goLight(LIGHTUNIT2) :=
                        not(goLight(LIGHTUNIT2))
                    stopLight(LIGHTUNIT2) :=
                        not(stopLight(LIGHTUNIT2))
                    phase := GO2STOP1
                endpar
            endif
        endif

    rule r_go2stop1_to_stop2stop1 = ...
```

```
rule r_stop2stop1_to_go1stop2 = ...

rule r_go1stop2_to_stop1stop2 =
    if phase=GO1STOP2 then
        if passed = ONEHUNDREDTWENTY then
            par
                goLight(LIGHTUNIT1) :=
                    not(goLight(LIGHTUNIT1))
                stopLight(LIGHTUNIT1) :=
                    not(stopLight(LIGHTUNIT1))
                phase := STOP1STOP2
            endpar
        endif
    endif

main rule r Main =
    par
        r_stop1stop2_to_go2stop1[]
        r_go2stop1_to_stop2stop1[]
        r_stop2stop1_to_go1stop2[]
        r_go1stop2_to_stop1stop2[]
    endpar

default init s0:
    function stopLight($l in LightUnit) = true
    function goLight($l in LightUnit) = false
    function phase = STOP1STOP2
```

Fig. 3. Example of AsmetaL model for a one-way traffic light

using the AsmetaL language, exploiting its features, and refine every model which can be visualized in a graphical manner and analyzed with several verification and validation tools.

3.1 Modeling

Starting from the functional requirements, ASMETA allows the user to model the system using, if needed, model composition and refinement.

3.1.1 Modeling Language
System requirements can be modeled in ASMETA by using the AsmetaL language and the AsmetaXt editor.

Figure 3 shows the AsmetaL model[2] of the *one-way traffic light*: two traffic lights (LIGHTUNIT1 and LIGHTUNIT2), equipped with a *Stop* (red) and a *Go* (green) light, that are controlled by a computer, which turns the lights go and stop, following a four phases cycle: for 50 s both traffic lights show *Stop*; for 120 s only LIGHTUNIT2 shows *Go*; for 50 s both traffic lights show again the *Stop* signal; for 120 s only LIGHTUNIT1 shows *Go*.

The model, identified by a *name* after the keyword asm, is structured into four sections:

- The *header*, where the signature (functions and domains) is declared, and external signature is imported (see Modularization below);
- The *body*, where transitions rules are defined (plus concrete domains and derived functions definitions, if any);
- A *main rule*, which defines the starting rule of the machine;
- The *initialization*, where a *default* initial state (among a set of) is defined.

---

[2] Note that $x denotes the variable x in the AsmetaL notation.

Each `AsmetaL` rule can be composed by using the set of *rule constructors* (see Sect. 2.2) to express the different machine action paradigms.

Modularization. ASMETA modeling supports the modularization and information-hiding mechanism, by the `module` notation. When requirements are complex or when separation of concerns is desired, users can organize the model in several ASM modules and join them, by using the `import` statement, into a single main one (also defined as *machine*), declared as `asm`, which imports the others and may access to functions, rules, and domains declared within the sub-modules. Every ASM module contains definitions of domains, functions, invariants, and rules, while the ASM machine is a module that additionally contains an initial state and the main rule representing the starting point of the execution.

3.1.2   Refinement

The modeling process of an ASM is usually based on model refinement [25]: the designer starts with a high-level description of the system and proceeds through a sequence of more detailed models each introducing, step-by-step, design decisions and implementation details. At each refinement level, a model must be proved to be a correct refinement of the more abstract one.

ASMETA supports a special case of *1−n refinement*, consisting in adding functions and rules in a way that one step in the ASM at a higher level can be performed by several steps in the refined model. We consider the refinement *correct* if any behavior (i.e., run or sequence of states) in the refined model can be mapped to a run in the abstract model.

To automatically prove the correctness of the model refinement process, users can exploit the `AsmRefProver` tool [11], which is based on a Satisfiability Modulo Theories (SMT) solver. With the execution of this software, one can specify two refinement levels and ensure that an ASM specification ASM_i is a correct refinement of a more abstract one ASM_{i-1}. Then, `AsmRefProver` confirms whether the refinement is correctly performed with two different outputs: `Initial states are conformant` and `Generic step is conformant`.

Figure 4 shows a refinement of the one-way traffic light model (see Fig. 3) in which pulsing lights (`rPulse` and `gPulse`) are introduced and a different management method for the time is used, based on a `timer` function mapping each phase to a timer duration. Thus, the behavior of the system modeled in Fig. 3 is preserved and expanded during the refinement process.

Modeling by refinement allows adding to the model requirements of increasing complexity only when the developer has gained enough confidence in the basic behaviors of the modeled system. This can be done by alternating modeling and testing activities, as presented in [21], with different refinement levels.

3.1.3   Visualization

Model visualization is a good means for people to communicate and to get a common understanding, especially when model comprehension can be threatened by the model size. ASMETA supports model visualization by a visual notation

```
asm oneWayTrafficLight_refined

import StandardLibrary

signature:
    enum domain LightUnit = {LIGHTUNIT1 | LIGHTUNIT2}
    enum domain PhaseDomain = { STOP1STOP2 | GO2STOP1
        | STOP2STOP1 | GO1STOP2 | STOP1STOP2CHANGING
        | GO2STOP1CHANGING | STOP2STOP1CHANGING
        | GO1STOP2CHANGING }
    dynamic controlled phase: PhaseDomain
    dynamic controlled stopLight: LightUnit —> Boolean
    dynamic controlled goLight: LightUnit —> Boolean
    static timer: PhaseDomain —> Integer
    dynamic monitored passed: Integer —> Boolean
    dynamic controlled rPulse: LightUnit —> Boolean
    dynamic controlled gPulse: LightUnit —> Boolean

definitions:
    function timer($p in PhaseDomain) = switch($p)
        case STOP1STOP2 : 50
        case GO2STOP1 : 120
        case STOP2STOP1 : 50
        case GO1STOP2 : 120
    endswitch

    rule r_switchToStop1 =
        par
            r_emit[rPulse(LIGHTUNIT1)]
            r_emit[gPulse(LIGHTUNIT1)]
        endpar

    rule r_switchToGo2 = ...
    rule r_switchToStop2 = ...
    rule r_switchToGo1 = ...

    rule r_stop1stop2_to_stop1stop2changing =
        if(phase=STOP1STOP2) then
            if(passed(timer(STOP1STOP2))) then
                par
                    r_switchToGo2[]
                    phase:=STOP1STOP2CHANGING
                endpar
            endif
        endif

    rule r_go2stop1_to_go2stop1changing = ...
    rule r_stop2stop1_to_stop2stop1changing = ...
```

```
rule r_go1stop2_to_go1stop2changing = ...
macro rule r_switch($l in Boolean) = $l := not($l)
macro rule r_emit($pulse in Boolean) = $pulse := true

rule r_pulses =
    forall $l in LightUnit with true do
        par
            if(gPulse($l)) then
                par
                    r_switch[goLight($l)]
                    gPulse($l) := false
                endpar
            endif
            if(rPulse($l)) then
                par
                    r_switch[stopLight($l)]
                    rPulse($l) := false
                endpar
            endif
        endpar

macro rule r_changeState =
    par
        if(phase=STOP1STOP2CHANGING) then
            phase := GO2STOP1
        endif
        if(phase=GO2STOP1CHANGING) then ... endif
        if(phase=STOP2STOP1CHANGING) then ... endif
        if(phase=GO1STOP2CHANGING) then ... endif
    endpar

main rule r_Main =
    par
        r_stop1stop2_to_stop1stop2changing[]
        r_go2stop1_to_go2stop1changing[]
        r_stop2stop1_to_stop2stop1changing[]
        r_go1stop2_to_go1stop2changing[]
        r_pulses[]
        r_changeState[]
    endpar

default init s0:
    function stopLight($l in LightUnit) = true
    function goLight($l in LightUnit) = false
    function phase = STOP1STOP2
    function rPulse($l in LightUnit) = false
    function gPulse($l in LightUnit) = false
```

Fig. 4. Example of a refined `AsmetaL` model for a one-way traffic light

defined in terms of a set of construction rules and schema that give a graphical representation of an ASM and its rules [4]. The graphical information is represented by a visual graph in which nodes represent syntactic elements (like rules, conditions, rule invocations) or states, while edges represent bindings between syntactic elements or state transitions. The `AsmetaVis` tool supports two types of visualization: *basic visualization*, which represents the syntactic structure of the model and returns a visual tree obtained by recursively visiting the ASM rules; *semantic visualization*, which introduces visual patterns that permit to capture some behavioral information as control states. An example of semantic visualization of the one-way traffic light case study (see Fig. 3) is shown in Fig. 5: it displays how the four macro rules in the model change the phase of the system.

3.2 Validation and Verification

Once the `AsmetaL` model is available, the user can perform validation and verification activities.

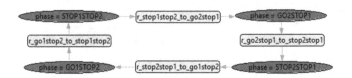

Fig. 5. `AsmetaVis` semantic visualization

Insert a boolean constant for passed(50):
true
<State 0 (monitored)>
passed(50)=true
</State 0 (monitored)>
<UpdateSet − 0>
goLight(lightUnit2)=true
phase=GO2STOP1
stopLight(lightUnit2)=false
</UpdateSet>
<State 1 (controlled)>
LightUnit={lightUnit1,lightUnit2}
goLight(lightUnit2)=true
phase=GO2STOP1
stopLight(lightUnit2)=false
</State 1 (controlled)>
Insert a boolean constant for passed(120):
false
<State 1 (monitored)>
passed(120)=false
</State 1 (monitored)>
<UpdateSet − 1>
</UpdateSet>

<State 2 (controlled)>
LightUnit={lightUnit1,lightUnit2}
goLight(lightUnit2)=true
phase=GO2STOP1
stopLight(lightUnit2)=false
</State 2 (controlled)>
Insert a boolean constant for passed(120):
true
<State 2 (monitored)>
passed(120)=true
</State 2 (monitored)>
<UpdateSet − 2>
goLight(lightUnit2)=false
phase=STOP2STOP1
stopLight(lightUnit2)=true
</UpdateSet>
<State 3 (controlled)>
LightUnit={lightUnit1,lightUnit2}
goLight(lightUnit2)=false
phase=STOP2STOP1
stopLight(lightUnit2)=true
</State 3 (controlled)>
Insert a boolean constant for passed(50):

Fig. 6. Simulation of one-way traffic light using `AsmetaS`

3.2.1 Simulation

This is the first validation activity usually performed to check the `AsmetaL` model behavior during its development and it is supported by the `AsmetaS` tool [13]. Given a model, at every step, the simulator builds the update set according to the theoretical definitions given in [27] to construct the model run. The simulator supports two types of simulation: *random* and *interactive*. In random mode, the simulator automatically assigns values to monitored functions choosing them from their codomains. In interactive mode, instead, the user inserts the value of monitored functions and, in case of input errors, a message is shown inviting the user to insert again the function value. In case of invariant violation or inconsistent updates, a message is shown in the console and the simulation is interrupted. In Fig. 6, we show the result of the simulation for the one-way traffic light `AsmetaL` model (see Fig. 3). When the desired time is passed, 50 or 120 s, the phase of the system changes.

3.2.2 Animation

The main disadvantage of the simulator is that it is textual, and this makes sometimes difficult to follow the computation of the model. For this reason, ASMETA has a model animator, `AsmetaA` [22], which provides the user with complete information about all the state locations, and uses colors, tables, and figures over simple text to convey information about states and their evolution. The animator helps the user follow the model computation and understand how the model state changes at every step.

	Type	Functions	State 0	State 1	State 2	State 3
☐ ⋀	M	passed(50)	true	true	true	
☐ ⋀	C	phase	STOP1STOP2	GO2STOP1	GO2STOP1	STOP2STOP1
☐ ⋀	C	stopLight(lightUnit2)	true	false	false	true
☐ ⋀	C	goLight(lightUnit2)	false	true	true	false
☐ ⋀	M	passed(120)		false	true	

Fig. 7. Animation of one-way traffic light using `AsmetaA`

Similarly to the simulator, the animator supports *random* and *interactive* animation. In the interactive animation, the insertion of input functions is achieved through different dialog boxes depending on the type of function to be inserted (e.g., in case of a Boolean function, the box has two buttons: one if the value is true and one if the value is false). If the function value is not in its codomain, the animator keeps asking until an accepted value is inserted. In random animation, the monitored function values are automatically assigned. With complex models, running one random step each time is tedious; for this reason, the user can also specify the number of steps to be performed and the tool performs the random simulation accordingly. In case of invariant violation, a message is shown in a dedicated text box and the animation is interrupted (as it also happens in case of inconsistent updates). Once the user has animated the model, the tool allows exporting the model run as a scenario (see Sect. 3.2.3), so that it can be re-executed whenever desired. Figure 7 shows the animation of the one-way traffic light model using the same input sequence of the simulator. The result is the same, but the tabular view makes it easier to follow the state evolution.

3.2.3   Scenario-Based Simulation

`AsmetaS` and `AsmetaA` tools require that the user executes the `AsmetaL` model step by step, each time the model has to be validated. Instead, in scenario-based simulation, the user writes a *scenario*, a description of external actor actions and reactions of the system [29], that can be executed whenever needed to check the model behavior. Scenarios are written in the `Avalla` language and executed using the `AsmetaV` tool. Each scenario is identified by its name and must `load` the ASM to be tested. Then, the user may specify different commands depending on the operation to be performed. The `set` command updates monitored or shared function values that are supplied by the user as input signals to the system. Commands `step` and `step until` represent the reaction of the system, which can execute one single ASM step and one ASM step iteratively until a specified condition becomes true. Then, the `check` command is used to inspect property values in the current state of the underlying ASM. Figure 8 shows an example of `Avalla` scenario for the one-way traffic light case study. The scenario reproduces the first two steps of the cycle: when 50 s are over, the second traffic light changes from *Stop* to *Go*; and only when 120 s are passed, the two traffic lights show *Stop* signal.

To simulate scenarios, `AsmetaV` invokes the simulator. During the simulation, `AsmetaV` captures any check violation and, if none occurs, it finishes with

scenario scenario1
load oneWayTrafficLight.asm
 	
set passed(50) := true;
step
check phase = GO2STOP1;
check goLight(lightUnit2) = true;
check goLight(lightUnit1) = false;
 	
set passed(120) := false;
	step
	check phase = GO2STOP1;
	check goLight(lightUnit2) = true;
	check goLight(lightUnit1) = false;
	
	set passed(120) := true;
	step
	check phase = STOP2STOP1;
	check goLight(lightUnit2) = false;
	check goLight(lightUnit1) = false;

Fig. 8. Example of `Avalla` scenario for the one-way traffic light case study

</State 1 (controlled)> check succeeded: phase = GO2STOP1 check succeeded: stopLight(lightUnit2) = false check succeeded: goLight(lightUnit2) = true <UpdateSet − 1>	</State 1 (controlled)> check succeeded: phase = GO2STOP1 CHECK FAILED: stopLight(lightUnit2) = true at step 1 check succeeded: goLight(lightUnit2) = true <UpdateSet − 1>

Fig. 9. `AsmetaV` output of one-way traffic light

a "PASS" verdict ("FAIL" otherwise). Moreover, the tool collects information about the coverage of the `AsmetaL` model, in particular, it keeps track of all the rules that have been called and evaluated, and it lists them at the end. Figure 9 shows the output of the validator upon executing the scenario in Fig. 8: in the first column, all the functions assume the expected value, while in the second column a check is failed because the function had a different value.

The user can exploit modularization also during scenario building. Indeed, it is possible to define `blocks`, i.e., sequences of `set`, `step`, and `check`, that can be recalled using the `execblock` when writing other scenarios that foresee the same sequence of `Avalla` commands.

3.2.4 Model Reviewing

When writing a formal model, a developer could introduce some errors that are not related to a wrong specification of the requirements but are just due to carelessness, forgetfulness, or limited knowledge of the formal method. For example, a developer could use a wrong function name, or could forget to properly guard an update, and so on. An error that is commonly done in ASM development is due to its computational model, where all possible updates are applied in parallel: if a location is simultaneously updated to two different values, this is known as *inconsistent update* [26], and it is considered as an error in ASMs. Such kind of error occurs quite frequently (especially in complex models) because the developer does not properly guard all the updates. Other types of errors done using ASMs are *overspecifying* the model, i.e., adding model elements that are not needed, or writing rules that can never be triggered.

All these types of errors can be captured automatically by doing a static analysis of the model. This is the aim of the `AsmetaMA` tool [7], which performs *automatic* review of ASM models. The tool checks the presence of seven types of errors by using suitable *meta-properties* specified in CTL and verified using the model checker `AsmetaSMV` (see Sect. 3.2.5). Figure 10a shows the selection of the seven meta-properties in `AsmetaMA`. For example, MP1 checks the presence of inconsis-

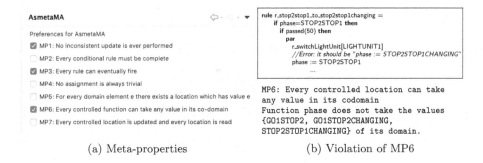

(a) Meta-properties   (b) Violation of MP6

Fig. 10. AsmetaMA

tent updates, and MP3 checks whether there are rules that can never be triggered. Figure 10b shows an example of a violation that can be found with the model review. It is an error that we discovered using AsmetaMA when writing the model of the traffic light; according to the requirements, when the **phase** is STOP2STOP1 and 50 time units are passed, the **phase** should become STOP2STOP1CHANGING in the next state; however, we wrongly typed the value as STOP2STOP1. Such error was discovered by MP6 that checks if there are possible values that are never assumed by a location: the violation of MP6 allowed us to reveal our mistake.

3.2.5 Model Checking

ASMETA provides classical model checking support by the tool AsmetaSMV [6]. The tool translates an ASM model into a model of the symbolic model checker NuSMV [30], which is used to perform the verification. Being NuSMV a finite state model checker, the only limitation of AsmetaSMV is on the finiteness of the number of ASM states: only finite domains can be used, and the *extend* rule (which adds elements to a domain) is not supported.

When using AsmetaSMV, the NuSMV tool is transparent to the user who can specify, directly in the ASM model, Computation Tree Logic (CTL) and Linear Temporal Logic (LTL) properties defined over the ASM signature. Moreover, also the output of the model checker is pretty-printed in terms of elements of the ASM signature. Figure 11a shows CTL and LTL properties specified for the traffic light case study.

The CTL property, for example, checks that if the second traffic light shows the stop light, it will show the go light in the future.

In order to better understand the verification results, the tool allows to simulate the returned counterexample. To this aim, a translator is provided that translates a counterexample into an Avalla scenario (see Sect. 3.2.3). Figure 11b shows the counterexample of the violation of the CTL property shown in Fig. 11a (in a faulty version of the ASM model); the corresponding Avalla scenario is reported in Fig. 11c.

AsmetaSMV has been used in several case studies to verify the functional correctness of the specified system. AsmetaSMV is also used as a back-end tool for other activities supported in ASMETA, e.g., model review (see Sect. 3.2.4).

```
CTLSPEC ag(stopLight(LIGHTUNIT2) implies ef(goLight(LIGHTUNIT2)))
LTLSPEC g(phase=STOP1STOP2 implies x(phase=GO2STOP1 or phase=STOP1STOP2))
```

(a) Specification of temporal properties in the **AsmetaL** model

```
-- specification AG (stopLight(LIGHTUNIT2) ->
                     EF goLight(LIGHTUNIT2)) is false
-- as demonstrated by the following execution sequence
Trace Description: CTL Counterexample
Trace Type: Counterexample
-> State: 7.1 <-
stopLight(LIGHTUNIT2) = true
goLight(LIGHTUNIT2) = false
rPulse(LIGHTUNIT2) = false
passed(50) = false
phase = STOP1STOP2
passed(120) = false
gPulse(LIGHTUNIT2) = false
-> State: 7.2 <-
passed(50) = true
-> State: 7.3 <-
rPulse(LIGHTUNIT2) = true
passed(50) = false
phase = STOP1STOP2CHANGING
gPulse(LIGHTUNIT2) = true
-> State: 7.4 <-
stopLight(LIGHTUNIT2) = false
goLight(LIGHTUNIT2) = true
rPulse(LIGHTUNIT2) = false
phase = GO2STOP1
passed(120) = true
gPulse(LIGHTUNIT2) = false
-> State: 7.5 <-
rPulse(LIGHTUNIT2) = true
phase = GO2STOP1CHANGING
passed(120) = false
gPulse(LIGHTUNIT2) = true
-> State: 7.6 <-
stopLight(LIGHTUNIT2) = true
goLight(LIGHTUNIT2) = false
rPulse(LIGHTUNIT2) = false
phase = STOP2STOP1
gPulse(LIGHTUNIT2) = false
```

```
scenario oneWayTrafficLight_refined.test
load oneWayTrafficLight_refined.asm

check stopLight(LIGHTUNIT2) = true;
check goLight(LIGHTUNIT2) = false;
check rPulse(LIGHTUNIT2) = false;
check phase = STOP1STOP2;
check gPulse(LIGHTUNIT2) = false;

set passed(50) := false; set passed(120) := false;
step

set passed(50) := true;
step
check rPulse(LIGHTUNIT2) = true;
check phase = STOP1STOP2CHANGING;
check gPulse(LIGHTUNIT2) = true;

set passed(50) := false;
step
check stopLight(LIGHTUNIT2) = false;
check goLight(LIGHTUNIT2) = true; check
rPulse(LIGHTUNIT2) = false;
check phase = GO2STOP1;
check gPulse(LIGHTUNIT2) = false;

set passed(120) = true;
step
check rPulse(LIGHTUNIT2) = true;
check phase = GO2STOP1CHANGING;
check gPulse(LIGHTUNIT2) = true;

set passed(120) := false;
step
check stopLight(LIGHTUNIT2) = true;
check goLight(LIGHTUNIT2) = false;
check rPulse(LIGHTUNIT2) = false;
check phase = STOP2STOP1;
check gPulse(LIGHTUNIT2) = false;
```

(b) Counterexample in **AsmetaSMV**   (c) Executable counterexample in **Avalla**

Fig. 11. AsmetaSMV

4   ASMETA@development time

Once the **AsmetaL** model is available, the user can automatically generate
abstract tests, C++ code, and C++ unit tests. Moreover, Behavior-Driven Development scenarios in C++ can be generated from **Avalla** scenarios.

4.1   Model-Based Test Generation

Model-based testing [46] is a popular testing approach in which formal models
are used for testing purposes, in particular test generation. Indeed, the model
is an abstract representation of the System Under Test (SUT), from which it is
possible to generate both the test inputs and the expected output (so, tackling
the *oracle problem* of software testing [18]). In offline test generation, *abstract
tests* are generated from the model, and then these are translated into *concrete
tests* for the SUT. Coverage criteria over the model are used to define the test
goals. A typical approach for generating tests achieving these goals is to use
model checkers [32]: a test goal is translated into a suitable temporal property

(called *trap property*), whose counterexample (if any) is the test covering the test goal.

In ASMETA, the `ATGT` tool [34] performs model-based test generation using both model checkers SPIN and NuSMV. The generation is guided by coverage criteria defined or adapted for ASMs [33], such as rule coverage, parallel rule coverage, MCDC, etc. For example, the *rule coverage* criterion requires that for every transition rule r_i there exists at least one state in a test in which r_i fires, and another state in a test in which r_i does not fire. The abstract tests generated with `ATGT` can be later translated into concrete test cases for the implementation, as described in Sect. 4.3.

4.2   Model-Based Code Generation

According to best practices of model-driven engineering, the implementation of a system should be obtained from its model through a systematic model-to-code transformation. Thanks to `Asm2C++`, given an `AsmetaL` model, the C++ code is automatically generated [24]. This is done through a series of steps: the `AsmetaL` model is parsed and its (internal) representation in terms of Java objects as an instance of the ASMETA metamodel (AsmM) is built; then, a *model-to-text* transformation, implemented in Xtext, is applied to translate the model into C++ code. The generated code is composed of two files: header (.h) and source (.cpp). The header file contains the interface of the source file and the translation of ASM domains declaration and definition, functions and rules declaration. The rules implementation, the functions/domains initialization, and the definitions of the functions are contained in the source file. The translation of the one way traffic light case study in C++ is shown in Fig. 12.

Since an ASM run step consists in the execution of the main rule and the update of the locations, in C++ the ASM step has been implemented by two methods: `mainRule()` and `fireUpdateSet()`. The former corresponds to the translation of the ASM main rule, while the latter updates the locations to the next state values. Moreover, we have addressed two semantic ASM concepts that do not have a direct implementation in C++: parallel execution and nondeterminism. More details on their implementation in C++ and the translation of ASM rules to corresponding C++ instructions can be found in [24].

Given the translation of an `AsmetaL` model in C++ code, it is easy to adapt the code generation process for a specific platform. We have chosen Arduino since it supports C++, it is cheap and it is easily accessible. After C++ code generation, three new steps are required: HW configuration and integration, ASM runner generation, and merging of all generated files. HW configuration contains the mapping between ASM functions and Arduino input/output, and other specific hardware settings. A first draft is automatically generated, and then the user links monitored and out functions to physical hardware pins. The ASM runner automatically generates a .ino file which contains the `loop()` function to run ASM on Arduino. The `loop()` function iteratively executes the following functions: `getInputs()`—reads the data from the input devices like sensors; `mainRule()`—contains the behavior described in the `AsmetaL`

```
#ifndef  ONEWAYTRAFFICLIGHT_H
#define  ONEWAYTRAFFICLIGHT_H

#include <set>
using namespace std;


/* DOMAIN DEFINITIONS */
namespace oneWayTrafficLightnamespace{
    class LightUnit;
  enum PhaseDomain {STOP1STOP2, GO2STOP1,
        STOP2STOP1, GO1STOP2};
}

using namespace
    oneWayTrafficLightnamespace;


class oneWayTrafficLightnamespace::
        LightUnit{
 public:
  static std::set<LightUnit*> elems;
  LightUnit(){elems.insert(this);}
};


class oneWayTrafficLight {



/* DOMAIN CONTAINERS */
const set<PhaseDomain> PhaseDomain_elems;

public:


  /* FUNCTIONS */
  PhaseDomain phase[2];
  std::map<LightUnit*, bool> stopLight[2];
  std::map<LightUnit*, bool> goLight[2];
  static int timer (PhaseDomain
        param0_timer);
  static LightUnit* lightUnit1;
  static LightUnit* lightUnit2;
  std::map<int, bool> passed;


  /* RULE DEFINITION */
  void r_switch (bool _l);
  void r_switchToGo2();
  void r_switchToStop2();
  void r_switchToGo1();
  void r_switchToStop1();
  void r_stop1stop2_to_go2stop1();
  void r_go2stop1_to_stop2stop1();
  void r_stop2stop1_to_go1stop2();
  void r_go1stop2_to_stop1stop2();
  void r_Main();


  oneWayTrafficLight();


  void initControlledWithMonitored();
  void getInputs();
  void setOutputs();
  void fireUpdateSet();

};

#endif
```

```
#include "oneWayTrafficLight.h"
using namespace oneWayTrafficLightnamespace;
/* Conversion of ASM rules in C++ methods */
void oneWayTrafficLight::r_switch (bool _l){
    _l = ! (_l);}
void oneWayTrafficLight::r_switchToGo2(){
    { r_switch (goLight[0][lightUnit2]);
      r_switch (stopLight[0][lightUnit2]);}}
void oneWayTrafficLight::r_switchToStop2(){
    { r_switch (goLight[0][lightUnit2]);
      r_switch (stopLight[0][lightUnit2]);}}
void oneWayTrafficLight::r_switchToGo1(){
    { r_switch (goLight[0][lightUnit1]);
      r_switch (stopLight[0][lightUnit1]);}}
void oneWayTrafficLight::r_switchToStop1(){
    { r_switch (goLight[0][lightUnit1]);
      r_switch (stopLight[0][lightUnit1]);}}
void oneWayTrafficLight::r_stop1stop2_to_go2stop1(){
    if ((phase[0] == STOP1STOP2)){
      if (passed[timer(STOP1STOP2)]){
        { r_switchToGo2();
          phase[1] = GO2STOP1;}}}}
void oneWayTrafficLight::r_go2stop1_to_stop2stop1(){
    if ((phase[0] == GO2STOP1)){
      if (passed[timer(GO2STOP1)]){
        { r_switchToStop2();
          phase[1] = STOP2STOP1;}}}}
void oneWayTrafficLight::r_stop2stop1_to_go1stop2(){
    if ((phase[0] == STOP2STOP1)){
      if (passed[timer(STOP2STOP1)]){
        { r_switchToGo1();
          phase[1] = GO1STOP2;}}}}
void oneWayTrafficLight::r_go1stop2_to_stop1stop2(){
    if ((phase[0] == GO1STOP2)){
      if (passed[timer(GO1STOP2)]){
        { r_switchToStop1();
          phase[1] = STOP1STOP2;}}}}
void oneWayTrafficLight::r_Main(){
    { r_stop1stop2_to_go2stop1();
      r_go2stop1_to_stop2stop1();
      r_stop2stop1_to_go1stop2();
      r_go1stop2_to_stop1stop2();}}
/* Static function definition */
int oneWayTrafficLight::timer(PhaseDomain _p){return [&](){
    if(_p==STOP1STOP2)
    return 50;
    else if(_p==GO2STOP1)
    return 120;
    else if(_p==STOP2STOP1)
    return 50;
    else if(_p==GO1STOP2)
    return 120; }();}
/* Function and domain initialization */
oneWayTrafficLight::oneWayTrafficLight(){
    //Static domain initialization
    PhaseDomain_elems:{STOP1STOP2,GO2STOP1,STOP2STOP1,GO1STOP2;};
    /* Init static functions Abstract domain */
    lightUnit1 = new LightUnit;
    lightUnit2 = new LightUnit;
    /* Function initialization */
    for(const auto& _l : LightUnit::elems){
        stopLight[0].insert({_l,true});
        stopLight[1].insert({_l,true}); }
    for(const auto& _l : LightUnit::elems){
        goLight[0].insert({_l,false});
        goLight[1].insert({_l,false}); }
        phase[0] = phase[1] = STOP1STOP2;}
void oneWayTrafficLight::initControlledWithMonitored(){}
/* Apply the update set */
void oneWayTrafficLight::fireUpdateSet(){}
/* init static functions and elements of abstract domains */
std::set< LightUnit*>LightUnit::elems;
LightUnit*oneWayTrafficLight::lightUnit1;
LightUnit*oneWayTrafficLight::lightUnit2;
```

Fig. 12. oneWayTrafficLight.h and oneWayTrafficLight.cpp

model; `fireUpdateSet()`—updates the state at the end of each loop; and `setOutputs()`—sets the output values like the current state of light-emitting diode (LED). The merging step takes care of merging all files.

```
BOOST_AUTO_TEST_SUITE( TestoneWayTrafficLight )
BOOST_AUTO_TEST_CASE( my_test_0 ){
  // instance of the SUT
  oneWayTrafficLight onewaytrafficlight;
  // state
  // set monitored variables
  onewaytrafficlight.passed[50]=false;
  ...
  BOOST_CHECK( onewaytrafficlight.phase[0]==STOP1STOP2);
  // call main rule
  onewaytrafficlight.r_Main();
  onewaytrafficlight.fireUpdateSet();
  ...
}
...
```

Fig. 13. C++ unit test

4.3 Unit Test Generation

If the C++ code is available (automatically generated or not) and the user wants to test it, C++ unit tests can be automatically generated given the AsmetaL model [23]. Unit tests are generated in two different ways. The first approach consists in running randomly the AsmetaS simulator for a given number of steps as requested by the tester, then the generated state sequence is translated into a C++ unit test. The second approach, instead, translates the abstract tests generated with ATGT (see Sect. 4.1) in C++ unit tests. In both cases, the C++ unit tests are written using the Boost Test C++ library.

A test suite is defined by using the BOOST_AUTO_TEST_SUITE(testSuiteName) macro; it automatically registers a test suite named testSuiteName. A test suite definition is ended using BOOST_AUTO_TEST_END(). Each test suite can contain one or more test cases. A test case is declared using the macro BOOST_AUTO_-TEST_CASE(testCaseName). An example of a test case in presented in Fig. 13.

4.4 Behavior-Driven Development Scenarios

In parallel to classical unit tests which focus more on checking internal functionalities of classes, developers and testers employ also Behavior-Driven Development (BDD) tests which should be examples that anyone from the development team can read and understand. Since the use of scenarios is common at code-level and at the level of the (abstract) model, and since there is a translator that automatically generates C++ code from AsmetaL model, we have introduced the AsmetaBDD tool which translates an abstract scenario written in the Avalla language to BDD code using the Catch2 framework [24]. The AsmetaBDD tool generates a C++ scenario that can be compiled together with the C++ code and executed. An example is shown in Fig. 14, where both scenarios check the correctness of the phase transition when 50 s are passed.

```
scenario scenario1

load oneWayTrafficLight.asm

check phase = STOP1STOP2;

set passed(50) := true;

step

check phase = GO2STOP1;
...
```

```
#include "catch.hpp"
#include "oneWayTrafficLight.hpp"
SCENARIO("oneWayTrafficLight starts") {
  GIVEN("The traffic lights are stopped") {
  oneWayTrafficLight trafficLight;
  REQUIRE(trafficLight.phase == STOP1STOP2);
  WHEN("passed 50 sec") {
    trafficLight.passed(50);
    THEN( "the traffic light is changing state " ) {
    REQUIRE(trafficLight.phase == GO2STOP1);}}
  ....
}}
```

(a) `Avalla` scenario   (b) BDD scenario using Catch2

Fig. 14. `AsmetaBDD` scenario example

5 ASMETA@operation time

Formal validation and verification techniques usually allow the identification and resolution of problems at design time. However, the state space of a system under specification is often too large or partially unknown at design time, such as for CPSs with uncertain behavior of humans in the loop and/or endowed with self-adaptation capabilities or AI-based components. This makes a complete assurance impractical or even impossible to pursue completely at design time. Runtime assurance methods take advantage of the fact that variables that are free at design time are bound at runtime; so, instead of verifying the complete state space, runtime assurance techniques may concentrate on checking the current state of a system.

Currently, ASMETA supports two types of runtime analysis techniques: runtime simulation described in Sect. 5.1, and runtime monitoring described in Sect. 5.2. Both approaches view the model as a *twin* of the real system and use the model as *oracle* of the correct system behavior. The former exploits the twin execution to prevent misbehavior of the system in case of unsafe model behavior, while the latter exploits the twin execution to check the correctness of the system behavior w.r.t. the model behavior.

5.1 Runtime Simulation

Recently, a runtime simulation platform [44] has been developed within ASMETA to check safety assertions of software systems at runtime and support on-the-fly changes of these assertions. The platform exploits the concept of executable ASM models and it is based on the `AsmetaS@run.time` simulator to handle an ASM model as a living/runtime model [47] and execute it in tandem with a prototype/real system. To this purpose, the runtime simulation platform operates between the system model and the real running system; it traces the state of the ASM model and of the system allowing us to realize a conceivable causal relation depending on the analysis scope and on low-level implementation details. This runtime simulation mechanism, for example, could be used in conjunction with an *enforcer* component tool to concretely sanitize/filter out input events for the running system or to prevent the execution of unsafe commands by the system – *input/output sanitization* [31].

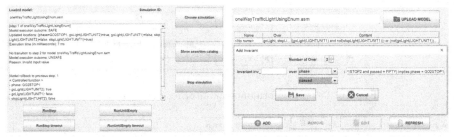

(a) Runtime simulation dashboard `SimGUI`   (b) Assertion catalog GUI

Fig. 15. `AsmetaS@run.time`

`AsmetaS@run.time` supports simulation *as-a-service* features of the `AsmetaS` simulator and additional features such as model execution with timeout and model roll-back to the previous safe state after a failure occurrence (e.g., invariant violations, inconsistent updates, ill-formed inputs, etc.) during model execution. `AsmetaS@run.time` allows also the dynamic adaptation of a running ASM model to add/change/delete invariants representing, for example, system safety assertions. This mechanism could be exploited to dynamically add new assertions and guarantee a safer execution of the system after its release, in case dangerous situations have not been foreseen at design time or because of unanticipated changes or situational awareness.

The runtime simulation platform includes also UI dashboards for dynamic *Human-Model-Interaction* (both in a graphical and in a command-line way) which allow the user to track the model execution and change safety assertions. Figure 15a shows the ASM model of the one-way traffic light model through the graphical dashboard `SimGUI`.

In particular, the central panel shows the ASM runs and the simulation results. The last one produced the verdict UNSAFE due to an invalid input value read by the ASM for the enumerative monitored function `passed`. Then, the model is rolled back to its previous safe state. The running ASM model can be adapted dynamically to incorporate new safety invariants or simply modify or cancel existing ones. This can be requested by an external client program or done manually by the user through the GUI `Assertion Catalog` to the simulator engine (see Fig. 15b). Model adaptation is carried out when the model is in a quiescent state, i.e., it is not currently in execution and no other adaptation activity of it is going on. Once adapted, the ASM model execution continues from its current state. A newly added safety invariant that would be immediately violated in the current state of the ASM model is forbidden.

5.2 Runtime Monitoring

ASMETA allows to perform runtime monitoring of a Java program using the tool `CoMA` (*Conformance Monitoring through ASM*) [8]. The approach is shown in Fig. 16 and described as follows:

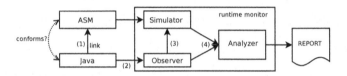

Fig. 16. CoMA: Conformance monitoring through ASM

```
import org.asmeta.monitoring.*;

@Asm(asmFile = "oneWayTrafficLight.asm")
public class OWTL {
  @FieldToLocation(func = "stopLight",
            args={"LIGHTUNIT1"})
  boolean redLight1;
  @FieldToLocation(func = "stopLight",
            args={"LIGHTUNIT2"})
  boolean redLight2;
  @FieldToLocation(func = "goLight",
            args={"LIGHTUNIT1"})
  boolean greenLight1;
  @FieldToLocation(func = "goLight",
            args={"LIGHTUNIT2"})
  boolean greenLight2;
  private boolean turn1;

  @StartMonitoring
  public OWTL() {
    redLight1 = true; redLight2 = true;
    greenLight1 = false; greenLight2 = false;
    turn1 = false;
  }
}
```

```
@RunStep
public void
updateLights(@Param(func = "passed") Time passedTime) {
  if((passedTime == Time.FIFTY
      && redLight1 && redLight2) ||
     (passedTime == Time.ONEHUNDREDTWENTY
      && greenLight1 != greenLight2)) {
    if(turn1) {
      greenLight1 = !greenLight1;
      redLight1 = !redLight1;
    }
    else {
      greenLight2 = !greenLight2;
      redLight2 = !redLight2;
    }
    if (redLight1 && redLight2) {
      turn1 = !turn1;
    }
  }
}

enum Time {FIFTY, ONEHUNDREDTWENTY, LESS;}
```

Fig. 17. CoMA – Java implementation of the one-way traffic light

- The Java program under monitoring and the ASM model are *link*ed by means of a set of Java annotations[3] (step ①). Some annotations are used to link the Java state with the ASM state; namely, they link class fields of the Java program with functions of the ASM model. Other annotations, instead, specify the methods of the Java program that produce state changes that must be monitored; Fig. 17 shows the Java implementation for the running case study, annotated for the linking with the ASM model shown in Fig. 3;
- the *observer* (step ②) monitors the Java program execution and, whenever a method under monitoring is executed, it performs a simulation step of the ASM model with the simulator (step ③);
- the *analyzer* (step ④) checks whether the Java state after the method execution is conformant with the ASM state after the simulation step. Details on the conformance definition can be found in [8].

CoMA can also check the conformance of nondeterministic systems in which multiple states can be obtained by executing a method under monitoring; namely, the tool checks whether there exists a *next* ASM state that is conformant with the obtained Java state. There are two implementations of this approach: by explicitly listing all the possible next ASM states [9], or by using a symbolic representation with an SMT solver [10].

---

[3] A Java annotation is a meta-data tag that permits to add information to code elements (class declarations, method declarations, etc.). Annotations are defined similarly as classes.

6   Conclusion and Outlook

This article provided an overview of the ASMETA model-based analysis approach and the associated tooling to the safety assurance problem of software systems using ASMs as underlying analysis formalism. ASMETA allows an open and evolutionary approach to safety assurance as depicted in Fig. 1.

ASMETA is an active open-source academic project. Over the years, it has been improved with new techniques and tools to face the upcoming new challenging aspects of modern systems. It has also been used as a back-end for system analysis of domain-specific front-end notations (as those for service-oriented and self-adaptive systems).

Recently, ASMETA has been extended to deal with model time features, and improvement to support the verification of *quantitative* system properties by means of probabilistic model checking is under development. Application domains under current investigations are those of IoT security, autonomous and evolutionary systems, cyber-physical systems, and medical software certification.

References

1. Al-Shareefi, F.: Analysing Safety-Critical Systems and Security Protocols with Abstract State Machines. Ph.D. thesis, University of Liverpool (2019)
2. Arcaini, P., Bonfanti, S., Gargantini, A., Mashkoor, A., Riccobene, E.: Formal validation and verification of a medical software critical component. In: 2015 ACM/IEEE International Conference on Formal Methods and Models for Codesign (MEMOCODE), pp. 80–89. IEEE, September 2015. https://doi.org/10.1109/MEMCOD.2015.7340473
3. Arcaini, P., Bonfanti, S., Gargantini, A., Mashkoor, A., Riccobene, E.: Integrating formal methods into medical software development: the ASM approach. Sci. Comput. Program. **158**, 148–167 (2018). https://doi.org/10.1016/j.scico.2017.07.003
4. Arcaini, P., Bonfanti, S., Gargantini, A., Riccobene, E.: Visual notation and patterns for abstract state machines. In: Milazzo, P., Varró, D., Wimmer, M. (eds.) Software Technologies: Applications and Foundations, pp. 163–178. Springer International Publishing, Cham (2016)
5. Arcaini, P., Bonfanti, S., Gargantini, A., Riccobene, E., Scandurra, P.: Modelling an automotive software-intensive system with adaptive features using ASMETA. In: Raschke, A., Méry, D., Houdek, F. (eds.) ABZ 2020. LNCS, vol. 12071, pp. 302–317. Springer, Cham (2020). https://doi.org/10.1007/978-3-030-48077-6_25
6. Arcaini, P., Gargantini, A., Riccobene, E.: AsmetaSMV: a way to link high-level ASM models to low-level NuSMV specifications. In: Frappier, M., Glässer, U., Khurshid, S., Laleau, R., Reeves, S. (eds.) ABZ 2010. LNCS, vol. 5977, pp. 61–74. Springer, Heidelberg (2010). https://doi.org/10.1007/978-3-642-11811-1_6
7. Arcaini, P., Gargantini, A., Riccobene, E.: Automatic review of Abstract State Machines by meta property verification. In: Muñoz, C. (ed.) Proceedings of the Second NASA Formal Methods Symposium (NFM 2010), NASA/CP-2010-216215, pp. 4–13. NASA, Langley Research Center, Hampton VA 23681–2199, USA, April 2010

8. Arcaini, P., Gargantini, A., Riccobene, E.: CoMA: conformance monitoring of Java programs by Abstract State Machines. In: Khurshid, S., Sen, K. (eds.) RV 2011. LNCS, vol. 7186, pp. 223–238. Springer, Heidelberg (2012). https://doi.org/10.1007/978-3-642-29860-8_17

9. Arcaini, P., Gargantini, A., Riccobene, E.: Combining model-based testing and runtime monitoring for program testing in the presence of nondeterminism. In: Proceedings of the 2013 IEEE Sixth International Conference on Software Testing, Verification and Validation Workshops, pp. 178–187. ICSTW 2013, IEEE Computer Society, Washington, DC, USA (2013). https://doi.org/10.1109/ICSTW.2013.29

10. Arcaini, P., Gargantini, A., Riccobene, E.: Using SMT for dealing with nondeterminism in ASM-based runtime verification. ECEASST **70**, 1–15 (2014). https://doi.org/10.14279/tuj.eceasst.70.970

11. Arcaini, P., Gargantini, A., Riccobene, E.: SMT-based automatic proof of ASM model refinement. In: De Nicola, R., Kühn, E. (eds.) SEFM 2016. LNCS, vol. 9763, pp. 253–269. Springer, Cham (2016). https://doi.org/10.1007/978-3-319-41591-8_17

12. Arcaini, P., Gargantini, A., Riccobene, E.: Rigorous development process of a safety-critical system: from ASM models to Java code. Int. J. Softw. Tools Technol. Transfer **19**(2), 247–269 (2015). https://doi.org/10.1007/s10009-015-0394-x

13. Arcaini, P., Gargantini, A., Riccobene, E., Scandurra, P.: A model-driven process for engineering a toolset for a formal method. Softw. Pract. Experience **41**, 155–166 (2011). https://doi.org/10.1002/spe.1019

14. Arcaini, P., Holom, R.-M., Riccobene, E.: ASM-based formal design of an adaptivity component for a Cloud system. Formal Aspects Comput. **28**(4), 567–595 (2016). https://doi.org/10.1007/s00165-016-0371-5

15. Arcaini, P., Mirandola, R., Riccobene, E., Scandurra, P.: MSL: a pattern language for engineering self-adaptive systems. J. Syst. Softw. **164**, 110558 (2020). https://doi.org/10.1016/j.jss.2020.110558

16. Arcaini, P., Riccobene, E., Scandurra, P.: Formal design and verification of self-adaptive systems with decentralized control. ACM Trans. Auton. Adapt. Syst. **11**(4), 25:1-25:35 (2017). https://doi.org/10.1145/3019598

17. ASMETA (ASM mETAmodeling) toolset. https://asmeta.github.io/

18. Barr, E.T., Harman, M., McMinn, P., Shahbaz, M., Yoo, S.: The oracle problem in software testing: a survey. IEEE Trans. Softw. Eng. **41**(5), 507–525 (2015). https://doi.org/10.1109/TSE.2014.2372785

19. Benduhn, F., Thüm, T., Schaefer, I., Saake, G.: Modularization of refinement steps for agile formal methods. In: Duan, Z., Ong, L. (eds.) ICFEM 2017. LNCS, vol. 10610, pp. 19–35. Springer, Cham (2017). https://doi.org/10.1007/978-3-319-68690-5_2

20. Bombarda, A., Bonfanti, S., Gargantini, A.: Developing medical devices from abstract state machines to embedded systems: a smart pill box case study. In: Mazzara, M., Bruel, J.-M., Meyer, B., Petrenko, A. (eds.) TOOLS 2019. LNCS, vol. 11771, pp. 89–103. Springer, Cham (2019). https://doi.org/10.1007/978-3-030-29852-4_7

21. Bombarda, A., Bonfanti, S., Gargantini, A., Radavelli, M., Duan, F., Lei, Yu.: Combining model refinement and test generation for conformance testing of the IEEE PHD protocol using abstract state machines. In: Gaston, C., Kosmatov, N., Le Gall, P. (eds.) ICTSS 2019. LNCS, vol. 11812, pp. 67–85. Springer, Cham (2019). https://doi.org/10.1007/978-3-030-31280-0_5

22. Bonfanti, S., Gargantini, A., Mashkoor, A.: AsmetaA: animator for Abstract State Machines. In: Butler, M., Raschke, A., Hoang, T.S., Reichl, K. (eds.) ABZ 2018. LNCS, vol. 10817, pp. 369–373. Springer, Cham (2018). https://doi.org/10.1007/978-3-319-91271-4_25

23. Bonfanti, S., Gargantini, A., Mashkoor, A.: Generation of C++ unit tests from Abstract State Machines specifications. In: 2018 IEEE International Conference on Software Testing, Verification and Validation Workshops (ICSTW), pp. 185–193, April 2018. https://doi.org/10.1109/ICSTW.2018.00049

24. Bonfanti, S., Gargantini, A., Mashkoor, A.: Design and validation of a C++ code generator from Abstract State Machines specifications. J. Softw. Evol. Process 32(2), e2205 (2020). https://doi.org/10.1002/smr.2205

25. Börger, E.: The ASM refinement method. Formal Aspects Comput. 15, 237–257 (2003)

26. Börger, E., Raschke, A.: Modeling Companion for Software Practitioners. Springer, Heidelberg (2018). https://doi.org/10.1007/978-3-662-56641-1

27. Börger, E., Stärk, R.: Abstract State Machines: A Method for High-Level System Design and Analysis. Springer Verlag, Heidelberg (2003). https://doi.org/10.1007/978-3-642-18216-7

28. Calinescu, R., Weyns, D., Gerasimou, S., Iftikhar, M.U., Habli, I., Kelly, T.: Engineering trustworthy self-adaptive software with dynamic assurance cases. IEEE Trans. Softw. Eng. 44(11), 1039–1069 (2018). https://doi.org/10.1109/TSE.2017.2738640

29. Carioni, A., Gargantini, A., Riccobene, E., Scandurra, P.: A scenario-based validation language for ASMs. In: Börger, E., Butler, M., Bowen, J.P., Boca, P. (eds.) ABZ 2008. LNCS, vol. 5238, pp. 71–84. Springer, Heidelberg (2008). https://doi.org/10.1007/978-3-540-87603-8_7

30. Cimatti, A., et al.: NuSMV 2: an opensource tool for symbolic model checking. In: Brinksma, E., Larsen, K.G. (eds.) CAV 2002. LNCS, vol. 2404, pp. 359–364. Springer, Heidelberg (2002). https://doi.org/10.1007/3-540-45657-0_29

31. Falcone, Y., Mariani, L., Rollet, A., Saha, S.: Runtime failure prevention and reaction. In: Bartocci, E., Falcone, Y. (eds.) Lectures on Runtime Verification. LNCS, vol. 10457, pp. 103–134. Springer, Cham (2018). https://doi.org/10.1007/978-3-319-75632-5_4

32. Fraser, G., Wotawa, F., Ammann, P.E.: Testing with model checkers: a survey. Softw. Test. Verif. Reliab. 19(3), 215–261 (2009)

33. Gargantini, A., Riccobene, E.: ASM-based testing: coverage criteria and automatic test sequence. J. Univers. Comput. Sci. 7(11), 1050–1067 (2001). https://doi.org/10.3217/jucs-007-11-1050

34. Gargantini, A., Riccobene, E., Rinzivillo, S.: Using spin to generate tests from ASM specifications. In: Börger, E., Gargantini, A., Riccobene, E. (eds.) ASM 2003. LNCS, vol. 2589, pp. 263–277. Springer, Heidelberg (2003). https://doi.org/10.1007/3-540-36498-6_15

35. Gargantini, A., Riccobene, E., Scandurra, P.: A semantic framework for metamodel-based languages. Autom. Softw. Eng. 16(3–4), 415–454 (2009). https://doi.org/10.1007/s10515-009-0053-0

36. Gargantini, A., Riccobene, E., Scandurra, P.: Ten reasons to metamodel ASMs. In: Abrial, J.-R., Glässer, U. (eds.) Rigorous Methods for Software Construction and Analysis. LNCS, vol. 5115, pp. 33–49. Springer, Heidelberg (2009). https://doi.org/10.1007/978-3-642-11447-2_3

37. Gaspari, P., Riccobene, E., Gargantini, A.: A formal design of the Hybrid European Rail Traffic Management System. In: Proceedings of the 13th European Conference on Software Architecture - Volume 2. pp. 156–162. ECSA 2019, Association for Computing Machinery, New York, NY, USA (2019). https://doi.org/10.1145/3344948.3344993
38. Gurevich, Y.: Evolving Algebras 1993: Lipari Guide, pp. 9–36. Oxford University Press Inc., USA (1995)
39. Harel, D., Rumpe, B.: Meaningful modeling: What's the semantics of "Semantics"? Computer 37(10), 64–72 (2004). https://doi.org/10.1109/MC.2004.172
40. Leveson, N.: Are you sure your software will not kill anyone? Commun. ACM 63(2), 25–28 (2020). https://doi.org/10.1145/3376127
41. Lutz, R.R.: Software engineering for safety: a roadmap. In: Proceedings of the Conference on The Future of Software Engineering, pp. 213–226. ICSE 2000, Association for Computing Machinery, New York, NY, USA (2000). https://doi.org/10.1145/336512.336556
42. Mirandola, R., Potena, P., Riccobene, E., Scandurra, P.: A reliability model for service component architectures. J. Syst. Softw. 89, 109–127 (2014). https://doi.org/10.1016/j.jss.2013.11.002
43. Riccobene, E., Scandurra, P.: A formal framework for service modeling and prototyping. Formal Aspects Comput. 26(6), 1077–1113 (2013). https://doi.org/10.1007/s00165-013-0289-0
44. Riccobene, E., Scandurra, P.: Model-based simulation at runtime with Abstract State Machines. In: Muccini, H., et al. (eds.) Software Architecture, pp. 395–410. Springer International Publishing, Cham (2020)
45. Schmidt, D.C.: Guest editor's introduction: model-driven engineering. IEEE Comput. 39(2), 25–31 (2006). https://doi.org/10.1109/MC.2006.58
46. Utting, M., Legeard, B., Bouquet, F., Fourneret, E., Peureux, F., Vernotte, A.: Chapter two - recent advances in model-based testing. Advances in Computers, vol. 101, pp. 53–120. Elsevier (2016). https://doi.org/10.1016/bs.adcom.2015.11.004
47. Van Tendeloo, Y., Van Mierlo, S., Vangheluwe, H.: A multi-paradigm modelling approach to live modelling. Softw. Syst. Model. 18(5), 2821–2842 (2018). https://doi.org/10.1007/s10270-018-0700-7
48. Vessio, G.: Reasoning about properties with Abstract State Machines. In: Gogolla, M., Muccini, H., Varró, D. (eds.) Proceedings of the Doctoral Symposium at Software Technologies: Applications and Foundations 2015 Conference (STAF 2015), L'Aquila, Italy, 20 July 2015. CEUR Workshop Proceedings, vol. 1499, pp. 1–10. CEUR-WS.org (2015). http://ceur-ws.org/Vol-1499/paper1.pdf
49. Weyns, D., et al.: Perpetual assurances for self-adaptive systems. In: de Lemos, R., Garlan, D., Ghezzi, C., Giese, H. (eds.) Software Engineering for Self-Adaptive Systems III. Assurances. LNCS, vol. 9640, pp. 31–63. Springer, Cham (2017). https://doi.org/10.1007/978-3-319-74183-3_2

Flashix: Modular Verification of a Concurrent and Crash-Safe Flash File System

Stefan Bodenmüller, Gerhard Schellhorn$^{(\boxtimes)}$, Martin Bitterlich, and Wolfgang Reif

Institute for Software and Systems Engineering, University of Augsburg, Augsburg, Germany
{stefan.bodenmueller,schellhorn,martin.bitterlich, reif}@informatik.uni-augsburg.de

Abstract. The Flashix project has developed the first realistic verified file system for Flash memory. This paper gives an overview over the project and the theory used. Specification is based on modular components and subcomponents, which may have concurrent implementations connected via refinement. Functional correctness and crash-safety of each component is verified separately. We highlight some components that were recently added to improve efficiency, such as file caches and concurrent garbage collection. The project generates 18K of C code that runs under Linux. We evaluate how efficiency has improved and compare to UBIFS, the most recent flash file system implementation available for the Linux kernel.

1 Introduction

Modular software development based on refinement has always been a core concern of our Formal Methods group.

One of the constant positive and inspiring influences on our work has always been Prof. Börger's research on the formalism of Abstract State Machines (ASMs) [7].

The earliest starting point of this has been the Prolog Compiler specified in [6] that describes compilation to the Warren Abstract Machine as stepwise ASM refinement. Mechanized verification of this refinement tower was posed by him as a challenge in a DFG priority program. Our solution to this case study led to a formalization of the ASM refinement theory [5] proposed there [33] and later on to a completeness proof [34]. Using this theory we managed to do a mechanized verification of the compiler [35] using our theorem prover KIV [15]. The work led to the PhD of one of the authors of this paper [32], with Prof. Börger being one of the reviewers.

Partly supported by the Deutsche Forschungsgemeinschaft (DFG), "Verifikation von Flash-Dateisystemen" (grants RE828/13-1 and RE828/13-2).

It was also Prof. Börger who pointed us to the Mondex challenge [10,41], which consists of mechanizing the proofs of a security protocol for Mondex electronic purses. Among other groups [25] we verified the original protocol [21], but also proposed an improvement that would avoid a weakness. We extended the case study to the development of a suitable cryptographic protocol and to the verification of a Java implementation [19]. The Java calculus [40] we used in KIV was influenced by the semantics of Prof. Börger's book [39]. The work also influenced our work on the development of a systematic development of security protocols using UML specifications [20].

Since then, we have tackled our most ambitious case study: Development of a fully verified, realistic file system for flash memory, called Flashix.

In a first phase of the project we had to develop the necessary theory that allowed to manage the complexity of such an undertaking. A concept of components and subcomponents was developed that are connected by refinement. This allowed modular software development as well as modular verification of proof obligations for each component. Together the proofs of this refinement tower guarantee functional correctness as well as crash-safety of the resulting implementation. An overview was given in [17]. The generated code from this first phase is a sequential implementation that can be run in Linux.

This paper gives an overview of the second phase of the project, where we tackled aspects crucial for efficiency: we enhanced the theory with a concept that allows to add concurrency incrementally to a refinement tower. We now also allow caches for files, which lead to a new crash-safety criterion called *write-prefix crash consistency*. We summarize the concepts and the theory in Sect. 2 and give an overview over the structure of Flashix in Sect. 3.

We then highlight two of the new features of Flashix. File content is now cached as described in Sect. 4, and write-prefix crash consistency has been proved [4]. Like wear leveling (described in [36]) garbage collection is now concurrent (Sect. 5).

The specifications of the Flashix file system implementation uses a language of abstract programs similar to the rules of Turbo-ASMs [7], although the concept for concurrency is based on interleaving [37]. We generate Scala- as well as C-Code from such abstract programs. Currently the generated C-Code has 18k loc, which can be used in Linux either with the Fuse library [42] or integrated in the kernel.

Section 6 evaluates the performance of our implementation. Since the concepts of UBIFS (the newest implementation of a file system in the Linux kernel) served as a blue-print for the concepts we used and verified in Flashix, we also compare efficiency to UBIFS.

Finally Sect. 7 gives related work, and Sect. 8 concludes the paper.

2   Methodology

This section gives an informal summary of the methodology. It consists of three core concepts that together establish that the top-level specification of POSIX is correctly implemented having crash-safe and linearizable behavior. The three concepts detailed in the following subsections are

- State-based components with specifications of operations. These are refined to implementations which are components, too. The implementations may call operations of other *subcomponent* specifications.
- Refinement from specifications comes in two flavors: (a form of) *data refinement* that allows to exchange abstract data structures (e.g. a set) with more efficient, concrete ones (e.g. a red-black tree) and *atomicity refinement*, which replaces an atomic operation with the non-atomic steps of a program, which allows concurrent execution.
- A concept for verifying crash-safety.

2.1 Components

A component is similar to an ASM. We distinguish between specification and implementation components, although they are specified using the same syntax.

A component specifies a number of state variables that store values of data types like numbers, lists, sets, arrays, maps, or tuples. Data types themselves are axiomatized using simply-typed lambda calculus. Most axioms use many-sorted first-order logic only, but there are exceptions like infinite sequences which use function variables (which represent dynamic functions as used in ASMs).

The operations of a component are given by a precondition, inputs and outputs, together with an imperative program that modifies the state. Programs contain assignments (function updates are possible), conditionals, loops and recursion. Using non-deterministic **choose** is allowed in specifications. Thereby an arbitrary postcondition can be established, simply by choosing a state that satisfies the predicate. Implementations allow specific versions only for which executable code can be generated. Two common cases are the allocation of a new reference for a heap-based data structure and choosing an element from a set.

There are two distinguished operations: Initialization, which computes initial states, and recovery, which is called after a crash when remounting to re-initialize the state.

Specification components are used to describe parts of the file system in an abstract and simple way, mainly by specifying functionality algebraically. Implementation components, on the other hand, implement functionality programmatically using low-level data structures.

For example, in Flashix we use a specification component to access a set of ordered elements. The component provides interface operations to add or delete an element. Another operation returns the greatest element below some given threshold. The precondition of this operation requires the set to be non-empty. The programs for these operations typically consist of a single assignment as the functionality is axiomatized over algebraic sets.

The corresponding implementation component gives an efficient realization of the interface using a red-black tree defined as a heap-based pointer structure. The separation into specification and implementation components allows to generate high-performance code from implementations while client components do not have to deal with their complexity but can rely on their abstract specification instead.

The semantics of a specification component is always that of a data type as in data refinement: it is a set of traces (also called histories). A trace is a sequence of matching pairs $[inv(in_1, op_1), ret(op_1, out_1), \ldots, inv(in_n, op_n), ret(op_n, out_n)]$ of *invoke* events $inv(in_i, op_i)$ and *return* events $ret(op_i, out_i)$. The first corresponds to invoking the operation op_i with input in_i, the second to the call returning with output out_i. Such a trace corresponds to a client sequentially calling operations $op_1, \ldots, op_n$ that execute atomically. Note that we immediately use a pair instead of a single call event to simplify the description of concurrency and crash-safety. The trace is observable (i.e. an element of the semantics) if there is a suitable sequence of states $[s_0, \ldots, s_n]$ (the run of the ASM) which is hidden from the client. State s_0 must be initial, and if calling operation op_i in state s_{i-1} with input in_i has a valid precondition, then it must have an execution that terminates and leads to state s_i with output out_i. Since the client is responsible for calling an operation only, when its precondition is satisfied, observations after calling an operation with violated precondition are arbitrary: the called operation should behave chaotically, yielding an arbitrary (even an illegal) successor state or none at all by non-termination.

Implementation components differ from specification components in two aspects. First, they may call operations from one or more subcomponents. Second, their semantics can be either *atomic* or *interleaved*. In the first case, its semantics is the same as the one of a specification component. In the latter case the semantics of the implementation are interleaved runs of the programs. The semantics then is similar to a control-state ASM, where the control-state is encoded implicitly in the program structure. To accommodate the fact, that we can have an arbitrary number of threads (or agents in ASMs), the events in traces are now generalized (as in [22]) to consist of matching pairs of $inv_t(in_i, op_i)$ and $ret_t(op_i, out_i)$ events that are indexed with a thread t. Matching pairs of different threads may now overlap in a history, corresponding to interleaved runs of the programs. The atomic steps of executing an operation are now: first, the invoking step (where the invocation event is observed), then the execution of individual instructions of the program like assignments, conditionals, and subcomponent calls, and finally a returning step (if the program terminates). The execution of operations of several threads are now interleaved. Formally, a concurrent history is legal if the projection to events of each thread t gives a sequence of matching pairs, possibly ending with an invoke event of a still running (pending) operation.

2.2 Refinement

Refinement of a specification component to an implementation component with atomic semantics is done using the contract approach to data refinement (see [13] for proof obligations in Z and the history of the approach) adapted to our operational specification of contracts, with proof obligations in wp-calculus, as detailed in [17].

All our refinements in the first phase of the Flashix project were such sequential refinements. Adding concurrency by replacing an atomic implementation with a thread-safe implementation with interleaved runs typically requires to add locks and ownership concepts, together with information which lock protects which data structure. Details of this process of 'shifting the refinement tower' are described in [36].

In the second phase of the Flashix project we have now added concurrency in all three places, where this is useful: The implementation of the top-level POSIX specification is now thread-safe, as mentioned in Sect. 3. Wear leveling and garbage collection (see Sect. 5) are now executed concurrent to regular POSIX operations called by the user of the file system.

Proving an implementation with interleaved semantics to be correct can be done in one step, proving linearizability [22] (progress aspects such as termination and deadlock-freedom must additionally be proved).

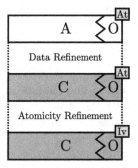

Fig. 1. Refinement to a concurrent implementation.

Our approach uses two steps, using the implementation with atomic semantics (denoted by $\boxed{\text{At}}$) as an intermediate level, see Fig. 1. This allows to have an upper refinement that (ideally) is the same as before and can be reused. In practice it is often necessary to add auxiliary operations that acquire/release ownership to specifications, as indicated by the **O** in the figure. These additional operations do not generate code, but they ensure that the client of the specification (the machine that uses this machine as a submachine) will not use arbitrary concurrent calls, but only ones that adhere to a certain discipline, as detailed in [36]. This leads to additional proof obligations for the refinement as well as for the client, which must call acquire/release to signal adherence to the discipline.

The lower *atomicity refinement* shows that the interleaved implementation (denoted by $\boxed{\text{Iv}}$) behaves exactly as if the whole code of the implementation would be executed atomically at some point in between the invoking step and the returning step. This point is usually called the linearization point. Correct atomicity refinement (and linearizability in general) can be expressed as reordering the events of the concurrent history H to a sequential history S (i.e. a sequence of matching pairs that do not interleave) that is correct in terms of the atomic semantics. The (total) order of matching pairs in the sequential history is determined from the order of linearization points. It preserves the partial order (called the *real-time order*) of operations in the concurrent history. If an operation is pending in the concurrent history, the corresponding sequential history may be *completed* in two ways: either the invoke event can be dropped, when the operation has not linearized, or a matching return can be added, when it has.

The proof method for proving atomicity refinement uses Lipton's theory of commuting operations [26] and borrows ideas from [14].

The proof has two phases, which may be alternated. In the first phase, we verify that specific assertions hold at all program points using a rely-guarantee approach. This proof also guarantees termination and deadlock freedom with a calculus similar to the one in [43]. Essentially, the steps must satisfy two conditions: First, assertions before and after a step must satisfy the usual conditions of Hoare's calculus for total correctness. Second, all assertions must be stable with respect to a rely condition that is proven to abstract the steps of other threads.

The second phase is to iteratively show that two sequential steps of one thread can be combined into a single atomic step. This is done by showing that the first step commutes (leaves the same final state, if it is a returning step it must also produce the same output) with every step of another thread to the right (the step is a *right mover*), or dually that the second step is a *left mover*.

We found that this proof technique is suitable for locking-based algorithms, where locking/unlocking instructions are simple cases of right/left movers. If a data structure is written in a section of the code, where the thread holds a suitable lock, then the operation is both a left as well as a right mover (a *both mover*).

Note that the assertions proven for program points play an essential role in proving such commutativity properties, since they are often incompatible, resulting in trivial commutativity. Usually writing a data structure does not commute with writing it in another thread. But if it is asserted that the updating thread holds a lock that protects it, then they trivially commute, since the two assertions that the lock is held by both threads contradict each other.

Combining steps into larger steps can be iterated. It typically leads to the innermost locking range to be contracted to one atomic instruction. Repeating the first phase, we can now prove that the lock is free at all times, which again allows new instructions to become left or right movers in phase two. Alternating phases ends, when all instructions of the program have been combined into a single, atomic step.

The approach so far guarantees *functional correctness*. For our instance this says that our concurrent implementation of the POSIX standard (which combines the code from all implementation components) has the same behavior (in the sense of linearizability [22]) as atomically executing POSIX operations. In particular, all operations terminate and there are no deadlocks.

To prove that this is the case, we have to show that refinement is compatible with the use of subcomponents: If C refines A, then implementation M that calls operations of A, is refined by machine M, calling their implementation from C, as shown in Fig. 2.

This is a folklore theorem ("substitutivity") that should hold for all meaningful definitions of refinement. For data refinement we are aware of the formal proof in [12], for linearizability it is informally stated in [22]. We have not given a detailed proof yet, the sequential case (including crash-safety, see below) is

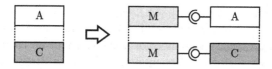

Fig. 2. Substitutivity of refinement.

proven in [17]. For the concurrent case we recently found an elegant proof in terms of combining IO-Automata [3], though this proof does not yet take into account non-terminating behavior of operations. Like for other refinement definitions (see e.g. [13] for data refinement or [34] for ASM refinement) the proof would have to be lifted to a scenario where states include a bottom element that represents non-termination.

2.3   Crash Safety

In addition to functional correctness, crash-safety is the second important aspect for a file system to work correctly. Informally it guarantees that when a crash happens (typically a power failure), the file system can be rebooted to a state that is consistent and does not contain any surprises: files and directories that have not been modified before the crash still should keep their content. Files where a write operation was running should not have modified content outside the range that was overwritten, and data within the range should be either old or new data, but nothing else.

A first observation relevant for crash-safety is that the only persistent state of the file system that is left unchanged by a crash is flash memory, which is the state of the lowest-level MTD interface. All other state variables are state, stored in RAM, that is deleted by a crash. Meaningful values for these states are constructed by running the recovery operations of all implementation components bottom-up.

A second, crucial observation is that if a running operation (on any level of the hierarchy) is aborted in the middle due to a crash, the resulting state can also be reached by crashing in a state after the operation has completed. The reason is that the flash hardware can always (nondeterministically) refuse any writes due to being worn out. Therefore, the alternative run that leads to the same state as the one with the crash is the one where all flash modifications fail after the point of the crash in the original one and the crash happens at the end. Proving this can be reduced to the simple proof obligation (expressible in wp-calculus) that all implementations of operations have a run such that running crash and recovery in the initial and final state yields the same state.

As a consequence, the question whether crashes at any time are guaranteed to lead to a consistent state can be reduced to the question whether crashes in between operations lead to a consistent state. Again, the latter gives a simple proof obligation in wp-calculus for the recovery program.

However, this does not specify how the final state looks in comparison to the final state of the original run, so we still might see an unexpected result.

A simple idea to specify the effect of a crash would be to specify a (total) *Crash* relation between states before and after the crash. However, this becomes intractable for higher levels of the refinement tower, due to the use of write-back caches. Such caches are used in all file system implementations for efficiency, since RAM access is significantly faster than writing or reading from persistent memory. In our flash file system such a cache, called the *write buffer*, is even necessary, since a page can only be written as a whole, and can not be overwritten. The write buffer therefore collects data until full page size is reached before writing the page to flash memory. The write buffer follows a queue discipline, it persists data first that was received first. We call such a buffer *order-preserving* and allow arbitrary use of such order-preserving caches on any level of an implementation.

The use of a cache makes it difficult to just specify a crash relation, since on higher level specifications, the information which part of the data is still in cache is no longer present. After all, the top-level POSIX specification specifies a directory tree and file contents with no information which parts are still cached. In principle, such information can be added as auxiliary data (used for verification only, deleted in the running code), but we found such an encoding to become intractable.

Instead we specify the effect of crashes mainly in an *operational* style, where the effect of a crash is to construct an alternative run that explains the resulting state after the crash. This alternative run mainly retracts a final piece of the original run with the intuition that the results of the retracted operations are still in the cache. We found that this is compatible with order-preserving caches, where losing the content of a queue corresponds to losing the effects of some of the final operations.

In addition to undoing part of the run it is however necessary that operations (one in a sequential setting, several in concurrent implementations) that are running at the time of the crash may be executed with a different result: when writing some bytes to a file crashes, an initial piece of the data may have persisted while the remaining bytes have not. Constructing an alternative run that writes fewer bytes is consistent with POSIX: A top-level write is allowed to return having only written a prefix of the data to the file (the number of written bytes is returned). The alternative run will therefore have a different completion of the write operation with fewer bytes written.

Therefore, in addition to undoing a final part of the run, we allow all running operations (which have a pending invoke in the shortened history) to be completed differently in the replacement run.

Two more considerations are necessary to ensure that crashes do not give surprising results. First, POSIX offers a *sync* operation that empties all cached data. This leads to synchronized states, which we specify on all levels of the refinement hierarchy. Operations that lead to a synchronized state are then forbidden to be retracted in the alternative run.

Finally, we still need a crash relation on all levels to specify an additional residual effect of the crash on RAM state. For the top-level of POSIX this is obvious, since even when no operation is running, a crash will at least close all open files and remove the resulting orphaned files (files that were removed in the directory tree, but are still open).

In summary, the effect of a crash after a run that went through states $(s_1, \ldots, s_n)$ will be a (consistent) final state s' of an alternative run, which executes an initial piece of the original run, say up to s_i, then completes operations that are running at this point to reach a state s''. Finally, the crash relation is applied and the recovery program is executed to reach s'.

The proof obligations resulting from this concept were formally verified to imply this crash-safety criterion for a sequential setting in [17]. However, it is applicable without changes in a concurrent setting, too. Note that it is again crucial that all operations that run at the point where a crash happens have an alternative run without any more changes to the persistent flash memory. Thus, when proving linearizability by reordering steps according to Lipton's theory, we can already consider a run with completed operations to show that an equivalent sequential execution exists where all programs execute atomically.

The theory given here must be extended when caching the data of individual files is considered as retracting a part of the run is no longer sufficient. We consider an appropriate extension in Sect. 4.

3   The Flashix File System – Overview

The Flashix file system is structured into a deep hierarchy of incremental refinements as shown in Fig. 3. Boxes represent formal models that can be connected via refinements (dashed lines) and can call operations of their subcomponents through a well-defined interface (–⊚–). We distinguish between specification components in white and implementation components in gray. Combining all implementation components then results in the final implementation of the file system.

The top layer of Fig. 3a is a formal specification of the POSIX standard [31]. It defines the interface and the functional correctness requirements of the file system. Here, the state of the file system is given by a directory tree where leaves store file identifiers, and a mapping of file identifiers to the corresponding file contents, represented by a sequence of bytes. An indirection between file identifiers and file content is necessary to allow hard links, where the same file is present in several directories. *Structural operations*, i.e. operations that modify the directory tree like creating/deleting directories or (un)linking files, are defined on paths. *Content operations*, such as reading or writing parts of the content of a file, work directly on file identifiers.

The bottom layer of the hierarchy in Fig. 3b is a formal specification of the Linux MTD Interface (Memory Technology Devices). It acts as a lower boundary of the file system and provides low-level operations to erase flash blocks and to read and write single pages within flash blocks. Preconditions ensure that calls

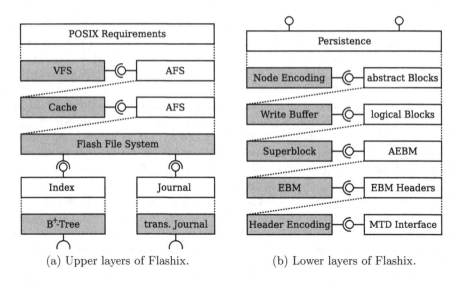

(a) Upper layers of Flashix.   (b) Lower layers of Flashix.

Fig. 3. Component hierarchy of the Flashix file system.

to these operations comply with the characteristics of flash memory, i.e. that pages are only written as a whole and that pages are only written sequentially within a block. Additionally, it formalizes assumptions about hardware failures or the behavior of the flash device in the event of a crash.

In a first refinement step, the POSIX model is refined by a Virtual Filesystem Switch (VFS) that uses an abstract specification of the core file system (AFS). Similar to the Linux Virtual Filesystem, the VFS component implements the resolution of paths to individual file system objects, permission checks, and the management of open files. Basically, the AFS provides an interface analogous to the POSIX interface but on the level of file system objects instead of paths. This specification abstracts completely from any flash-specific concepts and thus the VFS is not limited to be used exclusively with flash file systems. Details of the POSIX specification as well as the sequential refinement to VFS and AFS can be found in [18].

Recently, we worked on a locking concept for the VFS that allows concurrent calls to the file system interface. The approach taken focuses on enabling parallel access to file contents, in particular we want to allow arbitrary concurrent reads as well as concurrent writes to different files. Therefore, we chose a fine-grained locking strategy for files, whereas we applied a coarse-grained strategy for the directory tree. This means that each file is protected by an individual reader-writer lock while a single reader-writer lock is used for the entire directory tree. It should be noted, that parallel traversal of the directory tree is still possible as long as no *structural operation* is performed. Thus, we think this is a good trade-off between development or verification effort and performance gain. We augmented the existing sequential versions of VFS and AFS with locks and ownerships respectively and proved that the interleaved implementation of VFS is linearizable and deadlock-free using atomicity refinement as explained in Sect. 2. The verification showed that a strict order for acquiring and releasing locks is beneficial for our approach.

AFS is refined by the actual Flash File System (FFS). Additionally, AFS is refined by a Cache component that caches data structures used at the interface of the core file system. The Cache is integrated as a decorator, i.e. it wraps around the AFS in the sense that it uses AFS as a subcomponent and also implements the interface of AFS. This allows the file system to be used both with and without Cache. The main goal of this integration was to allow *write-back* caching of content operations. However, write-back caching can have significant effects on the crash behavior of a system. In [4] we presented a novel correctness criterion for this sort of file system caches and proved that Flashix complies with it. We sketch the most important concepts of this addition and the proof idea in Sect. 4.

The FFS was the layer at which we started the development of the Flashix file system in [38]. It introduces concepts specific to flash memory and to log-structured file systems. Updates to file system objects must be performed out-of-place and atomically. For this purpose, the FFS is built upon an efficient Index, implemented by a wandering B^+-Tree, and a transactional Journal. Both are specified abstractly in the component FFS-Core. New versions of file system objects are encapsulated in nodes and grouped into transactions that are then written to a log. To keep track of the latest versions of objects, the locations of them on the flash memory are stored in the Index. The Index exists in two versions, one persisted on flash and one in RAM. Updates on the Index are initially performed only in RAM in order to improve performance as these update are quite costly to perform on flash. Only during *commits* that are executed regularly, the latest version of the Index is written to flash. The transactional Journal ensures that, in the event of a crash, the latest version of the RAM Index can be reconstructed. This can be done by replaying the uncommitted entries in the log starting from the persisted Index on flash. In doing so, incomplete transactions are discarded to comply to the atomicity properties expected by the VFS.

Another crucial mechanism implemented in this layer is *garbage collection*. Due to their out-of-place nature, updates to the file system leave garbage data behind. This data must first be deleted before the storage space it occupies can be used again. But since flash blocks can only be erased as a whole, garbage collection chooses suitable blocks for deletion (preferably blocks with a high percentage of garbage), transfers remaining valid data of that block to another one, and finally triggers the erasure of the block. This mechanism is not triggered explicitly by calls to the file system, instead it must be performed periodically to ensure that the file system does not run out of space. Hence we extracted garbage collection into a separate thread, we give more details on this concurrency extension in Sect. 5.

Both the transactional Journal and the B^+-Tree write nodes on the flash device. The Node Encoding component is responsible for serializing these nodes to bytes before they can be written to flash. It also keeps track of the allocation of erase blocks and, for each block, the number of bytes still referenced by live data, i.e. by nodes of the index or nodes that store current versions of file system objects. This information is used to determine suitable blocks for garbage collection. Besides that, the layer ensures that writing of nodes appears to be atomic to the Journal and Index. It detects partially written nodes that may occur through crashes or hardware failures and takes care of them. A more in-depth view on these components and the garbage collection is given in [16].

All serialized nodes pass a Write Buffer. This buffer cache tackles the restriction that flash pages can only be written sequentially and as a whole. It caches all incoming writes and only issues a page write once a page-aligned write is possible, i.e. the write requests have reached the size of one flash page in total. Otherwise, padding nodes would have to be used in order to write partially filled pages, which both would increase the absolute number of writes to flash and the amount of wasted space on the flash device. Introducing such an *order-preserving* write-back cache (written data leaves the cache in the same order as it entered it) also affects the crash behavior of the file system. In [29] we give a suitable crash-safety criterion as well as a modular verification methodology for proving that systems satisfy this criterion.

The Superblock component is responsible for storing and accessing the internal data structures of the file system. A specific part of the flash device is reserved for this data. They are written during a *commit* only, since persisting each update would have a significant negative impact on the performance of the file system. A critical task of this layer is to ensure that commits are performed atomically using a data structure called *superblock*.

Finally, the Erase Block Manager (EBM) provides an interface similar to the one of MTD (read, write, erase). However, the EBM introduces an indirection of the *physical blocks* of the flash device to *logical blocks* and all of its interface operations address logical blocks only. These logical blocks are allocated on-demand and mapped to physical blocks. The indirection is used to move logical blocks transparently from one physical block to another one which is necessary to implement *wear leveling*. Wear leveling ensures that within some bounds all blocks are erased the same number of times. This is necessary to maximize the life time of the flash memory, as erasing a flash block repeatedly wears it out, making it unusable. To ensure a bound, the number of performed erases is stored in an *erase counter*. Wear leveling finds a logical block that is mapped to a physical block with low erase count and re-maps it to a block with high erase count. Since a logical block with low erase-count typically contains a lot of *stale data* that has not been changed for some time and therefore is not likely to change soon, the number of erases is kept at the same level and the lifetime of the flash device increases.

The EBM uses the Header Encoding component for the serialization and deserialization of administrative data, most important an inverse mapping stored in the physical blocks containing the numbers of the logical blocks they are mapped to.

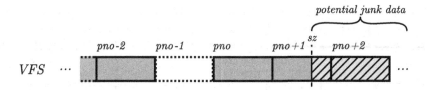

Fig. 4. Representation of file content in VFS.

A sequential version of the Erase Block Manager is explained in detail in [30]. But similar to garbage collection, wear leveling has to be performed regularly without being triggered by the user and so we adjusted the EBM to run wear leveling in a separate thread as well. Another thread is used to perform the erasure of blocks asynchronously, too. We illustrate this extension and the verification methodology for introducing concurrency to a refinement hierarchy on a simplified version of the EBM in [36].

4   Crash-Safe VFS Caching

A common technique to get a highly efficient file system implementation is the use of caching. Flashix features several caches in multiple layers: the B$^+$-Tree contains a *write-through* cache for the directory structure, the Write Buffer uses an *order-preserving* write-back cache for flash pages, and lately we added a *non-order-preserving* write-back cache for file contents to the VFS layer.

Since all in-memory data is lost in the event of a crash, crash-safety is a critical aspect when integrating caches. For write-through caches this is unproblematic as cached data is only a copy of persisted data on flash. In [29] we presented a crash-safety criterion for order-preserving caches: basically a crash has the effect of retracting a few of the last executed operations. But this criterion is too strong for non-order-preserving caches and so in [4] we proposed a more relaxed criterion and proved that our VFS caches comply with it. We will now give an overview over the crucial aspects of this latest extension.

While file contents are represented as a finite sequences of bytes in POSIX, VFS breaks this abstract representation down to a map of fixed-size byte arrays (*pages*) and an explicit file size as shown in Fig. 4. Each box depicts a page and is identified by a page number *pno*. The map offers the advantage of a sparse representation with the convention that missing pages are filled with zeros only. This is indicated by a white dashed box (*pno-1* in the figure). A important detail is the possibility of random data (hatched) beyond the file size *sz*, resulting from prior crashes or failed operation executions. This is especially relevant when the file size is not *page-aligned* and the page of *sz* (*pno+1* in the figure) contains actual garbage data (non-zero bytes) beyond *sz*.

When extending a file, such garbage data must not become visible as this would not match the POSIX requirements. There are two ways to change the file size: explicitly with a *truncation* or by writing content beyond the current

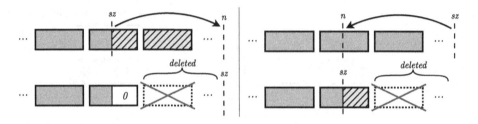

Fig. 5. Truncation to a larger size $sz \leq n$ (left) and to a smaller size $n < sz$ (right).

file size. A truncation crops the content of a file to a new size n and ensures that all data within the file is valid. Hence, in addition to updating the size, the actual content may also need to be updated. To increase the file size, possible junk data in the page of sz needs to be cleared with zeros and pages beyond the old file size are deleted (Fig. 5 on the left). On the other hand, junk data can remain in the page of the new size when shrinking the file (Fig. 5 on the right).

The VFS breaks down a write into three steps. First, possible junk data beyond the file size is removed. This is done by a truncation to the current file size ($n = sz$ in Fig. 5). Then the respective pages are written bottom up individually. If writing a page causes an error, writing stops and hence only a prefix of the requested bytes are written. Finally, the file size is adjusted to the position of the last written byte if data was written beyond the old file size.

By using the decorator pattern, VFS caching could be integrated into our refinement hierarchy without changing existing components by adding a single new component. The component Cache refines and uses AFS at the same time as shown by Fig. 3a. To cache content operations in a write-back manner, writes and truncations are aggregated in local page and size caches. Additionally, write-through caches for header nodes (containing meta data of files and directories) and directory entries are implemented. Page writes are stored in the page cache and truncations or size updates lead to updates in the respective size caches. We only cache the most current file size while multiple truncations are aggregated by caching the minimal truncation size since the last synchronization. Reading a page tries to find a corresponding cache entry. If it does not exist in the cache, the page is read from the Flash File System (FFS) and (if it exists) stored in the cache.

Updating the persisted content happens only if it is triggered by a call of the **fsync** operation for the respective file. A call to **fsync** starts the synchronization with a single truncation of the persisted content to the minimal truncation size. Then, similar to a VFS write, all dirty pages of the file in the cache are written to the FFS bottom up and finally the file size is adjusted if necessary.

Showing functional correctness of this addition can be done by a single data refinement and will not be considered further here. However, proving crash-safety is quite difficult. If a crash occurs, all data in the volatile state of the Cache component is lost such that unsynchronized writes and truncations have not taken place. To ensure crash-safety, it must be shown that each crash results

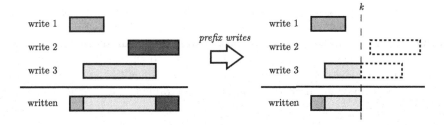

Fig. 6. Write-prefix crash consistency: prefix writes for a crash during the synchronization of a file just before syncing byte k.

in a consistent state. Normally, the effect of a crash would be described by an explicit change of state. But this is usually not practicable for write-back caches and the crash-safety criterion introduced in [29] is not suitable for non-order-preserving caches, too. So we use a new criterion called *Write-Prefix Crash Consistency (WPCC)*. It states that for any file a crash has the effect of retracting all write and truncate operations since the last synchronization of that file and re-executing them, potentially resulting in writing prefixes of the original runs [4].

This follows from the effects of a crash during persisting a cached file as shown in Fig. 6 for the POSIX data representation. On the left, there is shown how multiple overlapping writes combine to a sequence of written bytes. Since cached pages are written bottom up during synchronization, a crash in the middle of **fsync** results in a prefix of these bytes being persisted (on the bottom right in Fig. 6). If the crash occurs just before persisting the byte at position k, the resulting state can be explained by writing prefixes of the original instructions (namely those prefixes that have written exactly the bytes beyond k). This can result in complete writes (write 1 in Fig. 6), partial writes (write 3), or writes that are completely lost (write 2). To archive this behavior it is essential that VFS writes as well as synchronizations are performed bottom up and that writes can fail after writing a arbitrary number of bytes.

Informally, the criterion describes the effects of a crash by finding an alternative run where loosing cached data has no noticeable effect. These alternative runs may differ at most from their original runs in that writes since the last synchronization have written prefixes of their original runs. Because such an alternative run is a valid run and hence results in a consistent state, the original crashing run yields a consistent state as well.

The main effort for proving crash-safety is to show that such alternative runs exist for any possible occurrence of a crash. While finding suitable runs for crashes outside of **fsync** is unproblematic (if nothing was persisted, failed executions of cached operations can be chosen), this is especially hard for crashes within **fsync**. One particular challenge is to show that the aggregation of multiple truncate operations matches WPCC if the minimal truncation was executed but the final file size was not yet synchronized at the event of the crash. This can lead to slightly different junk data in the write-prefix run such that on the level of AFS the contents of the crashed run and the write-prefix run differ beyond the file size. However, this junk data is only visible in AFS as the abstraction to POSIX ignores all bytes beyond the file size. Hence an alternative run can be found on the level of POSIX, but this required to extend the proof work to another layer of abstraction. More details on the difficulties and the concrete proof strategy involving multiple commuting diagrams can be found in [4].

5   Concurrent Garbage Collection

Besides allowing concurrent calls to the file system interface as briefly outlined in Sect. 3, moving certain internal mechanisms into separate threads also introduces additional concurrency to the file system. Hence the affected models have to be modified in order to avoid conflicts resulting from parallel executions of operations.

The expansion with concurrent garbage collection ranges from the FFS layer to the Journal layer. In the FFS the concurrent operation for garbage collection is introduced (Fig. 8). This operation is not part of the interface (it refines skip, i.e. it has no visible effect for clients). Hence, it can not be called by any client components. Instead it will be repeated infinitely within its own thread. To ensure that garbage collection is not performed continuously, especially when no more space can be regained, a condition variable *gc_cond* is used[1]. At the beginning of each iteration the thread blocks at the **condition_wait** call until it is signaled by another thread to start. The concrete garbage collection algorithm is specified in the FFS-Core and implemented in the Journal component, so after being signaled the operation **ffs_core_gc** is called.

Signaling takes place in all FFS operations that may modify the file system state in the sense that either entries are written to the log and hence space on the flash device is allocated or garbage is introduced by invalidating allocated space. Such operations, as shown generically in Fig. 7, emit a signal to *gc_cond* after they have updated the index.

The implementation of **ffs_core_gc** in the Journal component then first checks whether there is a block which is suitable for garbage collection. If that is the case, all still referenced nodes of this block are collected, these nodes are then written to the journal, and their new addresses are updated in the index accordingly. Finally, if the referenced data was successfully copied, the block can be marked for erasure.

As an additional thread is introduced in the FFS, established ownerships in the VFS/AFS layer are not sufficient to prevent data races between the garbage collection thread and other threads. For this reason the reader-writer lock *core_lock* is added to the FFS component. It is used to acquire exclusive

---

[1] Note that condition variables are always coupled to a mutex. Here *gc_cond* is coupled to *gc_mutex*. Signaling a condition requires to hold the corresponding mutex. Starting to wait for a signal requires to hold the mutex as well, however, the mutex is released during waiting. As soon as a signal was emitted and the mutex is free, the waiting thread acquires the mutex and continues its execution.

```
ffs_operation(...) {
   ...
   nd₁ := inodenode(key₁, ...);
   ...
   rwlock_wlock( ; core_lock);
   ffs_core_wacquire();
   ffs_core_journal_add(nd₁, ... ; adr₁, ... ; err);
   if err = ESUCCESS then {
      ffs_core_index_store(key₁; adr₁);
      ...
      mutex_lock( ; gc_mutex);
      condition_signal( ; gc_cond, gc_mutex);
      mutex_unlock( ; gc_mutex);
   };
   ffs_core_release();
   rwlock_wunlock( ; core_lock);
   ...
}
```

```
ffs_gc() {
   mutex_lock( ; gc_mutex);
   condition_wait( ; gc_cond, gc_mutex);
   mutex_unlock( ; gc_mutex);

   rwlock_wlock( ; core_lock);
   ffs_core_wacquire();
   ffs_core_gc();
   ffs_core_release();
   rwlock_wunlock( ; core_lock);
}
```

Fig. 7. General operation scheme of modifying FFS operations.

Fig. 8. FFS garbage collection operation.

or shared ownership for the journal and index data structures. We did not head for a more fine-grained locking approach since usually updates affect nearly all parts of the state of FFS-Core anyway. However, using reader-writer locks still allows for concurrent read accesses to the Journal.

Modifying operations in the FFS as shown in Fig. 7 always follow the same scheme. First, all new or updated data objects are wrappend into nodes (nd_1, ...) with an unique key (key_1, ...). Depending on the concrete operation, nodes for inodes, dentries, and pages are created. Then these nodes are grouped into transactions and appended to the log using the **ffs_core_journal_add** operation. If successful, i.e. the operation returns the code ESUCCESS, the operation returns the addresses (adr_1, ...) where the passed nodes have been written to. Finally, the index is updated by storing the new addresses of the affected keys via the operation **ffs_core_index_store**[2]. It is cru-

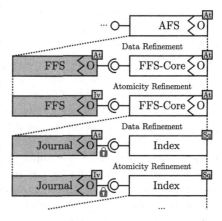

Fig. 9. Refinement hierarchy extended by concurrent garbage collection.

cial that garbage collection is never performed between these calls since this could result in a loss of updates (e.g. if garbage collection moves nodes updated by the operation), potentially yielding an inconsistent file system state. Hence, the locking range must include the **ffs_core_journal_add** as well as all **ffs_core_index_store** calls.

---

[2] Some operations also update the index by removing entries from it.

```
                          journal_operation(...) {
                              ...                              journal_operation(...) {
journal_operation(...) {      atomic {                             atomic {
    ...                           mutex_lock(; idx_lock);
    mutex_lock(; idx_lock);       index_operationᵢ(...);           ...
    index_operationᵢ(...);        mutex_unlock(; idx_lock);        mutex_lock(; idx_lock);
    mutex_unlock(; idx_lock);  }                                   index_operationᵢ(...);
                                                                   mutex_unlock(; idx_lock);
    ...                           ...                              ...
    mutex_lock(; idx_lock);       atomic {                         mutex_lock(; idx_lock);
    index_operationⱼ(...);            mutex_lock(; idx_lock);      index_operationⱼ(...);
    mutex_unlock(; idx_lock);         index_operationⱼ(...);       mutex_unlock(; idx_lock);
    ...                               mutex_unlock(; idx_lock);    ...
}                             }                                }
                                  ...                          }
                          }
```

Fig. 10. Reduction steps of a Journal operation (from left to right).

To prove that this locking strategy is in fact correct, i.e. that the interleaved components are linearizable, we again apply *atomicity refinement*. This results in the expansion of the refinement hierarchy shown in Fig. 9. Usually, atomicity refinement would have to be applied to all layers below Index, too, but we did not put any effort in making a interleaved version of the B$^+$-Tree yet. Instead we locked the interface of the Index (depicted by 🔒). This means that each call to an Index operations **index_operation** requires the current thread to be an exclusive owner of the Index component. In the Journal this is realized by surrounding these calls with a mutex *idx_lock* as shown in Fig. 10 on the left. Owning a subcomponent exclusively ensures that the subcomponent is only called sequentially and hence allows to directly use the unaltered sequential version of the subcomponent and its refinements (denoted by 🅢𝔮 in Fig. 9).

FFS-Core is augmented with ownership ghost state matching the reader-writer lock *core_lock* of FFS. The FFS operations acquire and release this ownership according to the locking ranges (see Fig. 7 and Fig. 8). While the ownership granularity of AFS (owned directory tree, owned files, ...) does not match the state of the FFS-Core or the Journal, the information about which files etc. are owned when an operation is called (encoded in the preconditions) is still relevant for the FFS in order to preserve functional correctness. For example, an owned file must not be removed from the index while its metadata is updated. Therefore, ownership ghost state is added to FFS analogously to AFS and corresponding ownership properties are established. This is sufficient to prove that the interleaved FFS can be reduced to an atomic FFS via atomicity refinement. The data refinement of the atomic AFS to the atomic FFS is basically identical to the original sequential refinement, in addition, it must only be shown that their respective ownerships match.

When proving the atomicity refinement of the Journal, it is apparent that Index operations together with their surrounding lock calls form atomic blocks like in the center of Fig. 10. But as most operations have multiple calls to the Index, this is not sufficient to reduce these operations to completely atomic ones. It remains to show that these blocks as well as statements that access the local state of the Journal move appropriately (usually they have to be both mover). To prove this, the ownership information of the FFS-Core component can be used. The Journal is augmented with ownership properties, operations and preconditions that match those of the FFS-Core and so accesses to the local state can be inferred to be both movers. The information that a certain ownership is acquired at the calls of Index operations and their associated locking operations allows to prove that these blocks in fact are movers and hence to further reduce the operations to be atomic (Fig. 10 on the right). Although the proofs are simple, this is quite elaborate since many commutations have to be considered. The data refinement of FFS-Core to the atomic Journal then again is basically identical to the sequential refinement.

6   Evaluation

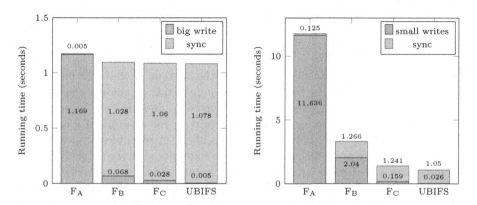

Fig. 11. Nano write benchmarks on Flashix and UBIFS: big write (left) and small writes (right). Flashix was used in three different configurations: sequentially without VFS cache (F_A), sequentially with VFS cache (F_B), and with VFS cache and concurrent wear leveling and garbage collection (F_C). (Color figure online)

To evaluate the performance of the Flashix file system we perform a collection of microbenchmarks. This gives us some insight in whether the expansions we have made, especially those described in Sect. 4 and Sect. 5 or in [4,36], have an impact on the performance. Furthermore, we want to compare the performance of Flashix with state-of-the-art flash file systems like UBIFS [23].

All benchmarks were run within a virtualized Linux Mint 19.3 distribution, using 3 Cores of a Intel Core i5-7300HQ CPU and 4, 8 GB of RAM. The flash device was simulated in RAM using the NAND simulator (nandsim) integrated into the Linux kernel [27]. The numbers shown in the following represent the mean of 5 benchmark runs in which the mean standard deviation across all runs is below 4.5% (this translates to a mean deviation in runtime of less than 0.16 s).

We chose some small workloads that represent everyday usage of file systems: copying and creating/extracting archives. Copying an archive to the file system

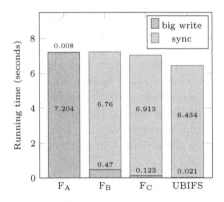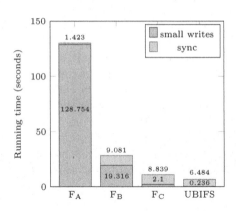

Fig. 12. Vim write benchmarks on Flashix and UBIFS: big write (left) and small writes (right). Flashix was used in three different configurations: sequentially without VFS cache (F_A), sequentially with VFS cache (F_B), and with VFS cache and concurrent wear leveling and garbage collection (F_C).

results in the creation of a file and writing the content of that one file. Analogously, copying an archive from the file system yields in reading the content of the file. Hence, we call these workloads *big write* and *big read* respectively. On the other hand, extracting an archive results in the creation of a directory structure containing many files. The contents of the created files are written as well, however, these are multiple smaller writes compared to the single big write when copying. Creating an archive from a directory structure on the file system requires to read all directories and files. Hence, we call such workloads *small writes* and *small reads* respectively. As sample data we used archives of the text editors Nano[3] and Vim[4].

Figure 11 shows the results of the write benchmarks with Nano. When comparing the uncached configuration (F_A) with the cached configuration (F_B) of Flashix, one can see that adding the VFS cache has indeed a significant impact on write times (depicted in blue). But as these times do not include persisting the cached data to flash, we enforced synchronization directly afterwards via *sync* calls (depicted in red). For big writes the combined runtime of the cached configuration is similar to the uncached one. For small writes though, the combined runtime of F_B is substantially faster since repeated reads to directory and file nodes during path traversal can be handled by the cache.

Moving wear leveling and garbage collection into separate threads (F_C) further improves the performance. This is especially noticeable in the small writes workload where the write time can be reduced by about one order of magnitude. In the sequential configurations (F_A and F_B), after each toplevel operation it was checked whether garbage collection or wear leveling should be performed. During these checks and potential subsequent executions of the algorithms, other

---

[3] nano-2.4.2.tar: approx. 220 elements, 6.7 MB.
[4] vim-7.4.tar: approx. 2570 elements, 40.9 MB.

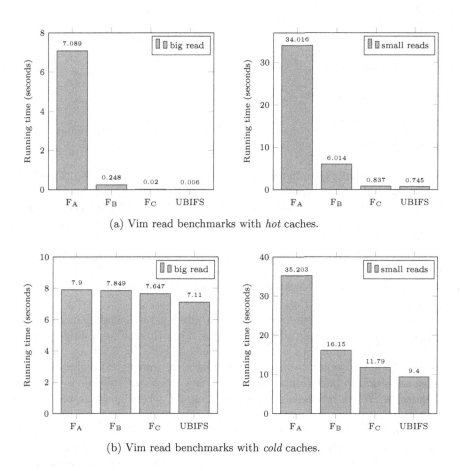

(a) Vim read benchmarks with *hot* caches.

(b) Vim read benchmarks with *cold* caches.

Fig. 13. Vim read benchmarks on Flashix and UBIFS: big read (left) and small reads (right). Flashix was used in three different configurations: sequentially without VFS cache (F_A), sequentially with VFS cache (F_B), and with VFS cache and concurrent wear leveling and garbage collection (F_C).

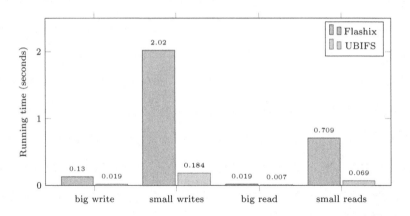

Fig. 14. Vim benchmarks on Flashix and UBIFS without flash delays.

POSIX operation calls were blocked. In the concurrent configuration (F_C) these blocked time can be eliminated for the most part since writing to the cache does not interfere with garbage collection or wear leveling. As small writes trigger considerably more toplevel operation calls, this effect is much more noticeable than with big write workloads.

Compared to UBIFS, the current version of Flashix performs as expected. Runtimes of the F_C configuration are always within the same order of magnitude of those of UBIFS. This also applies for running the benchmarks with a larger archive like Vim shown in Fig. 12.

Similar effects can be observed when considering read workloads as shown in Fig. 13. Adding caches significantly speeds up both reading a single big file and reading many small files when the caches are *hot* (Fig. 13a), i.e. when the requested data is present in the caches. Likewise, moving wear leveling and garbage collection to background processes brings down the runtime by an order of magnitude. When reading from *cold* caches, i.e. when no requested data is present in the caches, the speed up is much more subtle since the main delay results from reading data from flash. As shown in Fig. 13b, for big reads there is hardly any improvement from F_A to F_B or F_C. However, both expansions have an impact on the runtime of small read workloads for the same reasons as for small write workloads: repeated reads to the directory structure can be handled by the cache and blocked time for garbage collection and wear leveling can be eliminated. With these additions one can see that Flashix is competitive with UBIFS regarding read performance, too.

In future work we plan to further improve the performance of Flashix by improving our code generator as the generated code is not optimal in terms of allocating/deallocating and copying data structures. The optimization potential becomes apparent when comparing the raw in-memory runtimes of Flashix and UBIFS like in Fig. 14. Here we instructed nandsim to not simulate any delays for accessing the simulated NAND memory. The results show that UBIFS is still up to a factor of 10 faster than Flashix for the Vim microbenchmarks (the

Nano benchmarks yield similar results). First experiments show, that even simple routines can affect performance noticeably if they are generated inefficiently. For example, we found out that a simple optimization of a routine used in the Journal for calculating the required space of a node-list on flash improved the runtime by up to 30% compared to the generated code. Hence we plan to apply data flow analysis to identify this and other locations where such optimizations can be performed. We are optimistic that this will further close the gap to state-of-the-art handwritten file systems.

7   Related Work

There are some other projects related to verified file systems.

Damchoom et al. [11] develop a flash file system by using incremental refinement. Concurrency is verified on a similar level as AFS for reading and writing of file content and for wear-leveling as well. Synchronization between threads is implicit by semantics of Event-B [1] models. But this makes it difficult to derive executable code. Amani et al. [2] design the flash file system BilbyFS to research their tool Cogent for generating verified C code. The system can also derive specifications for Isabelle/HOL. BilbyFS has a similar but simpler structure as Flashix. For instance it builds on top of the EBM instead of MTD. It supports caching mechanism but not on the level of VFS. Crash-safety has not been considered so far.

(D)FSCQ [8,9] is a sequential implemented file system developed by Chen et al., which is targeted for regular disks with random access, not flash memory. Similar to our approach, structural updates to the file system tree are persisted in order. DFSCQ also uses a page cache, however, it does not specify an order in which cached pages are written to persistent store. Therefore, it is not provable that a crash leads to a POSIX-conforming alternate run. Instead a weaker crash-safety criterion is satisfied, called *metadata-prefix* specification: it is proved that a consistent file system results from a crash, where some subset of the page writes has been executed. Verification is done by using Crash Hoare Logic and Haskell code is derived from the specification.

Our crash-safety criterion for order-preserving caches is similar to buffered durable linearizability [24], though there are some differences: the criterion is purely history based, it allows to construct a prefix of the history, where pending operations can be completed anew, similar to our approach. It however allows to complete pending operations in the shortened history even after operations that have started after the crash. This is useful for a concurrent recovery routine, that may restart operations that crashed. It is disallowed in our approach, since not relevant for file systems. Buffered durable lineariability also disallows the effect of closing open files that we specify separately with a Crash predicate.

Verification of a sophisticated locking scheme that locks inodes hand-over-hand (lock coupling) has recently been done in the theorem prover Coq for a file system prototype called AtomFS that is directly programmed in C and stores data in RAM [44]. A particular challenge for the proof of linearizability solved there was the rename operation, that moves directories (whole subtrees). The operation has to lock both the source and target directory, but has to avoid deadlocks. It should be possible to port this locking scheme to our file system.

Other, older related work can be found in our prior work [4, 36].

8   Conclusion

The Flashix project has developed the first realistic verified file system using a refinement- and component-based approach that generates code at the end.

Being realistic however had the price that the individual components had to be intertwined carefully, which caused lots of effort and was therefore substantially harder than analyzing concepts individually.

We think that developing such a large system without suitable modularization and abstraction by verifying concrete C-code directly would have been an almost impossible task.

The use of abstract data types and components, that allows efficient verification comes at a price, however. Since abstract data types have the semantics of predicate logic, which is a value semantics that does not take sharing, allocation, or destructive updates into consideration, generating correct, efficient C-code is still a challenge.

Generating functional (non-destructive) code instead has long been done by theorem provers, but this would be hopelessly inefficient for a file system, where destructively updated arrays (buffers, pages, blocks) are crucial for efficiency: we tried using Scala's immutable type Vector once, but the generated code is slower by at least one order of magnitude.

It is possible to refine individual pieces of the abstract code to heap-based destructive code, and this is occasionally necessary (e.g. to represent search trees efficiently as pointer structures), and the verification task can be supported by using a library for separation logic, however refining *all* abstract data structures with pointer-structures would mean to analyze sharing manually, and to duplicate all code.

Another alternative is to enforce a linear type system on abstract specifications, for which a code generator could be proven correct [28].

Our current code generator follows the principle of *not* sharing data structures in the resulting C code to have definite allocation and deallocation points, and to allow destructive updates. This however, enforces copying x and all its substructures to y, when executing an assignment x := y. We already do a simple liveness check (if y is no longer used, then copying can be avoided), and some more ad-hoc optimizations to avoid unnecessary copying.

Still, a systematic data flow analysis, that allows sharing in places where it is harmless, and avoids copying wherever possible, should be able to close a large part of the still existing gap between the efficiency of our generated code and the hand-written code of UBIFS.

Implementation of such a data flow analysis is still future work, and we also want to tackle formalization and verification of such an approach, thereby establishing the correctness of the code generator.

References

1. Abrial, J.-R.: Modeling in Event-B - System and Software Engineering. Cambridge University Press, Cambridge (2010). https://doi.org/10.1017/CBO9781139195881
2. Amani, S., et al.: Cogent: verifying high-assurance file system implementations. In: Proceedings of the ASPLOS, pp. 175–188. ACM (2016)
3. Bila, E., Derrick, J., Doherty, S., Dongol, B., Schellhorn, G., Wehrheim, H.: Modularising verification of durable opacity. In: Logical Methods in Computer Science (2021). Submitted, draft available from the authors
4. Bodenmüller, S., Schellhorn, G., Reif, W.: Modular integration of crashsafe caching into a verified virtual file system switch. In: Dongol, B., Troubitsyna, E. (eds.) IFM 2020. LNCS, vol. 12546, pp. 218–236. Springer, Cham (2020). https://doi.org/10.1007/978-3-030-63461-2_12
5. Börger, E.: The ASM refinement method. Formal Aspects Comput. **15**(1–2), 237–257 (2003). https://doi.org/10.1007/s00165-003-0012-7
6. Börger, E., Rosenzweig, D.: The WAM–definition and compiler correctness. In: Beierle, C., Plümer, L. (eds.) Logic Programming: Formal Methods and Practical Applications. Studies in Computer Science and Artificial Intelligence 11, pp. 20–90. Elsevier, Amsterdam (1995)
7. Börger, E., Stärk, R.F.: Abstract State Machines – A Method for High-Level System Design and Analysis. Springer, Heidelberg (2003). https://doi.org/10.1007/978-3-642-18216-7
8. Chen, H.: Certifying a crash-safe file system. Ph.D. thesis, Massachusetts Institute of Technology, Cambridge, MA, United States (2016)
9. Chen, H., et al.: Verifying a high-performance crash-safe file system using a tree specification. In: Proceedings of the 26th Symposium on Operating Systems Principles (SOSP), pp. 270–286 (2017)
10. Cooper, D., Stepney, S., Woodcock, J.: Derivation of Z refinement proof rules: forwards and backwards rules incorporating input/output refinement. Technical report YCS-2002-347, University of York (2002). http://www-users.cs.york.ac.uk/susan/bib/ss/z/zrules.htm
11. Damchoom, K., Butler, M.: Applying event and machine decomposition to a flash-based filestore in event-B. In: Oliveira, M.V.M., Woodcock, J. (eds.) SBMF 2009. LNCS, vol. 5902, pp. 134–152. Springer, Heidelberg (2009). https://doi.org/10.1007/978-3-642-10452-7_10
12. de Roever, W., Engelhardt, K.: Data Refinement: Model-Oriented Proof Methods and Their Comparison, Volume 47 of Cambridge Tracts in Theoretical Computer Science. Cambridge University Press, Cambridge (1998)
13. Derrick, J., Boiten, E.: Refinement in Z and in Object-Z: Foundations and Advanced Applications. FACIT. Springer, London (2001). https://doi.org/10.1007/978-1-4471-0257-1. 2nd Revised Edition (2014)
14. Elmas, T., Qadeer, S., Tasiran, S.: A calculus of atomic actions. In: Proceeding of POPL 2009, pp. 2–15. ACM (2009)
15. Ernst, G., Pfähler, J., Schellhorn, G., Haneberg, D., Reif, W.: KIV: overview and VerifyThis competition. Softw. Tools Technol. Transfer (STTT) **17**(6), 677–694 (2015). https://doi.org/10.1007/s10009-014-0308-3
16. Ernst, G., Pfähler, J., Schellhorn, G., Reif, W.: Inside a verified flash file system: transactions and garbage collection. In: Gurfinkel, A., Seshia, S.A. (eds.) VSTTE 2015. LNCS, vol. 9593, pp. 73–93. Springer, Cham (2016). https://doi.org/10.1007/978-3-319-29613-5_5

17. Ernst, G., Pfähler, J., Schellhorn, G., Reif, W.: Modular, crash-safe refinement for ASMs with submachines. Sci. Comput. Program. **131**, 3–21 (2016). Abstract State Machines, Alloy, B, TLA, VDM and Z (ABZ 2014)
18. Ernst, G., Schellhorn, G., Haneberg, D., Pfähler, J., Reif, W.: Verification of a virtual filesystem switch. In: Cohen, E., Rybalchenko, A. (eds.) VSTTE 2013. LNCS, vol. 8164, pp. 242–261. Springer, Heidelberg (2014). https://doi.org/10.1007/978-3-642-54108-7_13
19. Grandy, H., Bischof, M., Stenzel, K., Schellhorn, G., Reif, W.: Verification of Mondex electronic purses with KIV: from a security protocol to verified code. In: Cuellar, J., Maibaum, T., Sere, K. (eds.) FM 2008. LNCS, vol. 5014, pp. 165–180. Springer, Heidelberg (2008). https://doi.org/10.1007/978-3-540-68237-0_13
20. Haneberg, D., Moebius, N., Reif, W., Schellhorn, G., Stenzel, K.: Mondex: engineering a provable secure electronic purse. Int. J. Softw. Inform. **5**(1), 159–184 (2011). http://www.ijsi.org
21. Haneberg, D., Schellhorn, G., Grandy, H., Reif, W.: Verification of Mondex electronic purses with KIV: from transactions to a security protocol. Formal Aspects Comput. **20**(1), 41–59 (2008). https://doi.org/10.1007/s00165-007-0057-0
22. Herlihy, M.P., Wing, J.M.: Linearizability: a correctness condition for concurrent objects. ACM Trans. Program. Lang. Syst. (TOPLAS) **12**(3), 463–492 (1990)
23. Hunter, A.: A brief introduction to the design of UBIFS (2008). http://www.linux-mtd.infradead.org/doc/ubifs_whitepaper.pdf
24. Izraelevitz, J., Mendes, H., Scott, M.L.: Linearizability of persistent memory objects under a full-system-crash failure model. In: Gavoille, C., Ilcinkas, D. (eds.) DISC 2016. LNCS, vol. 9888, pp. 313–327. Springer, Heidelberg (2016). https://doi.org/10.1007/978-3-662-53426-7_23
25. Jones, C., Woodcock, J. (eds.): Formal Aspects Comput. **20**(1) (2008). https://link.springer.com/journal/165/volumes-and-issues/20-1
26. Lipton, R.J.: Reduction: a method of proving properties of parallel programs. Commun. ACM **18**(12), 717–721 (1975)
27. Linux MTD: NAND and NAND simulator. http://www.linux-mtd.infradead.org/faq/nand.html
28. O'Connor, L., et al.: Refinement through restraint: bringing down the cost of verification. In: Proceedings of the 21st International Conference on Functional Programming Languages (ICFP), ICFP 2016, pp. 89–102. Association for Computing Machinery, New York (2016)
29. Pfähler, J., Ernst, G., Bodenmüller, S., Schellhorn, G., Reif, W.: Modular verification of order-preserving write-back caches. In: Polikarpova, N., Schneider, S. (eds.) IFM 2017. LNCS, vol. 10510, pp. 375–390. Springer, Cham (2017). https://doi.org/10.1007/978-3-319-66845-1_25
30. Pfähler, J., Ernst, G., Schellhorn, G., Haneberg, D., Reif, W.: Formal specification of an erase block management layer for flash memory. In: Bertacco, V., Legay, A. (eds.) HVC 2013. LNCS, vol. 8244, pp. 214–229. Springer, Cham (2013). https://doi.org/10.1007/978-3-319-03077-7_15
31. The Open Group Base Specifications Issue 7, IEEE Std. 1003.1, 2018 Edition. The IEEE and The Open Group (2017)
32. Schellhorn, G.: Verification of abstract state machines. Ph.D. thesis, Universität Ulm, Fakultät für Informatik (1999). https://www.uni-augsburg.de/en/fakultaet/fai/isse/prof/swtse/team/schellhorn/
33. Schellhorn, G.: Verification of ASM refinements using generalized forward simulation. J. Univ. Comput. Sci. (J.UCS) **7**(11), 952–979 (2001). http://www.jucs.org

34. Schellhorn, G.: Completeness of ASM refinement. Electron. Notes Theor. Comput. Sci. **214**, 25–49 (2008)
35. Schellhorn, G., Ahrendt, W.: The WAM case study: verifying compiler correctness for prolog with KIV. In: Bibel, W., Schmitt, P. (eds.) Automated Deduction – A Basis for Applications, Volume III: Applications, Chapter 3: Automated Theorem Proving in Software Engineering, pp. 165–194. Kluwer Academic Publishers (1998)
36. Schellhorn, G., Bodenmüller, S., Pfähler, J., Reif, W.: Adding concurrency to a sequential refinement tower. In: Raschke, A., Méry, D., Houdek, F. (eds.) ABZ 2020. LNCS, vol. 12071, pp. 6–23. Springer, Cham (2020). https://doi.org/10.1007/978-3-030-48077-6_2
37. Schellhorn, G., Tofan, B., Ernst, G., Pfähler, J., Reif, W.: RGITL: a temporal logic framework for compositional reasoning about interleaved programs. Ann. Math. Artif. Intell. **71**(1), 131–174 (2014). https://doi.org/10.1007/s10472-013-9389-z
38. Schierl, A., Schellhorn, G., Haneberg, D., Reif, W.: Abstract specification of the UBIFS file system for flash memory. In: Cavalcanti, A., Dams, D.R. (eds.) FM 2009. LNCS, vol. 5850, pp. 190–206. Springer, Heidelberg (2009). https://doi.org/10.1007/978-3-642-05089-3_13
39. Stärk, R.F., Schmid, J., Börger, E.: Java and the Java Virtual Machine-Definition, Verification, Validation. Springer, Heidelberg (2001). https://doi.org/10.1007/978-3-642-59495-3
40. Stenzel, K.: A formally verified calculus for full Java Card. In: Rattray, C., Maharaj, S., Shankland, C. (eds.) AMAST 2004. LNCS, vol. 3116, pp. 491–505. Springer, Heidelberg (2004). https://doi.org/10.1007/978-3-540-27815-3_37
41. Stepney, S., Cooper, D., Woodcock, J.: AN ELECTRONIC PURSE Specification, Refinement, and Proof. Technical Monograph PRG-126, Oxford University Computing Laboratory, July 2000. http://www-users.cs.york.ac.uk/susan/bib/ss/z/monog.htm
42. Szeredi, M.: File system in user space. http://fuse.sourceforge.net
43. Xu, Q., de Roever, W.-P., He, J.: The rely-guarantee method for verifying shared variable concurrent programs. Formal Aspects Comput. **9**(2), 149–174 (1997). https://doi.org/10.1007/BF01211617
44. Zou, M., Ding, H., Du, D., Fu, M., Gu, R., Chen, H.: Using concurrent relational logic with helpers for verifying the AtomFS file system. In: Proceedings of the SOSP, SOSP 2019, pp. 259–274. ACM (2019)

Computation on Structures
Behavioural Theory, Logic, Complexity

Klaus-Dieter Schewe[✉]

UIUC Institute, Zhejiang University, Haining, China
kd.schewe@intl.zju.edu.cn

Abstract. Over the last decades the field of computer science has changed a lot. In practice we are now dealing with very complex systems, but it seems that the theoretical foundations have not caught up with the development. This article is dedicated to a demonstration how a modernised theory of computation may look like. The theory is centred around the notion of algorithmic systems addressing behavioural theory, logic and complexity theory.

Keywords: Theory of computation · Behavioural theory · Computation on structures · Complexity theory · Logic · Abstract State Machines · Parallel algorithms · Insignificant choice · PTIME

Dear Egon,

The first time we met was at a Dagstuhl seminar in 1997 organised by Bernhard Thalheim. You were sceptical concerning my presentation on consistency enforcement in formal specifications due to the use of predicate transformers, but I think I could convince you that the existence proof (in infinitary logic) is possible, though the doubts concerning their usefulness remained. We also discussed about your 1985 monograph on computation theory, logic and complexity in the field of computer science, which in the very same year I used as a text in an introductory course on Theoretical Computer Science. Though some of the material was considered very demanding for the students, it was (and still is) one of the best texts describing the links between computation theory, logic and complexity theory, as it was handled until that time.

It took years to meet again, because Egon started to develop the very successful use of Abstract State Machines (ASMs) for rigorous software development, while I had turned my back on "formal methods" after discovering how little the FM community was interested in mathematical foundations. This changed again after getting to know ASMs better. I tentatively started putting Ph.D. students on the track, and in one case the intended quick exploitation of ASMs for database transformations became the basis of a convincing theory of parallel algorithms.

More than 35 years passed since the publication of your monograph on computation theory, logic and complexity, and over this period the field of computer

© Springer Nature Switzerland AG 2021
A. Raschke et al. (Eds.): Börger Festschrift, LNCS 12750, pp. 266–282, 2021.
https://doi.org/10.1007/978-3-030-76020-5_15

science has changed a lot. In practice we are now dealing with complex systems of systems, but though ASMs turned out very suitable to cover the developments, it seems that the theoretical foundations have not caught up with it. This birthday present is dedicated to a demonstration how a modernised theory of computation may look like. The theory is centred around the notion of algorithmic systems, which are harder to define than computable functions, in particular, when all developments in computing are to be taken into account.

I will argue that behavioural theories are key to the understanding, i.e. we require language-independent axiomatic definitions of classes of algorithmic systems that are accompanied by abstract machine models provably capturing the class under consideration. The machine models give further rise to tailored logics through which properties of systems in the considered class can be formalised and verified, and to fine-tuned classifications on the grounds of complexity restrictions. I will outline that all extensions will be (1) *conservative* in the sense that the classical theory of computation is preserved, (2) *universal* in the sense that all practical developments are captured uniformly, and (3) *practical* in the sense that languages associated with the abstract machine models can be used for rigorous high-level systems design and development, and the logics can be exploited for rigorous verification of desirable properties of systems. This links to your newer monographs focusing on the theory and application of the Abstract State Machine method.

1   Towards a Theory of Computation on Structures

In 1985 Egon Börger published his influential monograph on computation theory, logic and complexity (see the English translation in [8]), which focused on the concept of *formal language* as carrier of the precise expression of meaning, facts and problems, and the concept of *algorithm* or calculus, i.e. a formally operating procedure for the solution of precisely described questions and problems. At that time the text was at the forefront of a modern theory of these concepts, paving the way in which they developed first in mathematical logic and computability theory and later in automata theory, theory of formal languages and complexity theory.

Nonetheless, it became clear that the state of the theory left many open problems. Computing started to stretch out into many new application areas. Distributed computing over networks became possible, database systems facilitated concurrent computation, artificial intelligence ventured from a niche area to a useful technology enabling inferential problem solving in diagnosis, controlling machines through software became possible, etc. Now, only 35 years later the rapid progress in computing has led to a fascinating variety of interconnected systems that are used to support, manage and control many aspects of our life. There is hardly an area that has not yet been penetrated by computing, and still there are many open challenges for the continuation of this success story.

We are now dealing with systems of systems that are

- operating in *parallel* exploiting synchronously multiple processor cores and asynchronously computing resources distributed over networks,
- *hybrid* interacting with analogue systems with continuous behaviour,
- *adaptive* changing their own behaviour,
- *intelligent* reasoning about themselves and their environment,
- *interactive* communicating with their environment, and
- *random* depending on probability distributions.

All these developments require scientific foundations centred around computation theory, complexity and logic:

- Is there a theory of computation that faithfully covers all the aspects of systems of computing systems that occur in practice?
- Is there a methodology grounded in such a theory of computation that permits the definition and classification of complex systems and the provision of means for specification, systematic development, validation and verification?
- Is there a methodology that permits reasoning about problems and their solutions in terms of correctness and complexity?

In 1982 Chandra and Harel raised the problem, whether there exists a computation model over structures that captures the complexity class PTIME rather than Turing machines that operate over finite strings [17]. The problem reflects the typically huge gap between the abstraction level of an algorithm or more generally a system of algorithmic systems and the level of Turing machines. It is not sufficient to know that deep inside the core of systems we deal with computations that given a proper string encoding can be represented by Turing machines; instead, computation theory has to stretch to arbitrary Tarski structures that are omnipresent in all mathematical theories, and any extension should be conservative in the sense that the classical theory is preserved as a representation on the lowest level of abstraction.

A first answer was given in 1985 by Gurevich's "new thesis" [26], which was further elaborated in the 1995 Lipari guide [28]. The new theory emphasises Tarski structures (aka universal algebras) to capture abstract states of systems and evolving algebras, now known as Abstract State Machines (ASMs), as the abstract machines capturing the algorithms on arbitrary levels of abstraction. Egon Börger realised that these ideas do not only create a new paradigm for the foundations of computing subsuming the classical theory, but at the same can be exploited for rigorous systems engineering in practice thereby fulfilling the criteria of a "software engineering" discipline that deserves this name as envisioned in the 1968 meeting in Garmisch, where this notion was coined [32].

A remarkable success story started leading to proofs of compiler correctness for the Warren Abstract Machine for Prolog [13], the translation from Occam to transputers [10], the compilation of Java and the bytecode verifier [37], the development of the sophisticated theory of ASM refinements [9], and much more. The state of the theory and practice of ASMs is well summarised in Egon Börger's and Robert Stärk's monograph on ASMs [16]. More recent examples are found in the modelling companion by Börger and Raschke [12].

While the development proved that ASMs can take over the role of the formal languages in computation theory, it took until 2000 to develop the celebrated "sequential ASM thesis" [29], which is based on the observation that "if an abstraction level is fixed (disregarding low-level details and a possible higher-level picture) and the states of an algorithm reflect all the relevant information, then a particular small instruction set suffices to model any algorithm, never mind how abstract, by a generalised machine very closely and faithfully". On one hand the thesis provided a language-independent definition of the notion of *sequential algorithm* giving for the first time in history a precise axiomatic definition of the notion of "algorithm" (though restricted to sequential algorithms). On the other hand it contained the proof that all algorithms as stipulated by the defining postulates are faithfully captured by sequential ASMs. This justified further to establish another new notion: a *behavioural theory* comprises a machine-independent axiomatic definition of a class of algorithms (or more generally: algorithmic systems), an abstract machine model, and a proof that the machine model captures the class of computations.

Starting from the first behavioural theory, the theory of sequential algorithms, further success stories followed. Moschovakis's critical question how recursion could be captured was answered by the behavioural theory of recursive algorithms [15]. A first attempt to extend the theory to parallel algorithms was undertaken by Blass and Gurevich [5], but it was not well received due to the use of concepts such as mailbox, display and ken that were considered too close to the machine model, but another behavioural theory of parallel algorithms without these restrictions was then developed in [22]. This closed the case of synchronous parallel algorithms. A convincing behavioural theory for asynchronous algorithmic systems was developed in [14] with *concurrent ASMs* as the machine model capturing concurrent algorithms, i.e. families of sequential or parallel algorithms associated with agents that are oblivious to the actions of each other apart from recognising changes to shared locations. Recently, a behavioural theory of reflective algorithms was developed addressing the question how to capture algorithmic systems that can adapt their own behaviour [34].

The behavioural theories yield variants of Abstract State Machines that can be used for rigorous systems development. Furthermore, Stärk and Nanchen developed a logic for the reasoning about deterministic ASMs [36]. As discussed in [16] it was considered difficult to extend this logic to the case of non-deterministic ASMs[1]. This gap was closed in [23] by making update sets first-class objects in the theory and proving completeness with respect to Henkin semantics. It was also shown how the logic can be adapted to reason about concurrent ASMs [24]. An extension to reflective ASMs was approached in [35]. On one side it shows the tight connections between the classes of algorithmic systems handled in the behavioural theories. On the other side it shows that the development of the logical counterpart of the theories has not yet reached the same development state.

---

[1] Note a full behavioural theory of non-deterministic algorithms does not yet exist.

This applies even more so to complexity theory. One of the few studies trying to bring complexity theory to the theory of ASMs, which after all provide the theory of computations on structures as asked for by Chandra and Harel, is the theory of *choiceless polynomial time* (CPT) [6,7], which studies the choiceless fragment of PTIME using PTIME bounded deterministic Abstract State Machines. Though it was possible to show that CPT subsumes other models of computation on structures[2] such as relational machines [3], reflective relational machines [1] and generic machines [2], it is strictly included in PTIME. If the hope had been to exhaust PTIME the same as existential second-order logic captures NP [21], this failed. No systematic research trying to close the gap between CPT and PTIME followed, and Gurevich posted his conjecture that there is no logic capturing PTIME [27].

If true, it would doom all further attempts in this direction. This would further imply that complexity theory as a whole, in particular descriptive complexity theory [30] which is tighly coupled with finite model theory [20,31], could not be based on more abstract models of computations on structures. In particular, it would not be possible to avoid dealing with string encodings using Turing Machines. However, this consequence appears to be less evident in view of the ASM success stories. Various attempts have been undertaken to refute Gurevich's conjecture either by adding quantifiers such as counting [19] or by adding non-deterministic choice operators [4,25]. A comparison and evaluation is contained in [18].

All these attempts failed, and the main reason for the failure is the neglection of the computations understood as yielding sequences of abstract states with update sets defining the state transitions. Instead, only the functional relationship between the input structure and the Boolean output was emphasised. This restriction to Boolean queries blurs the subtle distinctions that become possible, when the behavioural theory and the associated logic are taken into account. A refutation of Gurevich's conjecture has been achieved in [33] exploiting *insignificant choice*[3] thus leading to *insignificant choice polynomial time* (ICPT). Based on the insight that choice is unavoidable to capture PTIME it is not too hard to see that PTIME problems can be solved by polynomial time bounded ASMs with insignificant choice, as it suffices to create an order on the set of atoms in the base set. This construction is rather specific, as it exploits to choose only atoms, and it permits to replace arbitrary insignificant choice ASMs by ASMs satisfying a *local insignificance condition*. This condition can be expressed in the logic of non-deterministic ASMs [23,24]. To show that the extension remains within PTIME it suffices to simulate PTIME ASMs with choices among atoms that satisfy the local insignificance condition by PTIME Turing machines with input strings given by the standard encoding of an ordered version of the input structure. Here the local insignificance permits to choose always the smallest atom,

---

[2] Strictly speaking, all these previous computational models are still based on Turing machines, which are coupled with queries on relational stores.

[3] Insignificant choice imposes two conditions on the update sets yielded by a choice. The first of these conditions is similar to semi-determinism [38].

and the PTIME bound results from the fact that local insignificance checking for choices among atoms can be done in polynomial time. With this logic capturing PTIME it then becomes possible to show that PTIME and NP differ [33].

In the remainder of this article I will further elaborate how behavioural theories, associated logics and complexity work together. The emphasis will be on parallel algorithms. In Sect. 2 I will start from the behavioural theory of parallel algorithm, which will be extended by insignificant choice. This does not alter the expressiveness, but justifies the use of choice rules in many practical examples using ASMs [12]. In Sect. 3 I will proceed with the logic of non-deterministic ASMs and outline how it needs to be modified to capture only insignificant choice. Finally, Sect. 4 brings in polynomial time, where the presence or absence of choice makes a significant difference. In fact, it is the difference between CPT and ICPT. I conclude with a brief outlook in Sect. 5 emphasising that this is just a brief demonstration of how a modernised theory of computation centred around the notion of algorithmic systems may look like.

2   Parallel Algorithms

Let us briefly review the parallel ASM thesis [22], and extend the theory by insignificant choice as in [33]. Note that different from classical computation theory the behavioural theory characterises the class of parallel algorithms by four postulates and then proves that the class is captured by the Abstract State Machines, which is more than just defining the semantics of ASMs.

2.1   The Parallel ASM Thesis

Deterministic algorithms proceed in steps, which is reflected in the sequential time postulate for sequential algorithms [29]. Parallel algorithms[4] do not make a change here; only the amount of updates characterising the transition from a state to its successor varies significantly.

Postulate 1 (Sequential Time Postulate). A *parallel algorithm* $\mathcal{A}$ comprises a non-empty set $\mathcal{S}$ of *states*, a non-empty subset $\mathcal{I} \subseteq \mathcal{S}$ of *initial states*, and a one-step transformation function $\tau : \mathcal{S} \rightarrow \mathcal{S}$.

Same as for sequential algorithms a *state* has to reflect all the relevant information, so we also preserve the abstract state postulate, which characterises states as Tarski structures over a fixed signature, i.e. a set of function symbols.

Postulate 2 (Abstract State Postulate). Every *state* $S \in \mathcal{S}$ of a parallel algorithm is a structure over a fixed finite signature Σ such that both $\mathcal{S}$ and $\mathcal{I}$ are closed under isomorphisms, the one-step transformation τ of $\mathcal{A}$ does not change the base set of any state, and if two states S and S' are isomorphic via $\zeta : S \rightarrow S'$, then $\tau(S)$ and $\tau(S')$ are also isomorphic via ζ.

---

[4] More precisely: unbounded parallel algorithms, as sequential algorithms algorithms already subsume bounded parallelism. The difference is that in the unbounded case the parallel branches of a computation depend on the state.

These two postulates alone give already rise to several decisive definitions. A *run* of a parallel algorithm $\mathcal{A}$ is a sequence $S_0, S_1, \ldots$ of states with $S_0 \in \mathcal{I}$ and $S_{i+1} = \tau(S_i)$ for all $i \geq 0$. A *location* of state S is a pair $(f, (a_1, \ldots, a_n))$ with a function symbol $f \in \Sigma$ of arity n and an n-tuple of elements a_i of the base set of S. The *value* $val_S(\ell)$ of a location ℓ in state S is $f_S(a_1, \ldots, a_n)$ using the interpretation f_S of f in S. An *update* in S is a pair (ℓ, v) comprising a location ℓ of S and an element v of the base set of S. An *update set* in S is a set of such updates.

An update set Δ is called *consistent* iff $(\ell, v_1), (\ell, v_2) \in \Delta$ imply $v_1 = v_2$. For a consuistent update set Δ in S we obtain a state $S' = S + \Delta$ with $val_{S'}(\ell) = v$ for $(\ell, v) \in \Delta$, and $val_{S'}(\ell) = val_S(\ell)$ otherwise. Any two states S, S' with the same base set define a unique minimal consistent update set $\Delta(S)$ with $S' = S + \Delta(S)$. In particular, we write $\Delta_{\mathcal{A}}(S)$ for the update set defined by S and its successor $\tau(S)$.

Update sets $\Delta_{\mathcal{A}}(S)$ must be determined by the parallel algorithm, which has an intrinsic finite representation. For sequential algorithms it suffices to assume that this finite representation contains a finite set of ground terms over the signature Σ such that the evaluation of these terms in a state S uniquely determines the updates in S. This gives rise to the *bounded exploration postulate*. For parallel algorithms this is slightly more complicated, as in every state the algorithm may execute an arbitrary number of parallel branches. However, these branches are determined by the state. As there must exist a finite representation, it is justified to assume that the branches are determined by terms, so it suffices to replace the ground terms by multiset comprehension terms[5].

Postulate 3 (Bounded Exploration Postulate). Every parallel algorithm $\mathcal{A}$ of signature Σ comprises a finite set W (called *bounded exploration witness*) of multiset comprehension terms $\{\!\{ t(\bar{x}, \bar{y}) \mid \varphi(\bar{x}, \bar{y}) \}\!\}_{\bar{x}}$ over signature Σ such that $\Delta_{\mathcal{A}}(S) = \Delta_{\mathcal{A}}(S')$ holds, whenever the states S and S' of $\mathcal{A}$ coincide on W.

Finally, each computation has a *background* comprising the implicit fixed values, functions and constructors that are exploited, but not defined in the signature. For sequential algorithms the background was kept implicit, as it merely requires the presence of truth values and the usual operators on them, a value *undef* to capture partial functions, and an infinite reserve, from which new values can be taken if necessary. Parallel algorithms must in addition require the presence of tuples and multisets as already used for bounded exploration. This leads to the *background postulate*.

Postulate 4 (Background Postulate). Each parallel algorithm $\mathcal{A}$ comprises a background class $\mathcal{K}$ defining at least a binary *tuple constructor* and a *multiset constructor* of unbounded arity, and a background signature Σ_B contains at least the following static function symbols:

---

[5] It must be multiset terms and not set terms, as there may be multiple branches doing the same.

- nullary function symbols `true`, `false`, `undef` and $\{\!\{\,\}\!\}$,
- unary function symbols `reserve`, `Boole`, $\neg$, `first`, `second`, $\{\!\{\cdot\}\!\}$, $\uplus$ and `AsSet`, and
- binary function symbols $=$, $\wedge$, $\vee$, $\rightarrow$, $\leftrightarrow$, $\uplus$ and $(\,,\,)$.

We assume general familiarity with Abstract State Machines [16], so we will not define them here. Then the key result in [22] is the following "parallel ASM thesis".

Theorem 1. *Abstract State Machines capture parallel algorithms as defined by the sequential time, abstract state, bounded exploration and background postulates.*

The proof that ASMs fulfil the requirement of the Postulates 1–4 is not very difficult. A bounded exploration witness can be constructed from an ASM rule; then showing the decisive property of Postulate 3 is rather straightforward.

The proof that every parallel algorithm as stipulated by the four postulates can be step-by-step simulated by an ASM with the same background and signature is complicated. The key argument is to show that if an update set $\Delta_{\mathcal{A}}(S)$ contains an update $((f, (a_1, \ldots, a_n)), a_0)$, then any $(n+1)$-tuple $(b_0, \ldots, b_n)$ with the same type as $(a_0, \ldots, a_n)$ also defines an update $((f, (b_1, \ldots, b_n)), b_0) \in \Delta_{\mathcal{A}}(S)$, where the *type* is defined by a bounded exploration witness W. Exploiting *isolating formulae* for types gives rise to a **forall**-rule r_S with $\Delta_{r_S}(S) = \Delta_{\mathcal{A}}(S)$. The extension to a single rule r with $\Delta_r(S) = \Delta_{\mathcal{A}}(S)$ for all states S uses the same ideas as the proof of the sequential ASM thesis with straightforward generalisations.

2.2 Parallel Algorithms with Choice

As shown by many examples in [12, 16] it is often useful to permit non-deterministic choice. We will therefore explore how to extend the parallel ASM thesis to a non-deterministic parallel ASM thesis, then restrict choice such that it becomes insignificant, i.e. the final result does not depend on the choice (up to isomorphism).

Clearly, the abstract state and background postulates can be preserved, but the sequential time postulate has to be replaced by a *branching time postulate*.

Postulate 5 (Branching Time Postulate). A *non-deterministic parallel algorithm* $\mathcal{A}$ comprises a non-empty set $\mathcal{S}$ of *states*, a non-empty subset $\mathcal{I} \subseteq \mathcal{S}$ of *initial states*, and a one-step transformation relation $\tau \subseteq \mathcal{S} \times \mathcal{S}$.

We continue to call each state $S' \in \tau(S)$ a *successor state* of S. Then S and $S' \in \tau(S)$ define a unique minimal consistent update set $\Delta(S, S')$ with $S + \Delta(S, S') = S'$. Let $\boldsymbol{\Delta}_{\mathcal{A}}(S) = \{\Delta(S, S') \mid S' \in \tau(S)\}$ denote the *set of update sets* in state S.

In the same way as the shift from sequential algorithms to parallel algorithms required multiset comprehensions, it seems plausible that also the shift from parallel algorithms to non-deterministic parallel algorithms will require multiset comprehensions. We therefore define a *witness term* as a term of the form

$$\{\!\{\{\!\{t(\bar{x}, \bar{y}) \mid \varphi(\bar{x}, \bar{y})\}\!\} \mid \psi(\bar{x})\}\!\} \ .$$

Then the bounded exploration postulate could be altered as follows:

Postulate 6 (Non-Deterministic Bounded Exploration Postulate). Every non-deterministic parallel algorithm $\mathcal{A}$ of signature Σ comprises a finite set W (called *bounded exploration witness*) of witness terms over signature Σ such that $\boldsymbol{\Delta}_{\mathcal{A}}(S) = \boldsymbol{\Delta}_{\mathcal{A}}(S')$ holds, whenever the states S and S' of $\mathcal{A}$ coincide on W.

It is again rather straightforward to show that non-deterministic ASMs satisfy the modified postulated for non-deterministic parallel algorithm.

Theorem 2. *Non-deterministic Abstract State Machines define non-deterministic parallel algorithms as defined by the branching time, abstract state, non-deterministic bounded exploration and background postulates.*

However, a proof that non-deterministic Abstract State Machines capture non-deterministic parallel algorithms has not yet been completed. This will be dealt with elsewhere. Our interest here is on a restricted version of non-determinism.

In a run $S_0, S_1, \ldots$ we call a state S_n *final* iff $\tau(S_n) = \{S_n\}$ holds, i.e. there is no more change to the state. Assuming that some function symbols in Σ have been declared as *output functions*. Let $out(S)$ denote the restriction of a final state S to its output locations. Then we call a non-deterministic parallel algorithm $\mathcal{A}$ a *insignificant choice algorithm* iff every run has a final state and for any two final states S_1 and S_2 the outputs $out(S_1)$ and $out(S_2)$ are isomorphic.

Next consider *locally insignificant choice* ASMs, i.e. non-deterministic ASMs with the following properties:

(i) For every state S any two update sets $\Delta, \Delta' \in \boldsymbol{\Delta}(S)$ are isomorphic, and we can write

$$\boldsymbol{\Delta}(S) \ = \ \{\sigma\Delta \mid \sigma \in G\} \,,$$

where $G \subseteq \mathcal{I}so$ is a set of isomorphisms and $\Delta \in \boldsymbol{\Delta}(S)$ is an arbitrarily chosen update set.

(ii) For every state S with $\boldsymbol{\Delta}(S) = \{\sigma_i\Delta_0 \mid 0 \le i \le k\}$ ($G = \{\sigma_0, \ldots, \sigma_k\} \subseteq \mathcal{I}so$) and the corresponding successor states $S_i = S + \sigma_i\Delta_0$ we have

$$\boldsymbol{\Delta}(S_i) \ = \ \sigma_i\boldsymbol{\Delta}(S_0) \,.$$

Theorem 3. *Locally insignificant choice Abstract State Machines define insignificant choice algorithms as defined by the branching time, abstract state, non-deterministic bounded exploration and background postulates and the insignificant choice restriction.*

We cannot yet provide a proof that locally insignificant choice Abstract State Machines capture insignificant choice algorithms, but it seems as plausible as Theorem 2.

3   The Logic of Non-deterministic ASMs

Let us now tend to associated logics. A logic for deterministic ASMs has been developed by Stärk and Nanchen [36] and proven to be complete. It is also described in [16]. We now look into the extension for non-deterministic ASMs [24] and how it can be adapted to capture the insignificant choice.

3.1   Unrestricted Logic

As the logic of non-deterministic ASMs has to deal with update sets, we let the signature contain a static constant symbol c_f for each dynamic function symbol $f \in \Sigma$, i.e. c_f is not dynamic and has arity 0. We also exploit that the base set contains elements that interpret c_f in every state. By abuse of notation we wrote $(c_f)_S = c_f$. Now let X be a second-order variable of arity 3. For a variable assignment ζ we say that $\zeta(X)$ *represents an update set* Δ iff for each $((f, \bar{a}), b) \in \Delta$ we have $(c_f, \bar{a}, b) \in \zeta(X)$ and vice versa. Here we write $\bar{a}$ for n-tuples, where n is the arity of f.

As for the syntax, with this extension the terms of $\mathcal{L}^{nd}$ are ASM terms. The formulae of the logic are defined inductively as follows:

- If t and t' are terms, then $t = t'$ is a formula.
- If X is an n-ary second-order variable and $t_1, \ldots, t_n$ are terms, then $X(t_1, \ldots, t_n)$ is a formula.
- If r is an ASM rule and X is a second-order variable of arity 3, then $\mathrm{upd}_r(X)$ is a formula.
- If φ and ψ are formulae, then also $\neg\varphi$, $\varphi \wedge \psi$, $\varphi \vee \psi$ and $\varphi \to \psi$ are formulae.
- If φ is a formula, x is a first-order variable and X is a second-order variable, then also $\forall x.\varphi$, $\exists x.\varphi$, $\forall X.\varphi$, $\exists X.\varphi$ are formulae.
- If φ is a formula and X is a second-order variable of arity 3, then $[X]\varphi$ is formula.

The semantics is defined for Henkin structures. A *Henkin prestructure* $\tilde{S}$ over signature Υ is a structure S over Σ with base set B together with sets of relations $D_n \subseteq \mathcal{P}(B^n)$ for all $n \geq 1$.

As the logic uses second-order variables we need extended variable assignments ζ into a Henkin prestructure. For first-order variables x we have $\zeta(x) \in B$ as usual, but for second-order variables X of arity n we request $\zeta(X) \in D_n$. Then with respect to a Henkin prestructure $\tilde{S}$ and such a variable assignment terms are interpreted as usual. The interpretation $[\![\varphi]\!]_{\tilde{S},\zeta}$ for formulae φ is mostly standard with the non-standard parts defined as follows:

- If φ has the form $\forall X.\psi$ with a second-order variable X of order n, then

$$[\![\varphi]\!]_{\tilde{S},\zeta} = \begin{cases} \mathbf{T} & \text{if } [\![\psi]\!]_{\tilde{S},\zeta[X \mapsto A]} = \mathbf{T} \text{ for all } A \in D_n \\ \mathbf{F} & \text{else} \end{cases}.$$

– If φ has the form $[X]\psi$, then

$$[\![\varphi]\!]_{\tilde{S},\zeta} = \begin{cases} \mathbf{F} & \text{if } \mathrm{val}_{S,\zeta}(X) \text{ represents a consistent update set } \Delta \\ & \text{with } [\![\psi]\!]_{\tilde{S}+\Delta,\zeta} = \mathbf{F} \\ \mathbf{T} & \text{else} \end{cases}.$$

While this interpretation is defined for arbitrary Henkin prestructures, it makes sense to restrict the collections D_n of n-ary relations to those that are closed under definability, which defines the notion of Henkin structure. We then say that a sentence is *valid* iff it is interpreted as 1 (i.e., true) in all Henkin structures.

A *Henkin structure* over signature Σ is a Henkin prestructure $\tilde{S} = (S, \{D_n\}_{n\geq 1})$ that is closed under definability, i.e. for every formula φ, every variable assignment ζ and every $n \geq 1$ we have

$$\{(a_1, \ldots, a_n) \in B^n \mid [\![\varphi]\!]_{\tilde{S},\zeta[x_1 \mapsto a_1, \ldots, x_n \mapsto a_n]} = \mathbf{T}\} \in D_n.$$

3.2 Capturing Insignificant Choice

We now approach a characterisation of the semantic insignificant choice restriction in the logic $\mathcal{L}^{nd}$ defined above. We use $\mathrm{isUSet}(X)$ to express that X represents an update set, and $\mathrm{conUSet}(X)$ to express that it is consistent—these are defined in [24].

Let us assume that the base set B is defined as the set of hereditarily finite sets $B = HF(A)$ over a finite set A of *atoms*. Then we can express that X is an isomorphism by

$$\mathrm{iso}(X) \equiv \forall x, y_1, y_2.(X(x, y_1) \wedge X(x, y_2) \to y_1 = y_2) \wedge$$
$$\forall x_1, x_2, y.(X(x_1, y) \wedge X(x_2, y) \to x_1 = x_2) \wedge \forall x \exists y.X(x, y) \wedge \forall y \exists x.X(x, y) \wedge$$
$$\bigwedge_{f \in \Upsilon_{dyn}} X(c_f, c_f) \wedge \forall x, y.\Big[X(x, y) \to (x \in \mathit{Atoms} \leftrightarrow y \in \mathit{Atoms}) \wedge$$
$$\forall u.(u \in x \to \exists v.v \in y \wedge X(u, v)) \wedge \forall v.(v \in y \to \exists u.u \in x \wedge X(u, v))\Big]$$

This leads to the following *insignificance constraint* for a rule r expressing that any two update sets yielded by r are isomorphic:

$$\forall X_1, X_2. \, \mathrm{upd}_r(X_1) \wedge \mathrm{upd}_r(X_2) \to$$
$$\exists X.(\mathrm{iso}(X) \wedge \mathrm{updIso}(X_1, X_2, X) \wedge \mathrm{updIsoSet}(X_1, X_2, X))$$

with

$$\mathrm{updIso}(X_1, X_2, X) \equiv \bigwedge_{f \in \Upsilon_{dyn}} [\forall \bar{x}_1, x_2, \bar{y}_1, y_2.$$

$$(X_1(c_f, \bar{x}_1, x_2) \wedge \bigwedge_{1 \leq i \leq ar(f)} X(x_{1i}, y_{1i}) \wedge X(x_2, y_2) \rightarrow X_2(c_f, \bar{y}_1, y_2)) \wedge$$

$$\forall \bar{x}_1, x_2, \bar{y}_1, y_2. (X_2(c_f, \bar{x}_1, x_2) \wedge \bigwedge_{1 \leq i \leq ar(f)} X(x_{1i}, y_{1i}) \wedge X(x_2, y_2) \rightarrow X_1(c_f, \bar{y}_1, y_2))]$$

and

$$\mathrm{updIsoSet}(X_1, X_2, X) \equiv \forall Y_1, Y_2. (\mathrm{isUSet}(Y_1) \wedge \mathrm{isUSet}(Y_2) \wedge \mathrm{updIso}(Y_1, Y_2, X))$$
$$\rightarrow ([X_1]\mathrm{upd}_r(Y_1) \leftrightarrow [X_2]\mathrm{upd}_r(Y_2))$$

We can use this characterisation of insignificant choice to modify the logic in such a way that a choice rule will either become an insignificant choice or interpreted as **skip**. For this recall the axiomatic definition of $\mathrm{upd}_r(X)$ from [24]. In order to express insignificant choice we introduce new formulae of the form $\mathrm{upd}_r^{ic}(X)$. If r is not a choice rule, we simply keep the definitions replacing upd by upd^{ic}. For a choice rule r of the form **choose** $v \in \{x \mid x \in Atoms \wedge x \in t\}$ **do** $r'(v)$ **enddo** we define

$$\mathrm{upd}_r^{ic}(X) \leftrightarrow \exists v. v \in Atoms \wedge v \in t \wedge \mathrm{upd}_{r'(v)}^{ic}(X) \wedge$$
$$\forall Y. (\exists x. x \in Atoms \wedge x \in t \wedge \mathrm{upd}_{r'(x)}^{ic}(Y)) \rightarrow$$
$$\exists Z. (\mathrm{iso}(Z) \wedge \mathrm{updIso}(X, Y, Z) \wedge \mathrm{updIsoSet}(X, Y, Z))$$

4 Complexity Restriction

Let us finally look at the link to complexity theory. We define PTIME restricted versions of parallel ASMs [6] and locally insignificant choice [33], which define choiceless polynomial time (CPT) and insignificant choice polynomial time. The former one is strictly included in PTIME; the latter one captures PTIME.

4.1 Choiceless Polynomial Time

In order to define a polynomial time bound on an ASM we have to count steps of a run. If we only take the length of a run, each step would be a macrostep that involves many elementary updates, e.g. the use of unbounded parallelism does not impose any restriction on the number of updates in an update set employed in a transition from one state to a successor state. So we better take the size of update sets into account as well. If objects are sets, their size also matters in estimating what an appropriate microstep is. This leads to the notion of PTIME bound from CPT [6].

A *PTIME (bounded) ASM* is a triple $\tilde{M} = (M, p(n), q(n))$ comprising an ASM M and two integer polynomials $p(n)$ and $q(n)$. A *run* of $\tilde{M}$ is an initial segment of a run of M of length at most $p(n)$ and a total number of at most $q(n)$ active objects, where n is the size of the input in the initial state of the run.

We say that a PTIME ASM $\tilde{M}$ *accepts* the input structure I iff there is a run of $\tilde{M}$ with initial state generated by I and ending in a state in which *Halt* holds and the value of *Output* is 1. Analogously, a PTIME ASM $\tilde{M}$ *rejects* the input structure I iff there is a run of $\tilde{M}$ with initial state generated by I and ending in a state in which *Halt* holds and the value of *Output* is 0.

A logic $\mathcal{L}$ can be defined by a pair (Sen,Sat) of functions satisfying the following conditions:

– *Sen* assigns to every signature Σ a recursive set $Sen(\Sigma)$, the set of $\mathcal{L}$-*sentences of signature* Σ.
– *Sat* assigns to every signature Σ a recursive binary relation Sat_Σ over structures S over Σ and sentences $\varphi \in Sen(\Sigma)$. We assume that $Sat_\Sigma(S, \varphi) \Leftrightarrow Sat_\Sigma(S', \varphi)$ holds, whenever S and S' are isomorphic.

We say that a structure S over Σ *satisfies* $\varphi \in Sen(\Sigma)$ (notation: $S \models \varphi$) iff $Sat_\Sigma(S, \varphi)$ holds.

If $\mathcal{L}$ is a logic in this general sense, then for each signature Σ and each sentence $\varphi \in Sen(\Sigma)$ let $K(\Sigma, \varphi)$ be the class of structures S with $S \models \varphi$. We then say that $\mathcal{L}$ is a *PTIME logic*, if every class $K(\Sigma, \varphi)$ is PTIME in the sense that it is closed under isomorphisms and there exists a PTIME Turing machine that accepts exactly the standard encodings of ordered versions of the structures in the class.

We further say that a logic $\mathcal{L}$ *captures PTIME* iff it is a PTIME logic and for every signature Σ every PTIME class of Σ-structures coincides with some class $K(\Sigma, \varphi)$.

4.2   Insignificant Choice Polynomial Time

An *insignificant choice ASM* (for short: icASM) is an ASM M such that for every run $S_0, \ldots, S_k$ of length k such that *Halt* holds in S_k, every $i \in \{0, \ldots, k-1\}$ and every update set $\Delta \in \boldsymbol{\Delta}(S_i)$ there exists a run $S_0, \ldots, S_i, S'_{i+1}, \ldots, S'_m$ such that $S'_{i+1} = S_i + \Delta$, *Halt* holds in S'_m, and *Output* = **true** (or **false**, respectively) holds in S_k iff *Output* = **true** (or **false**, respectively) holds in S'_m.

A *PTIME (bounded) insignificant choice ASM* (for short: PTIME icASM) is a triple $\tilde{M} = (M, p(n), q(n))$ comprising an icASM M and two integer polynomials $p(n)$ and $q(n)$ with runs such that whenever an input structure I is accepted by $\tilde{M}$ (or rejected, respectively) then every run on input structure I is accepting (or rejecting, respectively).

According to this definition whenever there exists an accepting or rejecting run, then all other runs on the same input structure, i.e. runs that result making different choices, are also accepting or rejecting, respectively.

Theorem 4. *ICPT captures PTIME on arbitrary finite structures, i.e. ICPT = PTIME.*

The full proof is given in [33]. In a nutshell, given a PTIME problem we simply use a non-deterministic ASM to first generate an order on the set of atoms, then create deterministically the standard encoding of the input structure with this order and finally simulate the PTIME Turing machine deciding the problem. Then it is clear that the choices in the ASM will only refer to atoms, and the local insignificance will be satisfied. This is then used to prove also the converse by creating a PTIME simulation by a Turing machine. The local insignificance condition implies global insignificance, i.e. any choice can be replaced by a fixed choice of the smallest element, and the fact that choices are restricted to atoms guarantees that the local insignificance condition can be checked on a Turing machine in polynomial time. The first part of the proof further shows that PTIME is included in a fragment of ICPT defined by ASMs satisfying the local insignificance condition.

Corollary 1. *PTIME is captured by the fragment $ICPT_{loc}$ of ICPT, where the separating icASM satisfies the local insignificance condition.*

Through Theorem 4 and Corollary 1 ICPT highlights the similarities and differences between classical computation theory on strings using Turing machines and computation theory on structures using ASMs. Not only does the shift to arbitrary Tarski structures lead to a theory on arbitrary level of abstraction, while at the same time enabling the proofs of long-standing open problems such as the refutation of Gurevich's conjecture and the separation of PTIME from NP [33], it shows that computation theory requires more than just functions from input to output. Furthermore, it helps closing the gap between the theory of computation and the developments in practice with the perspective to obtain a thorough theoretical penetration of practice, which is what actually was claimed by the term "Software Engineering".

5   Concluding Remarks

Monographs written or co-authored by Egon Börger provide cornerstones for the development of the theory of computation and its applications [8,11,16,37]. In this article I outlined bits of a modernised theory of computation on structures grounded in behavioural theories of classes of algorithmic systems, associated logics and complexity theory. The emphasis was on polynomial time computations. Starting from parallel algorithms I showed how to extend them by insignificant choice, which requires a modification of the logic of non-deterministic ASMs. Then I sketched the recently proven capture of PTIME. This shows how all parts of the theory fit neatly together. Nonetheless, there are still many open problems associated with the theory, which need to be addressed such as a theory of non-determinism and randomness. The next decisive monograph will be a consolidated theory of computations on structures.

Computations on structures give rise to specification of algorithmic systems on arbitrary levels of abstraction, i.e. they directly feed into rigorous system development. The logics associated with a particular class of algorithmic systems can be used in this context for the verification of desirable properties. Complexity classes enable fine-tuned classification with further insights how an algorithm solving a problem in a complexity class looks like. The various successful applications of the ASM method with or without choice, with single or multiple machines and with sophisticated refinement strategies show that computation theory on structures is well positioned to bridge the gap between the increasing structural complexity of modern software-centric systems and the foundational theory.

References

1. Abiteboul, S., Papadimitriou, C.H., Vianu, V.: The power of reflective relational machines. In: Proceedings of the Ninth Annual Symposium on Logic in Computer Science (LICS 1994), pp. 230–240. IEEE Computer Society (1994)
2. Abiteboul, S., Vardi, M.Y., Vianu, V.: Fixpoint logics, relational machines, and computational complexity. J. ACM **44**(1), 30–56 (1997). https://doi.org/10.1145/256292.256295
3. Abiteboul, S., Vianu, V.: Generic computation and its complexity. In: Koutsougeras, C., Vitter, J.S. (eds.) Proceedings of the 23rd Annual ACM Symposium on Theory of Computing (STOC 1991), pp. 209–219. ACM (1991)
4. Arvind, V., Biswas, S.: Expressibility of first order logic with a nondeterministic inductive operator. In: Brandenburg, F.J., Vidal-Naquet, G., Wirsing, M. (eds.) STACS 1987. LNCS, vol. 247, pp. 323–335. Springer, Heidelberg (1987). https://doi.org/10.1007/BFb0039616
5. Blass, A., Gurevich, Y.: Abstract State Machines capture parallel algorithms. ACM Trans. Comput. Logic **4**(4), 578–651 (2003)
6. Blass, A., Gurevich, Y., Shelah, S.: Choiceless polynomial time. Ann. Pure Appl. Logic **100**, 141–187 (1999)
7. Blass, A., Gurevich, Y., Shelah, S.: On polynomial time computation over unordered structures. J. Symbol. Logic **67**(3), 1093–1125 (2002)
8. Börger, E.: Computability, Complexity, Logic, Studies in Logic and the Foundations of Mathematics, vol. 128. North-Holland (1989)
9. Börger, E.: The ASM refinement method. Formal Aspects Comput. **15**(2–3), 237–257 (2003). https://doi.org/10.1007/s00165-003-0012-7
10. Börger, E., Durdanovic, I.: Correctness of compiling Occam to Transputer code. Comput. J. **39**(1), 52–92 (1996). https://doi.org/10.1093/comjnl/39.1.52
11. Börger, E., Grädel, E., Gurevich, Y.: The Classical Decision Problem. Perspectives in Mathematical Logic. Springer, Heidelberg (1997)
12. Börger, E., Raschke, A.: Modeling Companion for Software Practitioners. Springer, Heidelberg (2018). https://doi.org/10.1007/978-3-662-56641-1
13. Börger, E., Rosenzweig, D.: A mathematical definition of full Prolog. Sci. Comput. Program. **24**(3), 249–286 (1995). https://doi.org/10.1016/0167-6423(95)00006-E
14. Börger, E., Schewe, K.D.: Concurrent abstract state machines. Acta Informatica **53**(5), 469–492 (2016). https://doi.org/10.1007/s00236-015-0249-7
15. Börger, E., Schewe, K.D.: A behavioural theory of recursive algorithms. Fundamenta Informaticae **177**(1), 1–37 (2020)

16. Börger, E., Stärk, R.: Abstract State Machines. Springer, Heidelberg (2003). https://doi.org/10.1007/978-3-642-18216-7
17. Chandra, A.K., Harel, D.: Structure and complexity of relational queries. J. Comput. Syst. Sci. **25**(1), 99–128 (1982). https://doi.org/10.1016/0022-0000(82)90012-5
18. Dawar, A., Richerby, D.: Fixed-point logics with nondeterministic choice. J. Log. Comput. **13**(4), 503–530 (2003). https://doi.org/10.1093/logcom/13.4.503
19. Dawar, A., Richerby, D., Rossman, B.: Choiceless polynomial time, counting and the Cai-Fürer-Immerman graphs. Ann. Pure Appl. Log. **152**(1–3), 31–50 (2008). https://doi.org/10.1016/j.apal.2007.11.011
20. Ebbinghaus, H.D., Flum, J.: Finite Model Theory. Perspectives in Mathematical Logic. Springer, Heidelberg (1995). https://doi.org/10.1007/978-3-662-03182-7
21. Fagin, R.: Generalized first-order spectra and polynomial-time recognizable sets. In: Karp, R. (ed.) SIAM-AMS Proceedings, pp. 43–73, no. 7 (1974)
22. Ferrarotti, F., Schewe, K.D., Tec, L., Wang, Q.: A new thesis concerning synchronised parallel computing - simplified parallel ASM thesis. Theoret. Comput. Sci. **649**, 25–53 (2016). https://doi.org/10.1016/j.tcs.2016.08.013
23. Ferrarotti, F., Schewe, K.D., Tec, L., Wang, Q.: A complete logic for Database Abstract State Machines. Logic J. IGPL **25**(5), 700–740 (2017)
24. Ferrarotti, F., Schewe, K.D., Tec, L., Wang, Q.: A unifying logic for nondeterministic, parallel and concurrent Abstract State Machines. Ann. Math. Artif. Intell. **83**(3–4), 321–349 (2018). https://doi.org/10.1007/s10472-017-9569-3
25. Gire, F., Hoang, H.K.: An extension of fixpoint logic with a symmetry-based choice construct. Inf. Comput. **144**(1), 40–65 (1998). https://doi.org/10.1006/inco.1998.2712
26. Gurevich, Y.: A new thesis (abstract). Am. Math. Soc. **6**(4), 317 (1985)
27. Gurevich, Y.: Logic and the challenge of computer science. In: Börger, E. (ed.) Current Trends in Theoretical Computer Science, pp. 1–57. Computer Science Press (1988)
28. Gurevich, Y.: Evolving algebras 1993: Lipari Guide. In: Börger, E. (ed.) Specification and Validation Methods, pp. 9–36. Oxford University Press (1995)
29. Gurevich, Y.: Sequential abstract state machines capture sequential algorithms. ACM Trans. Comput. Logic **1**(1), 77–111 (2000)
30. Immerman, N.: Descriptive Complexity. Graduate texts in Computer Science. Springer, New York (1999). https://doi.org/10.1007/978-1-4612-0539-5
31. Libkin, L.: Elements of Finite Model Theory. Texts in Theoretical Computer Science. An EATCS Series. Springer, Heidelberg (2004). https://doi.org/10.1007/978-3-662-07003-1
32. Naur, P., Randell, B.: Software Engineering (1968). Report on a conference sponsored by the NATO Science Committee
33. Schewe, K.D.: Insignificant choice polynomial time. CoRR abs/2005.04598 (2021). http://arxiv.org/abs/2005.04598
34. Schewe, K.D., Ferrarotti, F.: Behavioural theory of reflective algorithms I: reflective sequential algorithms. CoRR abs/2001.01873 (2020). http://arxiv.org/abs/2001.01873
35. Schewe, K.-D., Ferrarotti, F.: A logic for reflective ASMs. In: Raschke, A., Méry, D., Houdek, F. (eds.) ABZ 2020. LNCS, vol. 12071, pp. 93–106. Springer, Cham (2020). https://doi.org/10.1007/978-3-030-48077-6_7
36. Stärk, R., Nanchen, S.: A logic for abstract state machines. J. Univ. Comput. Sci. **7**(11) (2001)

37. Stärk, R.F., Schmid, J., Börger, E.: Java and the Java Virtual Machine: Definition, Verification, Validation. Springer, Heidelberg (2001). https://doi.org/10.1007/978-3-642-59495-3

38. Van den Bussche, J., Van Gucht, D.: Semi-determinism. In: Vardi, M.Y., Kanellakis, P.C. (eds.) Proceedings of the Eleventh ACM SIGACT-SIGMOD-SIGART Symposium on Principles of Database Systems, pp. 191–201. ACM Press (1992). https://doi.org/10.1145/137097.137866

The Combined Use of the Web Ontology Language (OWL) and Abstract State Machines (ASM) for the Definition of a Specification Language for Business Processes

Matthes Elstermann[2], André Wolski[3], Albert Fleischmann[1], Christian Stary[4(✉)], and Stephan Borgert[3]

[1] InterAktiv Unternehmensberatung, Pfaffenhofen, Germany
albert.fleischmann@interaktiv.expert
[2] Karlsruhe Institute of Technology, Karlsruhe, Germany
matthes.elstermann@kit.edu
[3] Technische Universität Darmstadt, Darmstadt, Germany
andre.wolski@stud.tu-darmstadt.de
[4] Johannes Kepler University Linz, Linz, Austria
Christian.Stary@jku.at

Abstract. The domain of Subject-oriented Business Process Management (S-BPM) is somewhat outstanding due to its embracing of Subject Orientation. However, at the same time, it is also a classic BPM domain concerned with typical aspects like creation and exchange of diagrammatic process models, elicitation of domain knowledge, and implementing process (models) into organisations and IT systems. Nevertheless, the Abstract State Machine (ASM) concept, a formal and abstract specification means for algorithms, has been and is fundamental and an important cornerstone for the S-BPM community. The first formal specifications for S-BPM has been developed by Egon Börger using ASM means—namely a specification for an interpreter engine for the subject-oriented modeling language PASS, the Parallel Activity Specification Schema. However, for the sake of intuitive and comprehensive use, ASM can be enriched with defining the passive aspects of PASS, namely the (data) structure of process models and data object appearing in the processes. Here it is useful to complement ASM description means with concepts that are better suited for that tasks. This work analyzes how the S-BPM research community has combined ASM with the Web Ontology Language (OWL) to generate a precise, while comprehensible, system specification for the execution of formal, subject-oriented process models. Furthermore, it will be argued why this combination is worthwhile overcoming the weaknesses of both generic and technology independent specification approaches.

Keywords: S-BPM · PASS · OWL · ASM · Formal standards · Subject-orientation

© Springer Nature Switzerland AG 2021
A. Raschke et al. (Eds.): Börger Festschrift, LNCS 12750, pp. 283–300, 2021.
https://doi.org/10.1007/978-3-030-76020-5_16

1   Introduction

The specification and implementation of business processes is an inherently complex activity involving multiple stages and multiple actors. Normally, the involved actors are specialists in different domains, e.g. business analysts, process stakeholders, software developers. The collected information about a process is usually formalized in models of and ideally are understood and agreed upon by all involved parties. For that purpose, a multitude of methodologies have been proposed to manage the complexity of gathering and structuring the required information. B. Thalheim gives in [21] a comprehensive overview of aspects on conceptual models.

According to [11] a model should have following properties:

1. Aiding a person's own reasoning about a domain
2. Supporting the communication of domain details among stakeholders
3. Supporting the communication of domain details to developers
4. Documenting relevant domain details for future reference

Furthermore, models are often created for the purposes to plan changes in their considered domain, e.g. to make business processes more effective and/or efficient. Today the standard way to achieve more effectiveness and efficiency is the use of IT. In order to transform a model into reality using IT there is another requirement for models:

5. Transformation of Models into an IT-solution

Diagrammatic or graphical modeling languages are the preferred approach for describing processes. They are seen as more understandable and intuitive than textual specifications [15]. However, to meet the transformation requirements the syntax and semantics of a graphical language must be precisely defined.

Because several stakeholders from different domains are involved in defining a model, several different visualisation styles based on the same modeling philosophy can be useful. Business people may prefer pictographs whereas developers may prefer a rectangle for representing the same entity.

Furthermore, since networked organization for business processes cross their borders, and the various organisations involved in a process likely use different IT-platforms, different parts of a process are implemented in different systems. Ideally, various platforms automatically must align to, or at least the humans as developers must understand the semantics of a process model in the same way.

There cannot only be different graphical tools for specifying models, but also different platforms where these models are migrated for execution. Figure 1 shows the basic structure of a modeling and execution environment.

A language which meets the first four properties is the Parallel Activity Specification Schema (PASS) which is a schema based on the subject-oriented philosophy where the active entities are the focus of model. As a modeling language, PASS is informal and semantically open, and allows to specify any kind of executable process system. However, due to the prior mentioned requirements

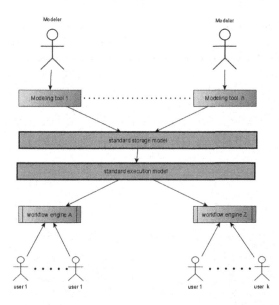

Fig. 1. Definition and execution of process models

caused by the possible involvement of a multitude of people, organisations, and IT systems, there are multiple editors and IT systems tailored to various stakeholders and their visual needs. Still or especially under these requirements, the exchange of process models between various tools and the identical execution on each platform needs to be assured. Consequently, model and execution semantics of any modeling complex with such a goal needs to be specified in a formal manner. We have done that for PASS using the Abstract State Machine (ASM) formalism—but only partially since, as will be argued, while being a proper tool for algorithmic definition, ASM can and possibly should be complemented by other specification means that better fit other modeling aspects.

In the first section of this work, the different aspects of modeling are considered in more detail. In the succeeding sections we describe the features of PASS and why we decided to use the ontological means of the web ontology language (OWL) to define the formal static model structure, and for formally specifying the dynamic aspects of PASS models Abstract State Machine (ASM).

2   Modeling in General

"You don't have to understand the world, you just have to orient yourself" ("Man muss die Welt nicht verstehen, man muss sich nur darin zurechtfinden."). According Internet source), this sentence is attributed to Albert Einstein (e.g. [7]). Who understands what is going on in the world? Who knows how it works? As we cannot answer these questions sufficiently detailed, we should take care of our world, namely the part of the world that is important to us in a specific situation. We should recognize that we create or construct our world on a daily basis. Any excerpt of reality is naturally determined by our subjective interests. Each of us decides which part of the world we want to consider and which aspects seem important to us. In doing so, we identify the artifacts and the relationships between them that are essential for us. Such an abstraction of a part of reality is called a model [20]. Figure 2 shows the general approach for creating models.

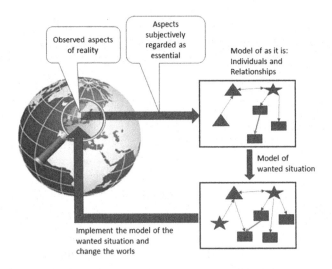

Fig. 2. Modeling and its implementation

This general approach is as follows: We consider a domain we want to change, e.g. we want to make a business process more effective and/or efficient. This means we create and adapt a model that fits our requirements. Now we have a model of the considered domain as we want it to be (SHOULD). This model represents what we implement in order to create a new world.

Models can never cover or represent all aspects of reality at once. They are always a reduction of reality made by and from the perspective of a specific interest group at a specific time for a certain purpose. In order to describe models in information science, ontologies can be applied. An ontology is a way of describing the attributes and properties of a domain and how they are related, by defining a set of concepts and categories that represent the considered part of the world.

A model has two major aspects: the static aspect which contains the considered entities and their relations, and the dynamic aspect which defines the behavioral part of the considered domain.

2.1 Static View

In the static view, *entities* of reality, which should become individuals of a model, are selected. This includes the identification of the *relations* between entities which become part of the model. The selected *entities* and the *relations* between them are called *attributes* of a static model.

2.2 Dynamic View

In a model, also dynamic aspects can be considered. Dynamic aspects can be concepts like chemical reaction processes, flow behaviors of fluids, the dynamics of a machines, or the observable behavior of an entity, or the observable behavior of a complex system. Describing these aspects has specific requirements that are not necessarily met by languages specialized for static aspects. Especially for state-based abstract system description, these requirements include the ability to describe concepts like concurrency, non-determinism, synchronization, and communication.

2.3 Modeling Languages

As stated, both, static structure and dynamic behavior are two major aspects of any system when we interpret the part of reality considered for modeling as system. Consequently, no system can be modeled without considering both in tandem [6].

However, not every modeling language, either graphical or textual, is equally sufficient to express both aspects and many of them tend to focus on one aspect.

For instance, entity relationship diagrams describe only data structures which means only static aspects are considered. UML, which is actually 14 different languages, considers both in respective diagram types, static aspects in, e.g., class diagrams and dynamic aspects in state transition diagrams. BPMN is a language for describing business processes. This means its focus is the specification of behavior.

3 The Parallel Activity Specification Schema: PASS

The aforementioned Parallel Activity Specification Schema (PASS) is special in this regard. It follows the paradigm of Subject-Orientation when describing models of process or rather process systems. In such models the term of "Subject" refers to active entities. These execute operations on objects and exchange data via messages to synchronize their operations.

PASS as a modeling language is used to specify dynamic concepts (processes). However, its SIDs (see later) as well as the Messages (Data)-Objects they contain are more akin to static structures. Furthermore, PASS models themselves are static structures that need to be described by a meta-model specialized for that. Finally, the execution of a PASS process model is, again, a dynamic concept and a specification for a correct execution would should be specialized for that.

In the following sections first we give an informal overview and after that we outline the precise and formal definition of PASS.

3.1   General Concept of Subject-Orientation

As any process model, subject-oriented process specifications are embedded in a context. A context is defined by the business organization and the technology by which a business process is executed. Subject-oriented system development has been inspired by various process algebras (see e.g. [10,13,17,18]), by the basic structure of nearly all natural languages (Subject, Predicate, Object) and the systemic sociology developed by Niklas Luhmann [2,16] and Jürgen Habermas [12,19]. In the active voice of many natural languages, a complete sentence consists of the basic components subject, predicate and objects. The subject represents the active element, the predicate (or verb) the action, and the object is the entity on which the action is executed. According to the organizational theory developed by Luhmann and Habermas, the smallest organization consists of communication executed between at least two information processing entities (Note, this is a definition by a sociologist, not by a computer scientist) [16]. Figure 3 summarizes the different inspirations of subject orientation.

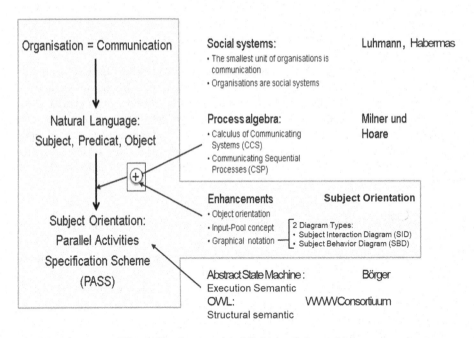

Fig. 3. Fundamentals of Subject-Orientation

Particularly, we describe how these various ingredients are combined in an orthogonal way to a modeling language for scenarios in which active entities play a prominent role like in Industry 4.0.

Structure of Models Described in PASS. A PASS model (system) consists of two separate but interconnected graph descriptions. First, the Subject Interaction Diagram (SID) that defines the existence of active entities (the *Subjects*) and their communication relationships, that they can use to exchange data in a process context. Furthermore, for each subject there can be an individual Subject Behavior Diagram (SBD) that defines its specific activities in a process.

A subject acts upon (data) objects that are owned by the subject and can not be seen by other subjects[1].

It is important to emphasize, that this specification is totally independent from the implementation[2] of subjects, objects, and the communication between subjects. This means subjects are abstract entities which communicate with each other and use their objects independent from possible implementations. The mapping to an actual implementation entity or technology is done in a succeeding step.

When an implementation technology is assigned (mapped) to a subject to execute its behavior (SBD) it becomes an actor/agent, e.g. a software agent. In Fig. 4 a model defined in PASS is shown. The upper part of that figure shows the graphical representation of a PASS model with SID (upper diagram) and SBDs (lower diagrams). In the example subject 'Customer' sends a message 'order' to the subject 'Companies' and receives the messages 'Delivery' or 'Decline'.

In principle, an SBD defines the behavior of a subject as a kind of state machine, interlinking three types of states (see lower part of Fig. 4). Send states (green nodes) represent the dispatch of messages to other subjects, Receive states (red nodes) represent the reception of messages from other subjects, and Function/Do states (yellow nodes) represent tasks that do not involve interaction with other subjects. States are connected using transitions representing their sequencing. The behaviour of a subject may include multiple alternative paths. Branching in PASS is represented using multiple outgoing transitions of a state, each of which is labelled with a separate condition. Merging of alternative paths is represented using multiple incoming transitions of a state. Within an SBD, all splits and merges in the process flow are always explicitly of the XOR type. There are no AND or OR splits!

Subjects are executed concurrently. Triggering and synchronizing concurrent behaviors is handled by the exchange of messages between the respective subjects.

For a subject-oriented process model to be complete and syntactically correct, all messages specified in the SID (and only those) must be handled in the SBDs of the two subjects involved. The SBD of the sending subject needs to include a Send state specifying the message and recipient name (see Fig. 4). Correspondingly, the SBD of the receiving subject needs to include a Receive state specifying the message and sender name. There is no explicit diagrammatic

---

[1] In an extended version of subject-orientation shared objects are also possible but are not considered here [9].

[2] Implementation of a subject = realization in form of a human, IT-System, or other technology.

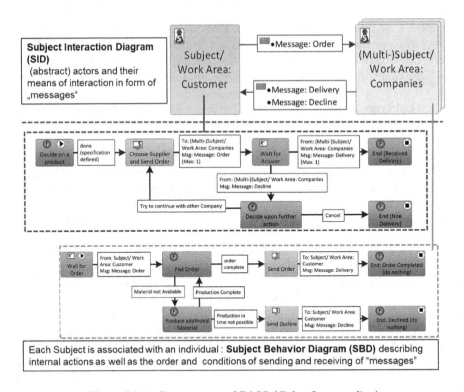

Fig. 4. Main Components of PASS (Color figure online)

association of the messages shown in the SID with the corresponding Send and Receive states in the SBDs.

At runtime, any incoming message is placed in the so-called input pool of the receiving subject, which can be thought of as a mailbox. When the execution of the subject has reached a Receive state that matches the name and sender of a message in the input pool, that message can be taken out of the input pool and behaviour execution can proceed as defined in the SBD. The default communication mode is asynchronous. Synchronous communication can be established by restricting the maximum number of messages that can be stored in the input pool. The input pool is not visualized in a diagram but is an important concept in order to understand how messages and behaviours are loosely coupled subjects in PASS.

3.2 Formal Description of PASS

In order to standardize any process modeling language, an informal description as described above is not sufficient for execution. In addition, formal specification for the static structure of the models as well as a specification of the execution dynamics is necessary. For PASS the static structure is defined formally using the Web Ontology Language (OWL see [1]. The dynamic aspects are specified with

ASM [5]. The following sections give an overview to both formal descriptions. More details can be found in [14].

Structure of PASS Models in OWL. In OWL, four principal (color corded) concepts exist, as shown here: Classes, relationships between classes, the so-called Object Properties, non-linking Data Properties that are attached to classes and allow attaching individual data values to instances of the classes, and finally the instances of classes themselves, the OWL **Individuals**.

Figure 5 shows the most important elements of PASS SIDs as they appear in the *Standard-pass-ont* ontology in OWL.

The central classes are Subject, MessageSpecification, and MessageExchange. Between these classes are defined the properties hasIncomingMessageExchange (in Fig. 5 number 217) and hasOutgoingMessageExchange (in Fig. 5 number 224). These properties define that subjects have incoming and outgoing messages. Each MessageExchange has a sender and a receiver (in Fig. 5 number 227 and number 225). Messages Exchanges also have a type. This is expressed by the property hasMessageType (in Fig. 5 number 222) linking it to a MessageSpecification. These Message Specifications are the actual existential definitions for Messages, while the model element of the Message Exchange is used to define that an existing message is indeed exchanged between specific subjects. Beyond that, message exchanges that have the same sender and receiver may be grouped into an MessageExchangeList that contains them.

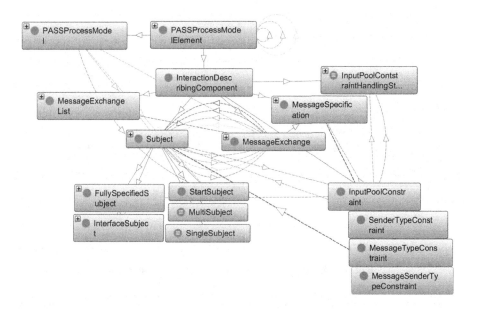

Fig. 5. Elements of PASS Process Interaction Diagram

Each Fully Specified Subject contains at least one Subject Behavior (containsBehavior), which is considered to be its Base Behavior (see property 202 in Fig. 6—containsBaseBehavior) and may have additional subject behaviors (see sub-classes of SubjectBehavior in Fig. 6) for Macro Behaviors and Guard Behaviors.

The details of all behaviors are defined as state transition diagrams (PASS behavior diagrams). These Behavior Diagrams themselves contain Behavior Describing Components (see Fig. 6). Inversely, the Behavior Describing Component have the relation belongsTo linking them to one Subject Behavior.

Execution Semantics of PASS Models with ASM. As stated, while the correct structure and syntax of Parallel Activity Specification Schema (PASS) are specified in the Web Ontology Language (OWL), its execution semantics is defined using the formalism of Abstract State Machines (ASM) as defined by Börger in [3]. The original PASS ASM execution specification has been formulated in [4] in 2011, but is now somewhat outdated in terms of PASS model elements covered by the specification as well as the used vocabulary.

While not yet covering all PASS elements, the current version of the PASS execution semantics is based on [23] and [22]. Note, that this is not a pure ASM specification, but rather a specification meant for *CoreASM*, an open-source ASM execution engine/software implemented for the JAVA Virtual Machine (JVM) [8]. Thereby, it is an execution specification as well as a reference implementation at the same time.

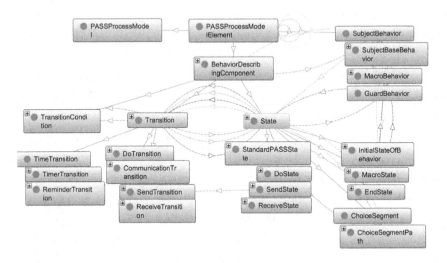

Fig. 6. Structure of subject behavior specification

The complete specification comprises all elements and aspects a workflow engine requires, including concepts to receive and send messages between various instances of subjects. However, at the core of each subject instance is what Börger had already defined in the original work: an interpreter for a Subject Behavior Diagram that defines the actual behavior possibilities of a subject instance within a process-instance.

In advanced PASS modeling, a subject may technically have more than one behavior, for example multiple MacroBehaviors. A FullySpecifiedSubject always has one main MacroBehavior given with containsBaseBehavior, which is loaded as 1st macro instance. Therefore, the rule SubjectBehavior simply calls the rule MacroBehavior for this 1st macro instance. Later on, the rule MacroBehavior can be called recursively from a MacroState with other instances of the MacroBehavior elements, the so-called *Additional Macros*.

```
rule SubjectBehavior = MacroBehavior(1)
```

Listing 1: SubjectBehavior

The rule MacroBehavior controls the repetitive evaluation of all States in the SBD of the given MacroBehavior instance.

```
rule MacroBehavior(MI) =
  let ch = channelFor(self) in
  choose stateNumber in activeStates(ch, MI) do
    Behavior(MI, stateNumber)
```

Listing 2: MacroBehavior

The evaluation of each individual State is structured into three main phases of the rule Behavior: *initialization*, the *state function*, and an optional *transition* behavior.

```
rule Behavior(MI, currentStateNumber) =
 let s = currentStateNumber,
     ch = channelFor(self) in
   if (initializedState(ch, MI, s) != true) then
     StartState(MI, s)
   else if (abortState(MI, s) = true) then
     AbortState(MI, s)
   else if (completed(ch, MI, s) != true) then
     Perform(MI, s)
   else if (initializedSelectedTransition(ch, MI, s) != true) then
     StartSelectedTransition(MI, s)
   else
    let t = selectedTransition(ch, MI, s) in
     if (transitionCompleted(ch, MI, t) != true) then
       PerformTransition(MI, s, t)
     else
       Proceed(MI, s, targetStateNumber(processIDFor(self), t))
```

Listing 3: Behavior

In the beginning the rule **Behavior** initializes the evaluation of a State with the rule **StartState**, which will also set **initializedState** to true .

Next, it is checked if the State has to be aborted, which is the case when a timeout or cancel is activated. Then the rule **AbortState** will reset the previous evaluation results of it.

Otherwise the rule **Perform** interprets the *state function*, which comprises the evaluation of the corresponding FunctionSpecification and Transition-Conditions[3], including the TimeTransitionCondition to supervise a possible timeout, until it is indicated that this phase is **completed** and an outgoing transition has been selected or determined.

Usually, the outgoing Transition will be selected by the environment (i.e. by the agent), which is for example the case for the FunctionSpecification **DefaultFunctionDo1_EnvironmentChoice**. However with auto-transitions it is possible, that such a transition is automatically selected by the *state function* as soon as it becomes enabled and as long as there are no other enabled transitions to choose from, as it is the case for the **DefaultFunctionReceive2_AutoReceiveEarliest** as it is specified in the rule **PerformReceive** (not shown here).

For the last phase the selected Transition will be initialized by the **StartSelectedTransition** and the transition behavior will be performed with the rule **PerformTransition**, until it is completed as well. As last step the rule **Proceed** updates the **activeStates** of the macro instance by removing the current State and adding the State that is indicated by the property hasTargetState on the selected Transition.

3.3   Combination of Structure and Behavior

As stated, the definitions of valid models, syntax and structure of a model, the static view, is given by the OWL specification. For that purpose, a model itself is a data object with a structure matching the ontology definition.

The concept is, that the ASM-engine accepts process model objects following the ontology definition as input and "produces" state sequences, i.e. the dynamic process flow, as an output. Naturally, this works only if the ASM interpreter can handle all concepts defined in ontology and therefore, has been created to match the ontology. In order for this to work, the process models have to be imported into the CoreASM environment—a transformation that links both technologies together.

Although the current ASM interpreter does not yet cover all PASS elements, it is shown in the previous section, that both the underlying concepts and nomenclature of both the specifications of the structure (OWL) and behavior (ASM) already fit well together. Such harmony cannot be taken for granted, as development of one specification has to be reflected in the other one in synchronization.

---

[3] Thereby the availability of each Transition can change, depending on the dynamic state of the TransitionCondition during execution. For example, the rule **CheckIP** enables or disables a Transition based on the availability of a Message in the Inputpool according to the ReceiveTransitionCondition.

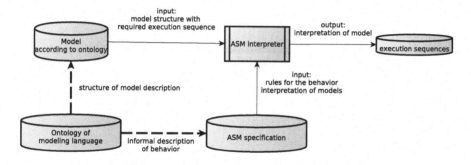

Fig. 7. Combination of OWL and ASM

The relationship between the OWL and ASM specifications is depicted in the bottom part of Fig. 7, which suggests to develop changes in the ontology first with an informal description of the behavior and later on to formally specify the behavior in the ASM specification and thereby to update the process model importer of the reference implementation.

The upper part of Fig. 7 summarises the run-time behavior: a process model, that adheres to the OWL PASS Ont, can be interpreted by the reference implementation, which is interpreting the ASM specification, to produce an interpretation of the process model with its execution sequences.

To detect errors in a process model, a reasoner can be used to detect inconsistencies within the process model and also against the standard-pass-ont (for example a ReceiveState must not have a SendTransition). However, not all problems can be found by a reasoner, since there is only a definition for a process model and not for an executable process model. It is deliberately allowed to store incomplete process models. This allows the exchange between various tools, for example to create a prototype in tool A and perform further refinement in tool B. Therefore, before or during the import of OWL process models to the interpreter, the process model has to be validated for executability. For example, all interactions defined on an SBD have to be defined on the SID, otherwise errors could occur during the execution.

4   Discussing the Combination

As stated, for subject-oriented modeling with PASS, this combination of OWL and ASM is used at the core of a standard architecture for PASS modeling and execution tools. The concept is that each modeling tool stores the models in the owl structure which in turn can be interpreted by execution tools. Various modeling tools can use different language variants or extensions for describing the same facts if they are derived from and therefore adhere to the PASS-OWL definition. Figure 8 shows the particular roles of OWL and ASM specifications.

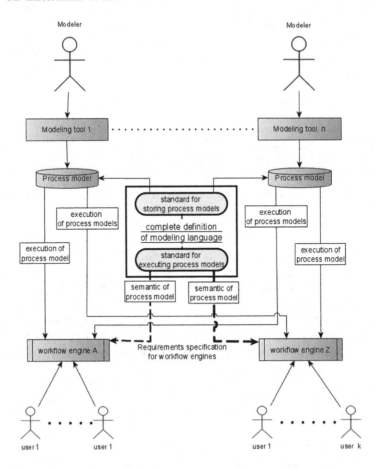

Fig. 8. Overview usage of OWL and ASM in tool suites

If done precisely, any CoreASM-Spec based interpreter is able to interpret the model in the right way. This conceptual environment allows creating process models with different graphical editors in an interchangeable way, and using any workflow engine ensuring interpretation in the same way.

To discuss a usage scenario (depicted in Fig. 9): Due to historical development, modeling tool A currently does not use the OWL structure to store models. The original ASM implementation had been developed before the OWL definition was complete. The data repository structure for storing specifications of modeling tool A (*.graphml files) has been directly transferred into a data structure (ScalaModel in PASSProcessModel) which is used as a high-level in-memory data structure for further transformation to the actual CoreASM data structure (ASM Map in PASSProcessWriterASM).

The intermediate data structure (ScalaModel in PASSProcessModel) is also capable to be stored in OWL and to be read from it, allowing the import of OWL process model definition into the CoreASM interpreter.

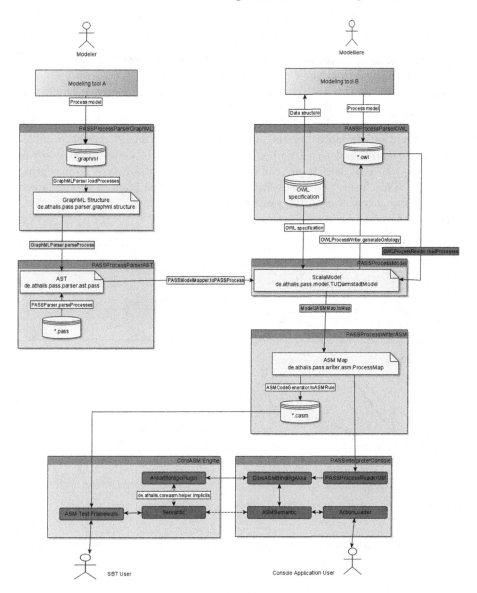

Fig. 9. Architecture of the reference implementation

Modeling tool B is based on Microsoft Visio. This modeling tool can store process models in the PASS standard OWL format. Due to adherence to OWL and ASM standards, both modeling tools can be used to generate valid, executable input for the same workflow engine.

Although the CoreASM execution engine works, it is only meant as a reference and far from having good performance and usability on a level where it would be applicable for industry applications. Moreover, a complete work-flow solution also needs an administration and management component. Consequently and as next steps, it is planned to implement a workflow engine in C# based on or derived from the CoreASM code.

Figure 10 shows the principal concept for usage of the reference implementation as a part of the requirement specification for a workflow engine. To do so, the programmers of the alternative workflow engines simply need to conform to the Execution Semantics of the ASM spec. The CoreASM reference implementation can be used as part of a test framework, which interprets a set of process models, that had been crafted to automatically converge to accepting end states. This forms test cases for the other workflow engines to validate if their executions converge to the same results.

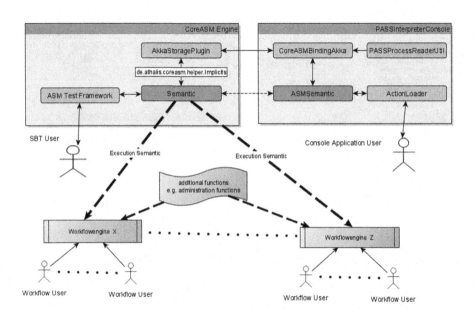

Fig. 10. Deriving a workflow engine from the reference implementation

5   Conclusion

In order to establish Subject-oriented Business Process Management (S-BPM) as behavior- and communication-centered system development approach, not only intuitive tools to (re)present models and specify contextual information are required, but, at the core, also stringent specifications of static and dynamic elements. Starting with its original formal ASM execution Specification by Egon Börger, the subject-oriented modeling language PASS (Parallel Activity Specification Schema), has become a semantically precise and formal instrument for behavior modeling. Complementing the ASM specification with the Web Ontology Language (OWL) enriches this capability with contextual system representations, and overcomes the weaknesses of both generic and technology-independent specification approaches.

In this paper we have demonstrated how this combination of ontology and an abstract state machine can be approached for specifying the structure and behaviour of systems in a mutually aligned way. We have exemplified in detail how these two concepts have been applied to PASS, and thus how a subject-oriented approach works for the benefit of specifying concurrent systems independent from their implementation. Utilizing this approach, a variety of modeling and execution tools can be developed to be used in an interchangeable way, while meeting different stakeholder requirements and preferences from a technological and user-centered perspective.

In a certain way it could be argued that having two different formalisms for static (OWL) and dynamic (ASM) aspects maybe somewhat unnecessary and double the work. Why the extra effort of having to create two different formal specification that are still sync with each other, regarding the vocabulary and concepts? Why the need to learn and understand both when wanting to extend specifications?

However, the experience during the work on the presented domain has shown that this combination is indeed the best approach to be advised. While being an official specification for S-BPM, the original Börger ASM interpreter rarely understood as being that, an interpreter, and partially people wondered where the Do States and Receive Transitions are? The ASM interpreter only implicated the existence of these elements indirectly. On the other side when defining the OWL specification, often model elements were proposed that came with assumptions about the execution, but without a precise execution semantics. Each on its own simply was always lacking precision and especially easy and direct comprehensibly in the other regard. Having both overcomes the shortcomings of both and furthermore allows learners that want to understand S-BPM and PASS modeling formally a subject-oriented approach in itself by allowing to make such simple statements as: "Here is an ASM interpreter. It should read PASS models that contain the following elements. Only a syntactically correct model can be executed. Here are the rules that tell you what is Okay and what not. When encountering an element during execution, the interpreter should behave in this specific way." It should be obvious where in the previous sentences each specification played its role and why both together form a great union.

References

1. Web ontology language (owl). https://www.w3.org/OWL/
2. Berghaus, M.: Luhmann leicht gemacht. Böhlau Verlag (2011)
3. Börger, E., Stärk, R.: Abstract State Machines: A Method for High-Level System Design and Analysis. Springer, Heidelberg (2003). https://doi.org/10.1007/978-3-642-18216-7

4. Börger, E.: A subject-oriented interpreter model for S-BPM. Appendix. In: Fleischmann, A., Schmidt, W., Stary, C., Obermeier, S., Börger, E. (eds.) Subjektorientiertes Prozessmanagement. Hanser-Verlag, München (2011)

5. Börger, E., Stärk, R.F.: Abstract State Machines: A Method for High-Level System Design and Analysis. Springer, Berlin (2003). https://doi.org/10.1007/978-3-642-18216-7

6. Dori, D.: Object-Process Methodology for Structure-Behavior Codesign. Springer, Heidelberg (2010). https://doi.org/10.1007/978-3-642-15865-0_7

7. Einstein: Zitat albert einstein. https://www.sasserlone.de/zitat/190/albert.einstein/

8. Farahbod, R.: CoreASM: an extensible modeling framework & tool environment for high-level design and analysis of distributed systems. Ph.D. Dissertation (2009)

9. Fleischmann, A., Stary, C.: Dependable data sharing in dynamic IoT-systems - subject-oriented process design, complex event processing, and blockchains. In: Betz, S., Elstermann, M., Lederer, M. (eds.) S-BPM ONE 2019, 11th International Conference on Subject Oriented Business Process Management. ICPC published by ACM Digital Library, Association of Computing Machinery (ACM) (2019)

10. Fleischmann, A., Schmidt, W., Stary, C., Obermeier, S., Boerger, E.: Subject-Oriented Business Process Management. Springer, Heidelberg (2012). https://doi.org/10.1007/978-3-642-32392-8

11. Gemino, A., Wand, Y.: A framework for empirical evaluation of conceptual modeling techniques. Requirements Eng. 9, 248–260 (2004). https://doi.org/10.1007/s00766-004-0204-6

12. Habermas, J.: Theory of Communicative Action Volume 1, Volume 2. Suhrkamp Paperback Science (1981)

13. Hoare, A.: Communicating Sequential Processes. Prentice Hall, Hoboken (1985)

14. I2PM: Standard document for pass (2019). https://github.com/I2PM/Standard-Documents-for-Subject-Orientation. Accessed 22 Jan 2020

15. Jolak, et al.: Software engineering whispers: the effect of textual vs. graphical software design descriptions on software design communication. Empir. Softw. Eng. 25, 4427–4471 (2020). https://doi.org/10.1007/s10664-020-09835-6

16. Luhmann, N.: Social Systems. Suhrkamp Verlag (1984)

17. Milner, R.: Communication and Concurrency. Prentice Hall, Hoboken (1989)

18. Milner, R.: Communicating and Mobile Systems: The Pi-Calculus. Cambridge University Press, Cambridge (1999)

19. Römpp, M.: Habermas leicht gemacht. Böhlau Verlag (2015)

20. Stachowiak, H.: Allgemeine Modelltheorie. Springer, Heidelberg (1973)

21. Thalheim B.: The theory of conceptual models, the theory of conceptual modelling and foundations of conceptual modelling. In: Embley, D., Thalheim, B. (eds.) Handbook of Conceptual Modeling. Springer, Heidelberg (2010). https://doi.org/10.1007/978-3-642-15865-0_17

22. Wolski, A., Borgert, S., Heuser, L.: A coreASM based reference implementation for subject oriented business process management execution semantics. In: Betz, S., Elstermann, M., Lederer, M. (eds.) S-BPM ONE 2019, 11th International Conference on Subject Oriented Business Process Management. ICPC published by ACM Digital Library, Association of Computing Machinery (ACM) (2019)

23. Wolski, A., Borgert, S., Heuser, L.: An extended subject-oriented business process management execution semantics. In: Betz, S., Elstermann, M., Lederer, M. (eds.) S-BPM ONE 2019, 11th International Conference on Subject Oriented Business Process Management. ICPC published by ACM Digital Library, Association of Computing Machinery (ACM) (2019)

Models and Modelling
in Computer Science

Bernhard Thalheim[✉][ⓘD]

Department of Computer Science, Christian-Albrechts University at Kiel,
24098 Kiel, Germany
bernhard.thalheim@email.uni-kiel.de
http://www.is.informatik.uni-kiel.de/~thalheim

Dedicated to Egon Börger

Abstract. Models are a universal instrument of mankind. They surround us our whole lifespan and support all activities even in case we are not aware of the omnipresence. They are so omnipresent that we don't realise their importance. Computer Science is also heavily using models as companion in most activities. Meanwhile, models became one of the main instruments. The nature and anatomy of models is not yet properly understood.

Computer Science research has not yet been properly investigating its principles, postulates, and paradigms. The well-accepted three dimensions are states, transformation, and collaboration. An element of the fourth dimension is abstraction. The fourth dimension is modelling. We review here the fourth dimension.

Keywords: Models · Art of modelling · Modelling theory · Computer science pitfalls · Model theory

1 Models Are the Very First Human Instrument

> Wavering forms, you come again;
> once long ago passed before my clouded sight.
> Should I now attempt to hold you fast? ...
> You bear the images of happy days,
> and friendly shadows rise to mind,
> With them, as almost muted tale,
> come youthful love and friendship.
>
> Goethe, *Faust I*

Remark: See too the presentations in https://vk.com/id349869409 *or in the youtube channel "Bernhard Thalheim" for the theory and practice of modelling.*

© Springer Nature Switzerland AG 2021
A. Raschke et al. (Eds.): Börger Festschrift, LNCS 12750, pp. 301–325, 2021.
https://doi.org/10.1007/978-3-030-76020-5_17

Models, Models, Models – We Are Surrounded with Models

Humans use many instruments and especially intellectual instruments. The very first intellectual instrument we use is a model. Babys quickly develop their own models of the 'mother' and 'father'. They cannot yet use a natural language but they know already models of their surroundings [22]. Later children realise that their models of 'father' or 'mother' are different from those used by other children although they use the same terms in their native language for referencing to mothers and fathers.

Later, we develop mental concepts and we learn to strengthen them to codified concepts. Concepts can be considered as the main and basic elements of mental models. Codified concepts are essential elements of conceptual models. Education is model-backed. Almost all scientific and engineering disciplines widely use modelling. Daily life and social interaction is also model-backed.

The wide use of models is caused by: models are often far simpler; models reduce complexity of systems; models support reasoning; models support interaction, communication, and collaboration; models are more focused, abstract, and truncated; models support perception; models must not be correct - they should however be coherent; models may be preliminary; models ease understanding; models can be understood on the fly; etc.

In general, models are and must be useful. Mental models are used by everybody in an explicit but often implicit form. They have an added value in life, society, science, and engineering, e.g. models in Ancient Egypt. Models have not to be named 'model'[1]. Their importance is far higher than often assumed.

Our discipline is not an exception for the misunderstanding of modelling. Moreover, modelling is a central activity due to the complexity of our systems and especially our software. There is a simple reason for the importance of models: systems are far more complex than a human can entirely understand. Models, however, allow to concentrate, scope and focus on certain aspects while neglecting others or coping with other through other models.

---

[1] The oldest mention we acknowledge is the usage in Ancient Egypt with the use of models as moulds, models as representations, and models of the right order ('maat'). The first explicit notion of model is 'metron' in Ancient Greece and 'modulus' in Roman time, i.e. at least 40BC. The wide use of this word came with engineering in the 16th century and with sciences in the 19th century.

Models, Everywhere, Anytime, for and by Everybody

Computer Science[2,3] (CS) and especially Applied Computer Science (ACS)[4] are unthinkable without modelling. Modelling has been already used when the history of CS started. We use very different names for artifacts that are models, e.g. specification, declaration, or description. We also know more that 60 different notions of conceptual model. As far we know, the variety of notions of model goes far beyond 450.

Analysing the state of art in CS&ACS we conclude that each branch widely uses models. Depending on the construction, description, prescription, prognosis, investigation, and application scenarios where models are used, models function as blueprint for system and software construction, as development companion, as starting or inspiration point, as means for systems analysis and prognosis, as means for learning and reflection, as reasoning instrument, as reflection and representation of some observed reality or system, as thought guide, as means for explanation and elaboration, as helper for integration and modernisation, etc. There is no branch in CS&ACS that does not use models.

Models are specific in dependence on the role they play as instruments in scenarios. If the model is dependable and especially of sufficient quality then it properly functions in those scenarios. Models are also used for sense-making in a more foundational setting, for delivering essential information about the system, and for representation of aspects that are currently of interest. Models are a good instrument for communication or more general interaction and collaboration. And they are used by everybody at the perception or conceptual or intelligible levels.

---

[2] A description of Computer Science has been given in [12]:
  "Computer Science and engineering is the systematic study of algorithmic processes – their theory, analysis, design, efficiency, implementation and application – that describe and transform information. The fundamental question underlying all of computing is, *What can be (efficiently) automated.*".

[3] Computer Science can be divided into kernel CS and applied CS. The first subdiscipline spans theoretical, practical and technical CS.

[4] We do not know a commonly accepted description of this subdiscipline. Essentially, applied CS has two branches: specific application-oriented CS and engineering of solutions in applications. The first branch has led to a good number of so-called 'hyphen' CS such as business informatics, biomedical informatics, and geoinformatics. The second branch is engineering of work systems [2], i.e. systems "in which humans or machines perform processes and activities using resources to produce specific products or services for customers". It spans topics such as information systems, information science, information theory, information engineering, information technology, information processing, or other application fields, e.g. studying the representation, processing, and communication of information in natural and engineered systems. It also has computational, cognitive and social aspects.

Inheritance of heritage knowledge is considered to be a really bad habit in CS. "Never cite a paper that is older than 5 years" and "never read and learn from old wise papers" are common approaches in modern research. Such revolutionary approach has never led to a real good outcome in history and disciplines. The body of knowledge of a science consists however of compiled achievements of research. Models and modelling is one of those. Modelling has its huge body of knowledge that waits its digestion, systematisation, and compilation.

2 The Fourth Dimension of Computer Science: Modelling

> What you inherit from your father,
> Earn it anew before you call it yours.
> What does not serve you is a heavy burden,
> What issues from the moment is alone of use.
>
> Goethe, *Faust I*

Towards an Understanding of our Disciplines

The great success of redevelopment of Mathematics by the Nicolas Bourbaki group [6] led to a structure-oriented redesign of Mathematics. This group found that Mathematics has three guiding dimensions: algebra, order, and topology. Following [47], we can distinguish three dimensions of computer science: states, transformation, and collaboration[5]. Computation is considered so far as state transformation. Systems are never completely monolithic. They consist of interacting components. The disciplinary orientation understands interaction as some kind of collaboration. So far we straiten human behaviour into this schema and squeeze together human behaviour and machine operating[6]. The fourth dimension is often underestimated but not less important: models. Figure 1 depicts the interrelation of the four dimensions and adds to the CS&ACS intext the outer context.

---

[5] A dimension that has not found its proper entry into our discipline is approximation. Approximation is the fifth dimension orthogonal to states, transformation, and collaboration.

[6] We expect that modern applications such as internet technology have to use humanisation as the sixth dimension in order to cope with modern interdisciplinary tasks.

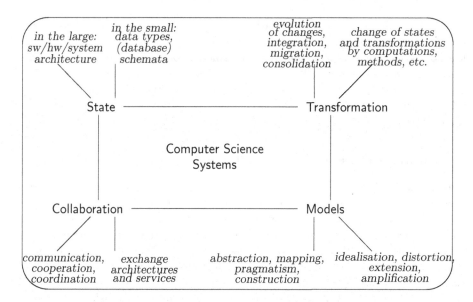

Fig. 1. The four dimensions of computer science: (1) the state dimension; (2) the transformation dimension; (3) the collaboration dimension; (4) the model dimension

The three dimensions are interwoven. For instance, central concerns in database management are states and state transformation. The fourth dimension is also interwoven with the other three. Models scope and focus to the essentials and allow us to restrict attention and to construct systems in a thoughtful way. Models allow us to avoid complexity, to concentrate first on the most important issues (pragmatism), to 'dream' on system extensions not yet observable or developed (amplification), and to abstract from the current state by an ideal state of affairs (idealisation). Humans are restricted in their mental abilities and cannot properly reason on complex systems. Instead, we order, categorise, generalise, abstract, and stereotype in a methodological mould in such a way that our models, macromodels, meso-models, and micro-models can be mapped to operating mechanisms.

CS&ACS have not yet developed a 'Bourbaki' collection of paradigms and principles. At present, Computer Science can be defined as the systematic study of computer and computation systems, "including their design (architecture) and their uses for computations, data processing, and systems control." (slightly changed from [32]). Following [12], we can assume that kernel CS is based on three main competencies: theory, abstraction, and construction (or composition; originally 'design'). We may consider modelling as a fourth competency. Modelling did not find a proper foundation in CS. It is however considered to be one – if not the main – of the core competencies in ACS.

What Means Modelling?

Models are used as instruments in our activities. Modelling thus includes mastering and manufacturing the instruments. Models should be useful. Modelling thus also includes deployment in preparation for work, utilisation with skilled acts of using, and well-organised application stories for models. We should not target for ideal or perfect instruments. The 3U macro-model considers usefulness, usability, and usage as a holistic quality for models. Modelling as mastering meets thus two challenges: appropriate[7] models with sufficient quality in use. Models are oriented on its use in application 'games' [46] by users within their context and usage culture.

CS Theoreticians, Researchers, and Developers Are Modelling

A common misbelief is that theoreticians, researchers, and developers do not need models. They can develop their ideas and thoughts without modelling. Code is then something like a plan and presentation of these ideas. What are then ideas and presentations? These are two mental states, the first one as perception or idea (in German 'Auffassung') and the second one as imagination (in German 'Vorstellung') [7,28,43]. These are then nothing else than perception models and their presentations.

A second claim is often made: the code is the model. We ask ourselves 'of what'? Of the operating of a machine from one side and of our understanding how the machine would work from the other side. So again, it is a model. Whether it is unconscious, preconscious, or subconscious does not matter. It is already a model mediating between expectations and operating.

A third claim sometimes asserted strongly is the belief that Theoretical Computer Science may sidestep models. What about their models of computation that became THE unshakable paradigm of computation? Nothing else than these models can be accepted. We may ask ourselves whether there are other useful models of operating.

CS Modelling is so far an Art or Handicraft

Modelling is so far treated as craft or handicraft, i.e. the skilled practice of a practical occupation. Modellers perform a particular kind of skilled work. Modelling is a skill acquired through experience in developing and utilising a model. At present it becomes an art. Art aims at creation of significant (here: appropriate) things (here: models)[8]. Artistry is a conscious use of skill and creative imagination that one can learn by study and practice and observation. Usually acquired proficiency, capability, competence, handiness, finesse, and expertise in modelling.

We envision that modelling is becoming a sub-discipline in CS due to its importance as the fourth dimension. It will be a branch of knowledge based on a controllable and manageable system of rules of conduct or methods of CS

---

[7] See below: adequate and justified, sufficient internal and external quality.

[8] Donald Knuth followed this meaning by calling his four volumes: 'The art of programming' [24].

practice. So far, a number of methodologies has been developed. Some of them became a mould[9] for modelling. Modelling is not yet a science[10] or culture[11].

Models in Applied Computer Science

Models are also used in manufacturing for instance as a template, pattern, reference, presentation, prototype, origin for production, master copy, and sample. We should note that models and modelling is different for engineering and also for ACS. It is far less well understood in CS. This observation is also true for software engineering [14] that makes heavy use of models but neither coherent, systematically, nor considering the engineering specifics. Software architecture [13] is not an exception.

Models can but don't have to be explicitly designed for usage. Engineering as the approach to create the artificial [36] is based on models as a starting point for construction, as a documentation of the construction, as a means for developing variations, as a thought and imagination instrument, and as an artifact within the creation process. There are, however, objects that became models at a far later stage of their existence[12].

3 Myths About Modelling

> I've studied now Philosophy and
> Jurisprudence, Medicine – and even, alas!
> Theology – from end to end with labor keen;
> and here, poor fool with all my lore I stand,
> no wiser than before.
>
> Goethe, *Faust I*

---

[9] A mould is a distinctive form in which a model is made, constructed, shaped, and designed for a specific function a model has in a scenario.
It is similar to mechanical engineering where a mould is a container into which liquid is poured to create a given shape when it hardens. In Mathematics, it is the general and well-defined, experienced framework how a problem is going to be solved and faithfully mapped back to the problem area.

[10] A model and modelling science consists of a system of knowledge that is concerned with models, modelling and their phenomena. It entails unbiased observations and systematic experimentation. It involves a pursuit of model and modelling knowledge covering general truths or the operations of fundamental laws.

[11] Culture combines approaches, attitudes, behaviour, conventional conducts, codes, traditions, beliefs, values, customs, thought styles, habits, the system comprising of the accepted norms and values, goals, practices, and manners that are favored by the community of practice. It is a characteristic of this community and includes all the knowledge and values shared by the community of practice.
Culture of modelling is a highly developed state of perfection that has a flawless or impeccable quality.

[12] Objects can be developed for usage. At a later stage, they were exhibited and become models used for explanation, e.g. about culture. See for instance, Ancient Egyptian objects in modern museums. We also may observe the opposite for the model-being of object, e.g. see [8].

Is modelling important? Should modelling become a professional skill? Or a profession? It seems that this is not (yet) and must not be the case. It is a side technique used by everybody. Modelling does not have a prominent role in kernel CS education. A separate module named 'Modelling' is a curiosity in curricula. Let us briefly consider some arguments against the existence of such modules in curricula. From the other side, all branches of CS including ACS make heavy use of models. So far we know, the earliest publication which uses the notion of model is 2.500 years old[13]. May be, the notion is even older since it seems that Ancient Egyptians already deliberately used models. CS and ACS have developed many novel approaches to modelling and are far away from considering their way of modelling as the sole right of representation[14]. We claim that modelling is one of the main activities in our field and one of our main competencies. The analysis [41] underpins this exceptional role of models in CS compared to other disciplines.

(1) There is no commonly acceptable notion of model

CS uses the term 'model' in a manifold of cases. Models have very different functions. It seems not to be fruitful to use a general singleton notion of model. Instead we may use a parameterised notion with parameters that can be refined and adapted according the specific function that a model plays in an application scenario, in a given context, for some community of practice, for a collection of origins which must be represented by the model[15].

(2) Modelling languages must be as expressive as only possible

After having their success stories, modelling languages become dinosaurs. Anything what is thinkable in the application domain of such languages is integrated. For instance, the entity-relationship modelling language started with 4

---

[13] The earliest source of systematic model consideration we know is Heraclitus (see [26] for a systematic and commented re-representation of Heraclitus fragments) with his concept of λόγος (logos).

[14] Mathematicians often claim that they are the only ones who know what is a model and what is modelling. We notice, however, that modelling is typically performed outside Mathematics.

[15] The notion

"A model is a simplified reproduction of a planned or real existing system with its processes on the basis of a notational and concrete concept space. According to the represented purpose-governed relevant properties, it deviates from its origin only due to the tolerance frame for the purpose." [44]

is a typical example of this parametrisation. The origin is the system and the inherited concept space. Analogy is essentially a mapping. Focus is simplification. Purpose is reproduction or documentation. The justification is inherited from the system and its processes. Sufficiency is based on tight mapping with some tolerated deviation. In a similar form we use parameters for the definition in [1]:

"A model is a mathematical description of a business problem.".

constructs (entity type, relationship type, attribute, cardinality constraints) and was quickly extended by more than 150 additional constructs [38]. Boiling down these extensions and reducing to essential elements we use nowadays around a dozen of constructs.

(3) Modelling languages should be primarily syntactic – Syntax first, semantics later, avoid pragmatics, disregard pragmatism

CS follows the mathematical or logical treatment of semiotics. A concept is introduced in a syntactic form and gets it meaning in a second semantification step. Pragmatics is often neglected as well as pragmatism. The two extremes are then full fledged syntactical constructions or rather simplistic and simple syntactical constructions that are then combined with heavy semantics. Natural languages use words and construction in a holistic semiotics form. For instance, the relational data structure language uses a very small syntactic language. Since the syntax for structures is insufficiently expressive for applications, this syntactical part is extended by more than threescore classes of integrity constraints.

(4) Each hype must have its modelling language

CS is an area of extensive and eager hype development. Since hypes come and vanish and since hypes are typically born due to some challenges to existing approaches, they use new vocabulary and thus new models. For instance, big data applications are considered to be schema-less and thus model-free. In reality they use models. Realising this we use a 'novel' language for their description. Essentially, big data modelling is based on the sixth normal form from relational database approaches.

(5) Implementable modelling languages should be minimalistic

The relational database modelling language is based on two syntactical constructors (relational type and attribute). Anything what is to be given in more detail can be semantically added. This treatment requires skills of a master or journeyman. It often results in introduction of artificial elements. For instance, RDF use only labelled binary relationships among labelled nodes. From predicate logic we know that any predicate can be expressed by associated binary predicates if we introduce a concept of identifier which is essentially a surrogate value and makes modelling a nightmare whenever the application is a bit more complex than the toy examples used for explaining RDF.

(6) A model reflects main aspects of applications and allows to derive all others

A model should holistically cover all aspects of a given application. This restricts development of a model or results in infeasibility or impossibility of model development. Database modelling is a typical example. A database schema is introduced in the global-as-design approach. All specific viewpoints of business users can be represented by views that are from the global schema. This philosophy results in assumptions such as unique-name, unique-granularity, and unique identifiability. New database applications needs then work-arounds or other languages.

(7) Models can be discarded after being used

A common programming approach is to use some kind of model for inspiration or for documentation of the negotiated agreement in a community of practice. These models are later thrown away since they seem not to be of any use anymore. For instance, UML models are used as inspiration models for programmers. They can be discarded after they have been used for coding. At the best, they are stored in the documentation. Revising, modernising, or migrating the code does not havde an impact on the model.

(8) At its best, models are used for programming and coding

Models often carry far more knowledge about an application domain than the program that has been derived from the model. For instance, language terms used in models have their own linguistic meaning and carry important semantics. The main function of models is the support for program construction. Programs might also be models to certain extend. But the model is used in this case mainly as the initial point for programming.

From the other side, modelling languages are often so well-developed that models written in these languages can be fully (or at least in form of templates) translated or compiled to code. The modelling-to-program initiative[16] matures the model-driven approaches to executable models, i.e. modelling is then professional programming.

(9) One model is sufficient for holistic representation

Systems have typically many sub-systems, are embedded, and are used with a good variety of features. Modelling often assumes that a singleton model allows to represent all these facets and aspects. The database schema approach mentioned above is a typical example of this belief. A weaker assumption is the existence of a model ensemble that uses separation of concern for development of models for each aspect. We better use model suites that consist of tightly and coherently associated models [10]. Models in a model suite reflect specific aspects and facets but are associated to each other in a way that allows to control and to manage coherence.

---

[16] See http://bernhard-thalheim.de/ModellingToProgram/.

(10) Modelling can be practised as handicraft or an art

Modelling uses cookbook approaches which can be used in a handicraft at the level of a journeyman who starts as apprentice with learning how to model on the basis of toy examples. Modelling is thus – at its best – an art and does not need a theory. We envision, however, that modelling must be performed at the level of a professional master craftsman. It will then develop its own modelling culture.

(11) Modelling languages should have a flexible semantics and fixed syntax

The UML community uses an approach with refinable semantics. Each member of the community of practice and tools may use their own interpretation. The result is a 'lost-in-translation' situation resembling a nightmare. The syntax allows then a variety of translations depending on a hidden intention of the language user. Instead, languages must support precise descriptions both in syntax and semantics. Moreover, the UML approach uses model suites with a number of somehow associated models each of it will have their variety of interpretations. Harmonisation becomes a nightmare. Integration, modernisation, and migration of systems cannot be properly supported. Instead, a model suite can be based on institutions for signatures. Each model in the model suite is expressed in a properly defined language. Models are associated by a fixed association schema. We can then uniquely translate one meaning to another one.

(12) Programs and algorithms first, structures later

Programmers are used to concentrate on an algorithmic problem solution. The lessons of first computer programming with bit and byte tricky realisation seem to be forgotten. At least with data mining we discovered that efficient algorithms have to incorporate supporting data structures. We should break with the bad habit to consider design of proper variable collection as nonessential for programming. CS solution development has to be based on a sophisticated integration of structures, functionality, interaction, and distribution. So, models have to be based on co-design approaches that guarantee this integration.

(13) There are no hidden deep models behind

Computer Science models often keep quiet about postulates, paradigms, and more general its basis or grounding. For instance, the basis of Turing machines includes a number of implicit principles such as compositionality, functional or relational state transformations, step-wise computation, and context-freeness. One of the guiding implicit postulates is the Von-Neumann-machine and its sequentiality and separation into computation, control, and data. An implicit and concealed deep model circumvents deliberate model utilisation. In order to overcome this pitfall, the abstract state machine approach explicitly uses the three guiding postulates for sequential computation (postulates of sequential time, of abstract state, of bounded exploration of the state space) [3].

(14) Models represent machines as state-transformers

Digital computers are based on a notion of state. Computation is controlled state transformation. Many technical devices like analogous computers are not built on this paradigm. Neural networks might be coarsened understood as state transformers. The first Leibniz multiplication machine integrated analog and digital computation. CS concentrated on state transformation. Models seem to follow the same paradigm although this is not necessary.

(15) All models are conceptual models

There is no common agreement which artifact should (not) be considered to be a (conceptual) model although the term 'conceptual model' is used for more than for five decades in computer science and for more than one century in science and engineering. It is often claimed that any model is a conceptual one. Conceptual models are, however, models with specific properties. Better we claim:
"A conceptual model is a function-oriented and consolidated model of origins that is enhanced by concept(ion) spaces of their origins, is formulated in a language that allows well-structured formulations, is based on mental/perception/domain-situation models with their embedded concept(ion)s and notions, and is oriented on a matrix that is commonly accepted." [20]

(16) Models are monolithic at the same level of abstraction

It seems that programming has to be based on the same level of abstraction, e.g. programs based on formal grammars with word generation at the same level of abstraction. Second-order grammars do not change this picture. Models representing UML model collections, OSI communication layering, and information systems are typically model suites (see myth (9)). Model suites use several levels of abstraction. The abstraction layer meta-model [38] for development of an information system is an example of integrated and controlled use of models in a model suite at several abstraction layers.

(17) CS is entirely based on programming

Computer application engineering is mainly treated as software production. CS is also concerned with hardware, with embedded systems, and proper system composition. Engineering has some common concerns with science. It is however a completely different activitiy[17]. Most of the practical and applied Computer Science is in reality engineering. We note that engineering and ACS are extensively using of models.

---

[17] Engineering is the art of building with completely different success criteria (see [33]: "Scientists look at things that are and ask 'why'; engineers dream of things that never were and ask 'why not'." (Theodore von Karman)).
"Engineers use materials, whose properties they do not properly understand, to form them into shapes, whose geometries they cannot properly analyse, to resist forces they cannot properly assess, in such a way that the public at large has no reason to suspect the extent of their ignorance." (John Ure 1998, cited in [33]).

(18) The inner structure of models must be static

Tools we use in daily life are constantly adapted. The structure and the composition of these tools change. So far, we assume that the structure of a system is static. For instance, ASM uses a static set of functions. The functions themselves may change. The category of the functions may change among being static, dynamic, private, and public. However, the set of functions is fixed. There is no reason beside convenience why the set of functions should not be dynamic. Practical systems may have functions that are incorporated depending on the state of a running system and especially on the context.

(19) Model-based reasoning is based on deductive systems

The reasoning approach is still often based on some kind of deductive system that allows to derive conclusions from rules, facts, premisses, presuppositions, etc. Model detection already uses inductive and evidential reasoning. Model application also uses plausible and approximative reasoning. For instance, abduction is used as a technique for detection of the best explanation in a given context. Although classical and non-classical mathematical logic has overwhelming success stories we should integrate reasoning systems that are completely different from deductive systems. Therefore, model-based reasoning is also based on a combination of various kinds of deductive, inductive, evidential, abductive or more general plausible, and approximative reasoning. Most of these reasoning mechanisms are based on coherence instead of on consistency. Moreover, deduction has to be handled beyond first-order predicate logic despite undecidability, incompleteness, and worse-case complexity of higher-order logics.

(20) AI models may cover human intelligence

Although AI tools are nudging, directing, and misleading us, they do not cover all sides of human intelligence. Neither emotional nor self-reflection or other kinds of human intelligence can be programmed with current approaches. As a matter of fact, human creativity in problem solution is far from being programmable. Mental models are essential for daily problem solution as well as for technical realisations. Therefore, modelling has to properly reflect this kind of intelligence.

4   Models - Towards a General Theory

> It's written here: 'In the Beginning was the Word!'
> Here I stick already! Who can help me? It's absurd,
> Impossible, for me to rate the word so highly.
> I must try to say it differently.
> If I'm truly inspired by the Spirit. I find
> I've written here: 'In the Beginning was the Mind'.
> Let me consider that first sentence,
> So my pen won't run on in advance!
> Is it Mind that works and creates what's ours?

> It should say: 'In the beginning was the Power!'
> Yet even while I write the words down,
> I'm warned: I'm no closer with these I've found.
> The Spirit helps me! I have it now, intact.
> And firmly write: 'In the Beginning was the Act!'
>
> Goethe, *Faust I*

We will not present approaches that allow to overcome all the myths. The development of a general theory of models and modelling is on the agenda for the next threescore years. Parts and pieces for this theory are already developed.

This agenda includes answers to a number of questions, e.g.:

- *What are the essential properties of models? Can modelling be systematised?*
- *Which artifacts and thoughts are models? Is there any demarcation for the model-being? How can the model-being can be characterised?*
- *What is the main orientation of a model? Can we separate aspects of model use?*
- *Are we using some model or are we using an ensemble of models?*
- *What are the properties of models that are used for interaction in social settings?*
- *What about myths on models that seems to be valid?*

Models are thoughts or artifacts. They are used as instruments at a certain point of time. At other points of time, they might not be models. I.e. thoughts and artifacts have their own journey in the model being. What is then the notion of a model?

The Notion of Model

Let us first introduce a general notion of model. This notion can be specialised to more specific in disciplines. The notion generalises almost all notions or prenotions used and known so far in general model theory [23,27,29,37]. More specific notions can be declined by parameter refinement and hardening from this notion.

"A **model** *is a well-formed, adequate, and dependable instrument that represents origins and that functions in utilisation scenarios.*
Its criteria of well-formedness[18], adequacy[19], and dependability[20] must be commonly accepted by its community of practice (CoP) within some context and correspond to the functions that a model fulfills in utilisation scenarios." [40]

---

[18] *Well-formedness* is often considered as a specific modelling language requirement.

[19] The criteria for *adequacy* are analogy (as a generalisation of the mapping property that forms a rather tight kind of analogy), being focused (as a generalisation of truncation or abstraction), and satisfying the purpose (as a generalisation of classical pragmatics properties).

[20] The model has another constituents that are often taken for granted. The model is based on a background, represents origins, is accepted by a community of practice, and follows the accepted context. The model thus becomes *dependable*, i.e. it is justified or viable and has a sufficient quality.

Most notions assume dependability either as a-priori given or neglect it completely.

This notion also allows consideration of the model-being of any instrument. We note that the instrument-being is based on the function that a model plays in some scenario. The generality of this notion and its parametric form provides a means for specialisation. For instance, models used in Computer Science are both based on 'normal' models that are adequate from one side, 'deep' models that determine justification, and canonical sufficiency frames for model quality from the other side. For instance, assuming the deep model, quality sufficiency, and the background as definitely given and as being unchangeable, we can restrict the determination of the model-being of an instrument to adequacy. Such assumptions are often made while considering models. Reasoning by models and deployment of models of models becomes problematic if the deep model, quality sufficiency, and the background are implicit.

The Logos, Archẽ, and Nous of Modelling

The quote we use for this Section is based on reasoning about four of the seven meanings of λóγος (logos): <u>word</u>, concept, judgement, <u>mind</u>, <u>power</u>, <u>deed</u>, and reason. The notion of logos characterises the model-being. Models must be well-formed (word), based on world-views (concepts), acceptable (judgement), understandable (mind), applicable (power), support their successful use (intended application), and allow to assess their utilisation (reason). We distinguish *perception* and idea (in German 'Auffassung'[21]; including thought, conception, opinion, view, understanding, apprehension, thinking,claim to truth) from imagination and *reflection* (in German 'Vorstellung'; including vision, image, (re)presentation, conceivability, belief, mental image or picture, speech, explanation, justification, accountability, justification, meaningful and -founding spoken/written/thought speech, verifiability, reasonableness, correctness). Essentially, the two kinds of mental models are *perception model* and the *imagination* or reflection *communication model.*

Logos is according to Heraclitus [26] the dynamic, creative, and ordering principle that is injected to perception and reflection. It is subordinated to Nous as 'intellectus', ordering spirit, reason, and rational soul. Archẽ is the third element in this triptych. It is the general principle of reality and the grounding one in the sense of a fundamental laws. According to the Platon's three analogies (cave, divided line, sun) we cannot fully access it.

Computer Science modelling uses archẽ as its suite of *domain-situation models* (see the encyclopedic side in the model triptych [30]). The model triptych has its linguistic or representation dimension. This dimension enables model expression. It also hinders it due to the obstinacy of each language. These languages heavily influence modelling due to their hidden grounding and basis. For instance, state-oriented languages with doubtful paradigm of computation-is-state-transformation have a hard time to represent events or continuously changing systems (see also [45] for Sapir-Whorf principle of linguistic relativity).

---

[21] The word fields in German and English languages are different.

ACS Needs Approaches Beyond Computation

ACS models and CS models substantially differ. Both use reflection and representation models. Prescription ACS models must be as precise and accurate as only possible if they are used as blueprint, plan, antetype, archetype, and realisation worksheet. They must also be more flexible due to adaptation to given environments. For documentation models, the analogy is a homomorphic mapping. ACS models are also used as steering, usage, and guiding models. On the other hand, CS models can also be description, conceptual, investigation, and explanation models. Due to the difference in their function, their adequacy and dependability follow different schemata, norms, and usages, i.e. different canons.

ACS models cannot be context-free or context-limited. They represent a variety of aspects and are thus typically model suites. The UML approach demonstrates this approach by separation of concern. CS models are typically more monolithic and follow the global-as-design approach. For instance, in CS we prefer to use UML class diagrams as the lead model from which the class aspects in other models can be derived as specific views. Additionally, engineering is a construction discipline and thus less – if at all – theory-oriented. However, it is more oriented on technological realisability. Engineering has to produce robust products with well-designed error-tolerance.

Proper Theory of Modelling Languages

Both CS and ACS make intensive use of formal or somehow narrative but orthonormalised languages. Formal languages follow a rigid syntax-first-semantic-second approach. The classical approach to define models within a modelling language that is again defined by a meta-model language is far to strict whenever essential constructions and constructors are not theoretically based (see the OMG approach with models M_1, meta-models M_2, meta-meta-models M_3 etc.). For instance, we know more than half a dozen different IsA relationships while database modellers are used to only one of them. Narrative languages are less rigid and less correct. A typical example of the last kind are standards, e.g. the BPMN standard. In this case, the definition of a formal semantics often becomes a nightmare and requires intensive foundational work in order to become rigid, e.g. [4, 5].

We do not know a theory of language development neither for conceptual languages nor for so-called domain-specific languages although parts and pieces are currently under development, e.g. [15, 42]. Languages must, however, be based on a proper theory whenever languages are used as an enabling tool for modelling in the model triptych approach. Nowadays, languages must also be supported by advanced tools and advanced guidance for their use.

5   Tasks for Next Decades

> Enough words have been exchanged,
> Let me finally see deeds;
> While you are paying compliments,
> Something useful can happen.
>
> Goethe, *Faust I*

Overcoming Limits, Obstinacies, and Problematic Solutions

We often hear a claim that ACS&CS are at their dead end. The claim is near to truth due to problematic paradigms and postulates that must be changed soon in the form as envisioned in [9, 25, 42][22]. The AI 4.0 hype illustrates the wild search for a solution of new paradigms. The first step will be a thorough reconsideration of the computation paradigms. For instance, data massives do not need algorithmic number crunching but rather sophisticated 'water supply systems'. Web computation is based on symbol crunching. Neural networks must be build on the real way how neurons work. We do not need optimal solutions. Instead we can live with approximative solutions near to the optimal. Programming-in-the-small was successful for the current infrastructure. Programming-in-the-large is not yet properly understood. Programming-in-the-world cannot be supported within the current approach. A path towards a solution is proper modelling and model-based reasoning. It might also solve problems of dehumanized algorithmic machines. Sophisticated systems such as AI systems operate without feelings, without heart, without compassion, without conscience, and without ethics. They are simply machines. Human infrastructures need however a different environment based on models that humans use.

Programming-in-the-large and programming-in-the-world have to use new kinds of models. The classical modelling approach to programming-in-the-small has to be incorporated. It needs, however, a deep reconsideration of modelling in the future. Model suites with a number of well-associated to each other models can be one of the solution. As a minimal requirement to a model suite, models representing human approaches and use have to become an integral part.

Modelling to Program and Modelling as Programming

The main usage of models and modelling in CS&ACS is models-for-programming (M4P). There are many initiatives that extend this usage towards modelling-to-program (M2P): Model-driven development, conceptual-model programming, models@runtime, universal applications, domain-specific modelling, framework-driven modelling, pattern-based development, round-trip engineering, model programming, inverse modelling, and reference modelling.

---

[22] The software crisis has been a crisis 1.0. Nowadays we have a data crisis, a (large and embedded) system crisis, an infrastructure crisis, and an energy crisis. For instance, it is estimated that one third of the world-wide produced electro energy is consumed by computers by 2025.

These initiatives are the starting points for a programme for true fifth generation programming that starts with models and then uses models as source code for a program beside of being useful for program specification, i.e. modelling-as-programming (MaP). It is similar to second and third generation programming where programmers are writing programs in a high-level language and rely on compilers that translate these programs to machine code. We propose to use models instead of programs and envision that models can be translated to machine code in a similar way. The approach has already been used in several projects, e.g. for ASMbacked C++ programming and for database development [21, 35].

Models will thus become *executable* while being as *precise* and *accurate* as appropriate for the given problem case, *explainable* and *understandable* to developers and users within their tasks and focus, *changeable* and *adaptable* at different layers, *validatable* and *verifiable*, and *maintainable*.

Humanised Systems

Web system development [34] taught us a really sad lesson: user are not judging the system by its functions but its form at the first step. They ask whether a system is intuitively understandable and usable. They judge on colours and other completely irrelevant features instead of functioning issues. Civil engineering has already properly learned this lessons long time ago and found a way how to cope with the problem: design of everyday things, e.g. [31]. With the broadband application we have to go away from the philosophy that the administrator is the only king and the user is at its best the slave.

Models of systems must, thus, be based on the cultures of users, especially on their models.

Application-Governed ACS

CS and ACS experts are used to be the ruler and the guidance for human behaviour. This orientation has been very useful for the first half century of CS existence. Computers became now an essential part of the everyday infrastructure. As such, they are services in applications. A fist step to primary consideration of application objectives instead of requirements analysis is the abstraction-layer macro-model [34] for web system information systems.

Applications bring in their own culture[23] including national and regional, community-backed and corporate, good and problematic habits, professional and educational, ethic and generational, and finally gender, class, and ideological ones. Therefore, models must be adaptable to the specific cultures.

---

[23] Culture is a "a collective phenomenon, which is shared with people who live or lived within the same social environment, which is where it was learned; culture consists of the unwritten rules of the social game; it is the collective programming of the mind that separates the member of one group or category of people from others." [19].

Applications found their way of incorporating very different people with their specific perception, reflection, and imagination. People come with their models that might vary a lot. The common binder is the business. The binder can be used as the starting point for an association schema among the models used. In this case, we have to properly and flexibly handle and adapt models in a model suite.

Applications continuously change and adapt to new ways of operating. That means that change management will be an essential feature for model suites too. Changes might be evolutionary changes, modernisation, migration, or partial replacement with partial cannibalisation. The change with a model suites must also be so robust and flexible.

Paradigms and Postulates for CS and ACS Models

Models, activities to model and to use models, and modelling (MMM) has not yet found its theoretical underpinning. Parts and pieces are already developed in different disciplines [41]. An MMM theory or MMM culture is one of the challenges. CS models can be stereotyped. Modelling can be systematised into a smaller set of canons and moulds by grouping MMM success stories according to those stereotypes. This approach is similar to the design science approach. The rigor cycle [11,18] aims at summarisation, generalisation, and systematisation of modelling experience. Exaptation is some kind of extrapolation of known solutions to new problems in other context.

From the other side, modelling can be based on separation into normal modelling and inheritance of already existing deep models from the same kind of application [39]. Deep models are the inner and inherent part of a model together with corresponding methodologies, techniques, and methods, i.e. its matrices. Modelling, model development, and model utilisation can be concentrated around the normal model.

Modelling themselves follows a number of paradigms, postulates, and principles. They are often implicitly assumed in CS. Reusable postulates (e.g. separation into state, evolution, and interaction for information systems), paradigms (e.g. global-as-design for database structures and views), and principles (e.g. incremental constructivity) support efficient development of normal models. Database structure and functionality modelling reuses and accepts a standardised matrix with development and utilisation methods and some robust methodology.

6   Finally

> A marshland flanks the mountain-side,
> Infecting all that we have gained;
> Our gain would reach its greatest pride
> If all noisome bog were drained.
> I work that millions may possess this space,
> If not secure, a free and active race.
>
> Goethe, *Faust II*

Towards Modelling 2.0 and Modelling 3.0

We may consider the current state of art as the first generation of modelling. The body of MMM knowledge is not yet synthesised into a subdiscipline of CS and ACS. The MMM discipline has to be based on its specific foundations, its specific reasoning methods, its methodologies, and its practices. Model suites are a good starting point for multi-view, multi-abstract, multi-level, and multi-culture modelling. This approach has to be generalised, well-founded for all its extensions, and properly supported by tools and workbenches. Modelling in our discipline is mainly language-based. We, thus, need an extensible but well-integrated collection of languages. Some of them can be oriented on general purpose while most will be domain-specific. Models should be easy to modify and to modernise. Model quality management includes also model analysis depending on functioning in scenarios. This BoK should also include means for certification and industrial licensing.

Models are almost useless if they are one-way-products. Currently, some modelling platforms (e.g. ADOxx) support generation of programs from models, at least for some of the modelling approaches. Modelling 2.0 could be considered as modelling-in-the-world. Modellers come with various cultures, commonsense, thought schools according to the branches of CS and ACS, and various languages. Collaborative model development cannot be based on a uniform language environment. It can use some kind of generic approach that allows to derive specific languages from some source in such a way that the specific adaptation can be reversed and then mapped back to the other partner, i.e. we use an approach with model institutions. Shared model usage is another real challenge. Modelling 3.0 could use model spaces based on solution libraries from which a generic and adaptable model can be extracted similar to the concept of namespace for injection of the meaning of words in Web x.0 .

As stated, models are the fourth dimension of modern CS and ACS. The associations to the other dimensions will become explicit. For instance, MMM 2.0 or 3.0 will be then based on MMM for collaboration or more specifically for communication, for cooperation, and for coordination in various forms and adaptable to various kinds of applications.

Modelling Becomes an Education Subdiscipline in CS and ACS

Modeling is currently an art. It should become a craftsmanship with integration into CS and ACS education. So far, modelling is taught as a side-method for system development. Each subdiscipline has its approach to integrate models and modelling. Some of them use models in an implicit form, some of them in an explicit form. Each subdiscipline founds its specific approach and way of acting. Best practices are used for demonstrating the potential of models. A compilation of this experience is one of the lacunas of current CS and ACS.

Systematic and well-guided modelling should be taught in a specific and separate module in any CS and ACS programme. Modelling will become an object in education and continuing education. A discipline of modelling needs its basis, its methods, its methodologies, and its theory. All this must be adaptable to specific application cases. A modelling discipline includes also proper quality management and handling of errors.

Each subdiscipline has its conception, its concept spaces, its methods, its terminologies, its success and failure stories, its construction approach, its accepted practices, its educational demonstration models, its competencies and skills, its approaches, and its theoretical underpinning. MMM as a module will be not different.

Deploying the Full Capacity and Potential of Models

The utility, potential and capacity of models is not yet fully understood and deployed. Models are used for complexity reduction, for communication within the communities of practice, for efficiency increase during development, and for mediating during development. We may launch out model suites into risk minimisation, generation of solutions directly from models, and handling interoperability among different systems. Models should become an integral component of delivery software and hardware.

Models will become a proper companion during all development steps for integrated team collaboration, for quality management, for forecasting performance and bottlenecks, for generating of neatly integrable partial solution. Model suites have to reflect a large variety of viewpoints in a well-associated form since applications become more and more multi-disciplinary and interdisciplinary. Modelling-in-the-world should allow to tolerate heterogeneity in organisations that are outside the control and understanding of the modelling team.

Heritage (currently often called 'legacy') models and reference models can be used for experience propagation. Componentisation of models shall support re-usage and recycling of already existing high-quality solutions. Domain-specific languages may be used for development of inheritable and adaptable domain-specific model suites and their integration into solutions. Models in a model suite also represent a variety of viewpoints within a community of practice. Tight association of models in suites allows to concentrate on certain issues and aspects without losing the coherence in the model suite. Model suites have to support multi-abstraction, multi-levels, and multi-cultures among all partners.

In order to serve the variety of viewpoints and cultures within a community of practice, a model suite should integrate a number of informative models for each member as well a number of representation models in a variety of languages. These different models may represent the same aspects and features but in different forms due to the differences in the languages used and preferred by some members in the community of practice. Models for theoreticians might be qualitative. Models might also concentrate around quantitative aspects for data practitioners. We need then a 'matching' theory for the two orientations.

Call for Contribution and Research

Wolfgang Hesse (see also [16,17]) summarised the consternation about modelling in the Modellierung 2009 workshop:

- ... but they do not know what they do ...;
- Babylonian language confusion and muddle;
- "It's not a bug, it's a feature", de-facto-standards and lobbyists;
- Why I should cope with what was the state of art yesterday;
- Each day a new wheel, new buzzwords without any sense, and a new trend;
- Without consideration of the value of the model;
- Competition is a feature, inhomogeneity;
- Laokoon forever;
- Dreams about a sound mathematical foundation;
- Take but don't think - take it only without critics;
- Academia in the ivory tower without executable models;
- Where is the Ariadne thread through?

This lead directly to a number of research and development issues: Can we develop a simple notion of adequateness that still covers the approaches we are used in our subdiscipline? Do we need this broad coverage for models? Or is there any specific treatment of dependability for subdisciplines or specific deployment scenarios? Which modelling methods are purposeful within which setting? Which model deployment methods are properly supporting the function of a model within a utilisation scenario? How does the given notion of model match with other understandings and approaches to modelling in computer science and engineering? What is the background of modelling, especially the basis that can be changed depending on the function that a model plays in some utilisation scenario? Language matters, enables, restricts and biases. What is the role of languages in modelling? Which modelling context results in which modelling approach? What is the difference between the modelling process that is performed in daily practice and systematic and well-founded modelling? Are we really modelling reality or are we only modelling our perception and our agreement about reality? What is the influence of the modeller's community and schools of thought?

The current situation is not really different from 2009. We need a lot of good research contributions.

> What you put off today will not be done tomorrow;
> You should never let a day slip by
> Let resolution grasp what's possible
> And seize it boldly by the hair;
> Then you will never lose you grip,
> But labor steadily, because you must.
> Goethe, *Faust I*

References

1. Abts, D., Mülder, W.: Grundkurs Wirtschaftsinformatik: Eine kompakte und praxisorientierte Einführung. Vieweg (2004)
2. Alter, S.: Work system theory and work system method: a bridge between business and IT views of IT-reliant systems in organizations. In: Proceedings of the ISEC 2017, p. 211. ACM (2017)
3. Börger, E., Stärk, R.: Abstract State Machines - A Method for High-level System Design and Analysis. Springer, Heidelberg (2003). https://doi.org/10.1007/978-3-642-18216-7
4. Börger, E., Thalheim, B.: A method for verifiable and validatable business process modeling. In: Börger, E., Cisternino, A. (eds.) Advances in Software Engineering. LNCS, vol. 5316, pp. 59–115. Springer, Heidelberg (2008). https://doi.org/10.1007/978-3-540-89762-0_3
5. Börger, E., Thalheim, B.: Modeling workflows, interaction patterns, web services and business processes: the ASM-based approach. In: Börger, E., Butler, M., Bowen, J.P., Boca, P. (eds.) ABZ 2008. LNCS, vol. 5238, pp. 24–38. Springer, Heidelberg (2008). https://doi.org/10.1007/978-3-540-87603-8_3
6. Bourbaki, N.: Foundations of mathematics for the working mathematician. J. Symb. Log. **14**(1), 1–8 (1949)
7. Brentano, F.: Psychologie vom empirischen Standpunkte. Dunker & Humblot, Leipzig (1874)
8. Chadarevian, S., Hopwood, N. (eds.): Models - The Third Dimension of Science. Stanford University Press, Stanford (2004)
9. Cohen, R.S., Schnelle, T. (eds.): Cognition and fact: Materials on Ludwik Fleck. Boston Studies in the Philosophy of Science, vol. 87. Springer, Dortrecht (1986). https://doi.org/10.1007/978-94-009-4498-5
10. Dahanayake, A., Thalheim, B.: Co-evolution of (Information) system models. In: Bider, I., Halpin, T., Krogstie, J., Nurcan, S., Proper, E., Schmidt, R., Ukor, R. (eds.) BPMDS/EMMSAD -2010. LNBIP, vol. 50, pp. 314–326. Springer, Heidelberg (2010). https://doi.org/10.1007/978-3-642-13051-9_26
11. Dahanayake, A., Thalheim, B.: Development of conceptual models and the knowledge background provided by the rigor cycle in design science. In Models: Concepts, Theory, Logic, Reasoning, and Semantics, Tributes, pp. 3–28. College Publications (2018)
12. Denning, P., et al.: Computing as a discipline. Computer **22**(2), 63–70 (1989)
13. Engelschall, R.S.: Schönheit und Unzulänglichkeit von Software-Architektur. In Software Engineering 2021, LNI, vol. P-310, p. 19. Video accessible via https://www.youtube.com/watch?v=AMOmvtsCRoU (2021). Gesellschaft für Informatik e.V
14. Engelschall, R.S.: Manifesto for true software engineering (2021). http://true-manifesto.org/. Accessed 8 Mar 2021
15. Frank, U.: Domain-specific modeling languages: requirements analysis and design guidelines. In: Reinhartz-Berger, I., Sturm, A., Clark, T., Cohen, S., Bettin, J. (eds.) Domain Engineering, Product Lines, Languages, and Conceptual Models, pp. 133–157. Springer, Heidelberg (2013). https://doi.org/10.1007/978-3-642-36654-3_6
16. Hesse, W.: Modelle - Janusköpfe der Software-Entwicklung - oder: Mit Janus von der A- zur S-Klasse. In: Modellierung 2006, LNI, vol. 82, pp. 99–113. GI (2006)

17. Hesse, W., Mayr, H.C.: Modellierung in der Softwaretechnik: eine Bestandsauf-nahme. Informatik Spektrum **31**(5), 377–393 (2008)
18. Hevner, A., Chatterjee, S.: Design Research in Formation Systems. Springer, Boston (2010). https://doi.org/10.1007/978-1-4419-5653-8
19. Hofstede, G., Hofstede, G.J., Minkow, M.: Cultures and Organizations: Software of the Mind: Intercultural Cooperation and Its Importance for Survival. McGraw-Hill, New York (2010)
20. Jaakkola, H., Thalheim, B.: Cultures in information systems development. In: Information Modelling and Knowledge Bases XXX, Frontiers in Artificial Intelligence and Applications, vol. 312, pp. 61–80. IOS Press (2019)
21. Jaakkola, H., Thalheim, B.: Models as programs: the envisioned and principal key to true fifth generation programming. In: Proceedings of 29'th EJC, Lappeenranta, Finland, pp. 170–189 (2019). LUT, Finland
22. Kangassalo, M., Tuominen, E.: Inquiry based learning environment for children. In: Information Modelling and Knowledge Bases XIX, Frontiers in Artificial Intelligence and Applications, vol. 166, pp. 237–256. IOS Press (2007)
23. Kaschek, R.: Konzeptionelle Modellierung. PhD thesis, University Klagenfurt (2003). Habilitationsschrift
24. Knuth, D.E.: The Art of Programming I-III. Addison-Wesley, Reading, 1968–1973
25. Kuhn, T.: The Structure of Scientific Revolutions. University of Chicago Press, Chicago (1962)
26. Lebedev, A.V.: The Logos Heraclitus - A Reconstruction of Thoughts and Words; Full Commented Texts of Fragments (In Russian). Nauka, Moskva (2014)
27. Mahr, B.: Information science and the logic of models. Softw. Syst. Modeling **8**(3), 365–383 (2009)
28. Mahr, B.: Intentionality and modeling of conception. In: Judgements and Propositions - Logical, Linguistic and Cognitive Issues, pp. 61–87. Berlin (2010)
29. Mahr, B.: Modelle und ihre Befragbarkeit - Grundlagen einer allgemeinen Modelltheorie. Erwägen-Wissen-Ethik (EWE) **26**(3), 329–342 (2015)
30. Mayr, H.C., Thalheim, B.: The triptych of conceptual modeling. Softw. Syst. Modeling **20**(1), 7–24 (2020). https://doi.org/10.1007/s10270-020-00836-z
31. Norman, D.A.: The Design of Everyday Things. Doubleday, New York (1990)
32. Safra, J.E., Aquilar-Cauz, J., et al. (eds.): Encyclopædia Britannica Ultimate Reference Suite, chapter Computer Science. Encyclopædia Britannica, Chicago (2015)
33. Samuel, A., Weir, J.: Introduction to Engineering: Modelling. Synthesis and Problem Solving Strategies. Elsevier, Amsterdam (2000)
34. Schewe, K.-D., Thalheim, B.: Design and development of web information systems. Springer, Heidelberg (2019). https://doi.org/10.1007/978-3-662-58824-6
35. Schmid, J.: Compiling abstract state machines to C++. J. Univ. Comput. Sci. **11**, 1069–1088 (2001)
36. Simon, H.: The Sciences of the Artificial. MIT Press, Cambridge (1981)
37. Stachowiak, H.: Allgemeine Modelltheorie. Springer, Cham (1973)
38. Thalheim, B.: Entity-Relationship Modeling - Foundations of Database Technology. Springer, Heidelberg (2000). https://doi.org/10.1007/978-3-662-04058-4
39. Thalheim, B.: Normal models and their modelling matrix. In: Models: Concepts, Theory, Logic, Reasoning, and Semantics, Tributes, pp. 44–72. College Publications (2018)
40. Thalheim, B.: Conceptual models and their foundations. In: Schewe, K.-D., Singh, N.K. (eds.) MEDI 2019. LNCS, vol. 11815, pp. 123–139. Springer, Cham (2019). https://doi.org/10.1007/978-3-030-32065-2_9

41. Thalheim, B., Nissen, I. (eds.): Wissenschaft und Kunst der Modellierung: Modelle, Modellieren. Modellierung. De Gruyter, Boston (2015)
42. Thomas, O., et al.: Global crises and the role of BISE. Bus. Inf. Syst. Eng. **62**(4), 385–396 (2020)
43. Twardowski, K.: Zur Lehre vom Inhalt und Gegenstand der Vorstellungen: Eine psychologische Untersuchung. Hölder, Wien (1894)
44. Wenzel, S.: Referenzmodell für die Simulation in Produktion und Logistik. ASIM Nachrichten **4**(3), 13–17 (2000)
45. Whorf, B.L.: Lost generation theories of mind, language, and religion. Popular Culture Association, University Microfilms International, Ann Arbor, Mich (1980)
46. Wittgenstein, L.: Philosophical Investigations. Basil Blackwell, Oxford (1958)
47. Zimmermann, W., Thalheim, B.: Preface. In: ASM 2004, LNCS, vol. 3052, pp. V–VII. Springer, Heidelberg (2004)

A Framework for Modeling the Semantics of Synchronous and Asynchronous Procedures with Abstract State Machines

Wolf Zimmermann[✉] and Mandy Weißbach

Institut für Informatik, Martin-Luther-Universität Halle-Wittenberg,
06099 Halle/Saale, Germany
{wolf.zimmermann,mandy.weissbach}@informatik.uni-halle.de

Abstract. We present a framework for modeling the semantics of frequent concepts in the context of procedures and functions. In particular, we consider the concepts of recursive and non-recursive procedures, parameter passing mechanisms, return values, procedures as values (with local and non-local procedures), synchronous and asynchronous procedures, and for the latter, synchronization mechanisms. The concepts are mostly modeled in a combinable manner. Hence the concepts for procedures for many popular programming languages are covered by the framework.

1 Introduction

A formal programming language semantics is used in many applications. An important application is compiler verification [7]. Another one is model-checking approaches where the language semantics is abstracted e.g. to a finite-state machine or a pushdown machine, and the correctness of the abstraction must be proven [17,18]. The semantics specifications using Abstract State Machines (ASM) is an operational semantics that specifies an abstract execution model for a programming language. Hence, executable ASM models for the semantics of programming languages enable the generation of interpreters. In addition high-level debugging is also possible, i.e., debuggers may also be generated from the ASM semantics of a programming language. Furthermore, it is possible to prove formally the type-safety of a programming language by proving that any valid program has the invariant that each value of an expression belongs to its type, i.e., a type error can never occur, see e.g. [13].

In order to specify the semantics of real-world programming languages, it is helpful to have a toolkit that specifies state transitions of the different variants of language concepts. Therefore, in contrast to many works on ASM-semantics of programming languages [3], this work focuses on the basic concepts in programming languages rather than modeling the semantics of a concrete programming language. The main goal of the paper is to provide a framework for formalization of procedure/function calls and returns. We only consider typed languages with

© Springer Nature Switzerland AG 2021
A. Raschke et al. (Eds.): Börger Festschrift, LNCS 12750, pp. 326–352, 2021.
https://doi.org/10.1007/978-3-030-76020-5_18

static scoping. The idea is that a set of ASM rules is introduced for different concepts in this context.

Informally, a procedure has a name p and a body b which is a statement sequence. If p is being called, the execution after the call is starting with the execution of b, and if the control returns from p (e.g. after the execution of b or a return-statement is being executed), then the execution continues with the statement after the call. A procedure may have parameters. Parameter passing mechanisms specify the interaction between the caller and the callee at the procedure call as well at the procedure return. Furthermore, a function returns a value and a function call is an expression that after returns this value. We further consider recursive procedures, i.e., a procedure can call itself either directly or indirectly. The latter is forbidden for non-recursive procedures and functions. Procedures and functions as values allow the assignment of procedures and functions to variables or passing them to parameters of other procedures and functions. Each procedure and function has a context which contains all bindings and visible declarations. In languages such as C, procedures have a global scope, i.e., it is possible to declare procedures and functions in a global scope only. Hence, all bindings to names in a global scope are visible. Therefore, procedure values are solely the procedure with its parameters and body. In contrast, languages such as Ada allow local declarations of procedures, too. Hence, the visible declarations depend on the context of a procedure declaration and the context needs to be part of a procedure value.

The classical execution of procedure calls is described above: the caller waits until the callee returns. For asynchronous procedures, the caller and the callee are executed concurrently. Without explicit synchronization, the caller can only return iff all callees have been returned. However, the question remains how asynchronous functions return their values. For this, an explicit barrier synchronization with the called function is needed, and the value is returned upon this synchronization point.

The paper is organized as follows: Sect. 2 introduces Abstract State Machines as required for the purpose of this paper. Section 3 introduce the general idea behind programming language semantics. The principal modeling is independent of the concrete statements in the programming language and required for the modeling of procedures/functions. Section 4 discusses the modeling of non-recursive and recursive procedures, Sect. 5 discusses parameter passing, Sect. 6 discusses functions, and Sect. 7 procedure parameters. These four sections only consider synchronous procedures. For asynchronous procedures in Sect. 8 we discuss (possibly recursive) asynchronous procedures, asynchronous functions, and barrier synchronizations.

Remark 1. The concepts for modelling the semantics of procedures stems from the experience of the Verifix project [7] on construction correct compilers. The part of asynchronous procedures has its foundation on the work of protocol conformance [9] and deadlock analysis of service-oriented systems [16].

2   Abstract State Machine

We introduce Abstract Machines as it is useful for defining programming language semantics. It is based on [4], but the definition is based on many-sorted, partial order-sorted signatures (similar to [19]) instead of one sorted signatures.

Remark 2. The usual definitions of Abstract State Machines [4] is based on one-sorted signatures. This is the only extension to [4]. In particular, updates are defined as in [4] and no updates on the interpretation of sort symbols are possible. There are several reasons for introducing sort. In particular, it turns out that is convenient using sorts for the different concepts in programming languages, and sub-sorts if these concepts are extended (e.g. procedure return and function return).

Definition 1 (Partial ordered signature, variable). *A* partial order-sorted signature *is a tuple* $\Sigma \triangleq (S, \sqsubseteq, F, F')$ *where*

i. S *is a finite set of* sorts
ii. $\sqsubseteq \subseteq S \times S$ *is a partial order on sorts*
iii. $F \triangleq (F_{w,s})_{w \in S^*, s \in S}$ *is a family of sets of total function symbols,*[1]
iv. *and* $F' \triangleq (F_{w,s})_{w \in S^+, s \in S}$ *is a family of sets of partial function symbols.*

A set of variables *for* Σ *is a family of pairwise disjoint sets* $X \triangleq (X_s)_{s \in S}$.

Notation: $f : s_1 \times \cdots s_n \to s \in F$ *denotes* $f \in F_{s_1, \ldots, s_n, s}$, $f : s_1 \times \cdots s_n \to ?s \in F'$ *denotes* $f \in F'_{s_1, \ldots, s_n, s}$, *and* $x : s \in X$ *denotes* $x \in X_s$.

Definition 2 (Terms, Equations, Formulas). *Let* $\Sigma \triangleq (S, \sqsubseteq, F, F')$ *be a signature and* X *be a set of variables. The set of* Σ*-terms of sort* s *over* X *is the smallest family of sets* $(\mathcal{T}_s(\Sigma, X))_{s \in S}$ *satisfying*[2]

i. $X_s \subseteq \mathcal{T}_s(\Sigma, X)$ *for all* $s \in S$,
ii. $F_{\varepsilon, s} \subseteq \mathcal{T}_s(\Sigma, X)$ *and* $F'_{\varepsilon, s} \subseteq \mathcal{T}_s(\Sigma, X)$ *for all* $s \in S$,
iii. $\{f(t_1, \ldots, t_n) \mid t_1 \in \mathcal{T}_{s_1}(\Sigma, X), \ldots, t_n \in \mathcal{T}_{s_n}(\Sigma, X)\} \subseteq \mathcal{T}_s(\Sigma, X)$ *for all*
   $f : s_1 \times \cdots \times s_n \to s \in F$ *and all* $f : s_1 \times \cdots \times s_n \to ?s \in F'$
iv. $\mathcal{T}_s(\Sigma, X) \subseteq \mathcal{T}_{s'}(\Sigma, X)$ *iff* $s \sqsubseteq s'$

Furthermore $\mathcal{T}(\Sigma, X) \triangleq \bigcup_{s \in S} \mathcal{T}_s(\Sigma, X)$. *A* Σ*-ground term is* Σ*-term* $t \in \mathcal{T}(\Sigma, \emptyset)$ *without variables.* $\mathcal{T}(\Sigma) \triangleq \mathcal{T}(\Sigma, \emptyset)$ *denotes the set of all ground terms.*

   A Σ*-equation of sort* $s \in S$ *is a pair of two* Σ*-terms* $t, t' \in \mathcal{T}_s(\Sigma, X)$. *The index* s *is omitted if it is clear from the context. A sort check for sort* s *is an unary predicate over* $\mathcal{T}(\Sigma, X)$, *denoted by* t **is** s, $t \in \mathcal{T}(\Sigma, X)$, *and an unary definedness predicate* D_s.

---

[1] S^* denotes the set of all sequences over S and S^+ denotes the set of non-empty sequences over S.
[2] ε denotes the empty sequence.

The set of Σ-formulas over x is the smallest set $\mathcal{F}(\Sigma, X)$ satisfying

v. $\{t_1 \doteq t_2 \mid t_1, t_2 \in \mathcal{T}_s(\Sigma, X)\} \subseteq \mathcal{F}(\Sigma, X)$ for all $s \in S$.
vi. $\{t \text{ is } s \mid t \in \mathcal{T}(\Sigma, X)\} \subseteq \mathcal{F}(\Sigma, X)$ for all $s \in S$.
vii $\{D_s(t) \mid t \in \mathcal{T}_s(\Sigma, X)\} \subseteq \mathcal{F}(\Sigma, X)$
viii. $\{\neg\varphi \mid \varphi \in \mathcal{F}(\Sigma, X)\} \subseteq \mathcal{F}(\Sigma, X)$
ix. $\{\varphi \wedge \psi \mid \varphi, \psi \in \mathcal{F}(\Sigma, X)\} \subseteq \mathcal{F}(\Sigma, X)$
x. $\{\forall x : s \bullet \varphi \mid \varphi \in \mathcal{F}(\Sigma, X), x : s \in X\} \subseteq \mathcal{F}(\Sigma, X)$

The Σ-formulas in (v), (vi) an (vii) are called atomic. *We use the following abbreviations: $e \triangleq e \doteq true$ if e is a term of sort* BOOL

$$\varphi \vee \psi \quad \triangleq \neg(\neg\varphi \wedge \neg\psi)$$
$$\varphi \Rightarrow \psi \quad \triangleq \neg\varphi \vee \psi$$
$$\exists x : s \bullet \varphi \triangleq \neg\forall x : s \bullet \neg\varphi$$

The operator $\wedge$ has a higher priority than $\vee$ and both operators are associative. The operator $\vee$ has higher priority than $\Rightarrow$ and $\Rightarrow$ is right-associative. Furthermore the scope of the variable of a quantor $\forall$ or $\exists$ is the whole formula (except it is hidden by another quantor over the same variable name).

Definitions 1 and 2 define syntax only. Their interpretation is based on the notion of algebras.

Definition 3 (Σ-Algebra, Interpretation of Terms). *Let $\Sigma \triangleq (S, \sqsubseteq, F, F')$ a partial order-sorted signature. A Σ-algebra is a pair $\mathfrak{A} \triangleq (A, [\![\cdot]\!]_A)$ where*

i. $A \triangleq (A_s)_{s \in S}$ *is a family of sets such that $A_s \subseteq A_{s'}$ iff $s \sqsubseteq s'$ (the* carrier sets*)*
ii. *For each $f : s_1 \times \cdots \times s_n \to A_s \in F$, $[\![f]\!]_A : A_{s_1} \times \cdots \times A_{s_n} \to s$ is a total function*
iii. *For each $f : s_1 \times \cdots \times s_n \to A_s \in F'$, $[\![f]\!]_A : A_{s_1} \times \cdots \times A_{s_n} \to s$ is a partial function with domain $dom(f) \subseteq A_{s_1} \times \cdots \times A_{s_n}$.*

Sometimes we denote $A \triangleq \bigcup_{s \in S} A_s$ A valuation is a family of total functions $\beta \triangleq (\beta_s : X_s \to A_s)_{s \in S}$. The interpretation of terms w.r.t a Σ-algebra $\mathfrak{A}$ and a valuation β is a function $[\![\cdot]\!]_{\mathfrak{A}}^{\beta} : \mathcal{T}(\Sigma, X) \to A$ inductively defined as follows:

iv. $[\![x]\!]_{\mathfrak{A}}^{\beta} \triangleq \beta_s(x)$ iff $x \in X_s$
v. $[\![c]\!]_{\mathfrak{A}}^{\beta} \triangleq [\![c]\!]_A$ iff $c :\to s \in F$
vi. $[\![f(t_1, \ldots, t_n)]\!]_{\mathfrak{A}}^{\beta} \triangleq [\![f]\!]_A([\![t_1]\!]_{\mathfrak{A}}^{\beta}, \ldots, [\![t_n]\!]_{\mathfrak{A}}^{\beta})$ for all $t_1 \in \mathcal{T}_{s_1}, \ldots, t_n \in \mathcal{T}_{s_n}$ iff $f : s_1 \times \cdots \times s_n \to s \in F$
vii. *For $f : s_1 \times \cdots \times s_n \to s \in F'$, $[\![f(t_1, \ldots, t_n)]\!]_{\mathfrak{A}}^{\beta}$ is undefined iff $(t_1, \ldots, t_n) \notin dom(f)$ or at least one t_i is undefined. Otherwise $[\![f(t_1, \ldots, t_n)]\!]_{\mathfrak{A}}^{\beta}$ is defined and $[\![f(t_1, \ldots, t_n)]\!]_{\mathfrak{A}}^{\beta} \triangleq [\![f]\!]_A([\![t_1]\!]_{\mathfrak{A}}^{\beta}, \ldots, [\![t_n]\!]_{\mathfrak{A}}^{\beta})$ for all $(t_1, \ldots, t_n) \in dom(f)$.*

Notation: We omit the indices if they are obvious from the context.
Now we define the interpretation of formulas.

Definition 4 (Interpretation of Formulas). *Let $\Sigma \triangleq (S, \sqsubseteq, F, F')$ be a signature, X be a set of variables over Σ, $\mathfrak{A} \triangleq (A, [\![\cdot]\!]_A)$ be a Σ-algebra, and $\beta : X \to A$ be a valuation. The pair $\mathfrak{A}, \beta$ is a* model *of formula $\varphi \in \mathcal{F}(\Sigma, X)$, denoted by $\mathfrak{A}, \beta \models \varphi$ that inductively satisfies the following properties:*

i. $\mathfrak{A}, \beta \models t_1 \doteq t_2$ *iff* $[\![t_1]\!]_{\mathfrak{A}}^{\beta} = [\![t_2]\!]_{\mathfrak{A}}^{\beta}$ *and* $[\![t_1]\!]_{\mathfrak{A}}^{\beta}, [\![t_2]\!]_{\mathfrak{A}}^{\beta}$ *are both defined*

ii. $\mathfrak{A}, \beta \models t$ **is** s *iff* $[\![t]\!]_{\mathfrak{A}}^{\beta}$ *is defined and* $[\![t]\!]_{\mathfrak{A}}^{\beta} \in A_s$

iii. $\mathfrak{A}, \beta \models D_s(c)$ *for* $c :\to s \in F$, $\mathfrak{A}, \beta \models D_s(x)$ *for* $x : s \in X$,
 $\mathfrak{A}, \beta \models D_s(f(t_1, \ldots, t_n))$ *iff* $\mathfrak{A}, \beta \models D_{s_1}(t_1), \ldots, \mathfrak{A}, \beta \models D_{s_n}(t_n)$ *and* $f : s_1 \times \cdots \times s_n \to s \in F$, *and*
 $\mathfrak{A}, \beta \models D_s(f(t_1, \ldots, t_n))$ *iff* $\mathfrak{A}, \beta \models D_{s_1}(t_1), \ldots, \mathfrak{A}, \beta \models D_{s_n}(t_n)$, $f : s_1 \times \cdots \times s_n \to ?s \in F'$, *and* $([\![t_1]\!]_{\mathfrak{A}}^{\beta}, \ldots, [\![t_n]\!]_{\mathfrak{A}}^{\beta}) \in dom(f)$.

iv. $\mathfrak{A}, \beta \models \neg\varphi$ *iff* $\mathfrak{A}, \beta \not\models \varphi$

v. $\mathfrak{A}, \beta \models \varphi \wedge \psi$ *iff* $\mathfrak{A}, \beta \models \varphi$ *and* $\mathfrak{A}, \beta \models \psi$

vi. $\mathfrak{A}, \beta \models \forall x : s \bullet \varphi$ *iff* $\mathfrak{A}, \beta' \models \varphi$ *for all* $a \in A_s$ *and all valuations* β' *with*

$$\beta'(y) = \begin{cases} \beta(y) & \text{if } y \neq x \\ a & \text{if } y = x \end{cases}$$

The pair $\mathfrak{A}, \beta$ is a model *of a set of formulas $\Phi \subseteq \mathcal{F}(\Sigma, X)$, denoted by $\mathfrak{A}, \beta \models \Phi$ iff $\mathfrak{A}, \beta \models \varphi$ for all $\varphi \in \Phi$. Algebra $\mathfrak{A}$ is a model of $\varphi \in \mathcal{F}(\Sigma, X)$ iff $\mathfrak{A}, \beta \models \varphi$ for all valuations β. Finally, algebra $\mathfrak{A}$ is a model of a set of formulas $\Phi \subseteq \mathcal{F}(\Sigma, X)$, denoted by $\mathfrak{A} \models \Phi$ iff $\mathfrak{A}, \beta \models \Phi$ for all valuations β.*

We now define the notion of an Abstract State Machine. Informally, an abstract state machine specifies a state transition system by a set of rules where the states are algebras over a partial order-sorted signature. The rules induce a set of updates which change the interpretations of function symbols in the signature.

Definition 5 (Update, Rule, Abstract State Machine).
Let $\Sigma \triangleq (S, \sqsubseteq, F, F')$ be a signature and X be a set of variables. A Σ-update is a pair of two ground-terms $f(t_1, \ldots, t_n)$, $t \in \mathcal{T}_s(\Sigma)$, denoted by $f(t_1, \ldots, t_n) := t$. A Σ-rule has the form **if** *φ* **then** *u where u is a Σ-update (φ is the* guard *of the rule), it has the form* **forall** *$x : s \bullet \varphi$* **do** *u where u is a Σ-update which may contain the variable $x : s \in X$ and φ may contain a free variable[3] $x : s$, or it has the form* **choose** *$x : s \bullet \varphi$* **in** *$u_1, \ldots, u_n$ where $u_1, \ldots, u_n$ are Σ-updates which may contain $x : s$ and φ may contain the free variable $x : s$. A Σ-Abstract State Machine is a tuple $\mathcal{A} \triangleq (\mathbb{I}, X, R, Ax)$ where*

i. $\mathbb{I}$ *is a set of Σ-algebras (the* initial *states)*

ii. X *is a set of variables*

iii. R *is a set of Σ-rules*

iv. $Ax \subseteq \mathcal{F}(\Sigma, X)$ *is a set of axioms where no axiom contains a dynamic function symbol.*

---

[3] A variable $x : s$ is *free* in φ iff there is an occurrence of x:s that is not bound to a quantor $\forall$ or $\exists$.

Any top-level function symbol f of a left-hand side of an update in a rule of R is called a dynamic function symbol. *Static function symbols are function symbols that are not dynamic.*

Notation: if φ **then** u_1 is an abbreviation for   **if** φ **then** u_1

$\cdots$ $\cdots$

u_n **if** φ **then** u_n

If φ does not contain free variables $x : s$, then **if** φ **then forall** $x : s \bullet \psi$ **do** u is an abbreviation for **forall** $x : s \bullet \psi \wedge \varphi$ **do** u.

For **choose**, we use analogous abbreviations.

We now define the semantics of ASMs by state transitions. The states of a Σ-ASM are Σ-algebras that satisfy Ax

Definition 6 (Activated Rule, Update Sets, State Transition). *Let* $\Sigma \triangleq (S, \sqsubseteq, F, F')$ *be a signature and* $\mathcal{A} \triangleq (\mathcal{I}, X, R, Ax)$ *be a* Σ*-Abstract State Machine. A state* $q \triangleq (A, \llbracket \cdot \rrbracket_q)$ *is a* Σ*-algebra such that* $q \models Ax$. *A rule* **if** φ **then** $u \in R$ *is activated in a state* q *iff* $q \models \varphi$. *A rule* **forall** $x : s \bullet \varphi$ **do** u *is activated for all valuations* $\beta : \{x : s\} \to A$ *such that* $q, \beta \models \varphi$, *and a rule* **choose** $x : s \bullet \varphi$ **in** $u_1, \ldots, u_n$ *is activated for a valuation* $\beta : \{x : s\} \to A$ *such that* $q, \beta \models \varphi$.

A semantic update is a triple $(f, (a_1, \ldots, a_n), a)$ *where* $f : s_1 \times \cdots \times s_n \to s \in F \cup F'$ *is a dynamic function symbol,* $a_1 \in A_{s_1}, \ldots, a_n \in A_{s_n}$, *and* $a \in A_s$. *Two semantic updates* $(f, (a_1, \ldots, a_n), a), (g, (a'_1, \ldots, a'_n), a')$ *are conflicting iff* $f = g$, $a_i = a'_i$ *for* $i = 1, \ldots, n$, *and* $a \neq a'$. *A set of* U *of updates is consistent iff it does not contain two conflicting updates. A possible set* U_q *of semantic updates in state* q *is a consistent set of semantic updates defined as follows:*

$$U_q \triangleq \{(f, (\llbracket t_1 \rrbracket_q, \ldots, \llbracket t_n \rrbracket_q), \llbracket t \rrbracket_q) \mid \text{ if } \varphi \text{ then } f(t_1, \ldots, t_n) := t \in R \text{ is activated}\} \cup$$
$$\{(f, (\llbracket t_1 \rrbracket_q^\beta, \ldots, \llbracket t_n \rrbracket_q^\beta), \llbracket t \rrbracket_q^\beta) \mid \text{ forall } x : s \bullet \varphi \text{ do } f(t_1, \ldots, t_n) := t \in R$$
$$\text{is activated for valuation } \beta\} \cup$$
$$\{(f, (\llbracket t_1 \rrbracket_q^{\beta_r}, \ldots, \llbracket t_n \rrbracket_q^{\beta_r}), \llbracket t \rrbracket_q^{\beta_r}) \mid r \triangleq \text{ choose } x : s \bullet \varphi \text{ in } u_1, \ldots, u_m \in R$$
$$\text{and there is a valuation that activates } r$$
$$f(t_1, \ldots, t_n) := t \in \{u_1, \ldots, u_m\}\}$$

where β_r *is a valuation that activates the choose-rule* r. *Let* $\mathbb{U}(q)$ *be the set of possible sets of semantic updates in state* q. *ASMs non-deterministically chooses an* $U \in \mathbb{U}(q)$ *and the successor state successor state* q' *w.r.t* U *is defined for each dynamic function symbol* $f : s_1 \times \cdots \times s_n \to s \in F \cup F'$ *by:*

$$\llbracket f \rrbracket_{q'}(a_1, \ldots, a_n) \triangleq \begin{cases} a & \text{if } (f, (a_1, \ldots, a_n), a) \in U \\ \llbracket f \rrbracket_q(a_1, \ldots, a_n) & \text{otherwise} \end{cases}$$

If $\mathbb{U}(q)$ *is empty then* q *has no successor.*

A run of $\mathcal{A}$ *is a sequence of states* $\langle q_i \mid 0 \leq i < n \rangle$ *where* $n \in \mathbb{N} \cup \infty$, $q_0 \in \mathbb{I}$ *is an initial state, and* q_i *is a successor state of* q_{i-1} *for all* $1 \leq i < n$.

Remark 3. A possible update set takes into account each activated rule. Informally, any choose-rule non-deterministically chooses a valuation that activates the rule. Each update set $U \in \mathbb{U}(q)$ is consistent according to the above definition.

In contrast to [4], Definition 5 excludes nested updates on **forall** and **choose**. Hence, Definition 6 does not consider this case. However, it is straightforward to extend the definition by defining update sets inductively. For this work, it is not necessary to consider this nesting as the ASM rules in the paper don't use it.

Note that Definition 5 implies that there are no axioms on dynamic function symbols. Hence, axioms cannot be violated due to state transitions. Static function symbols have the same interpretation in all states of a run.

Remark 4. The initial state can be specified as set of rules of the form **if** φ **then** $f(t_1, \ldots, t_n) := t$ and **forall** $x : s \bullet \varphi$ **do** $f(t_1, \ldots, t_n) := t$ where the formula ϕ and the terms $t_1, \ldots, t_n, t$ do not contain a dynamic function symbol. Thus, an initial state q_0 satisfies $q_0 \models \varphi \Rightarrow f(t_1, \ldots, t_n) \doteq t$ and $\forall x : s \bullet \varphi \Rightarrow f(t_1, \ldots, t_n) \doteq t$, respectively.

3   Programming Language Semantics

Programming language definitions distinguish between lexical elements, concrete syntax, abstract syntax, static semantics, and dynamic semantics. Lexical elements are identifiers, constants, keywords, special symbols and comments. They form the vocabulary of the concrete syntax which is usually defined by a context-free grammar. The non-terminals of this context-free grammar represent language concepts, i.e., they represent the abstract syntax. The abstract syntax abstracts from all unnecessary details of the concrete syntax. For example, the keywords of a while loop are not required; it is just necessary to know that it has a condition (which is an expression) and loop body (which is statement sequence). Similarly, special symbols are not needed. Another example are parantheses since e.g. $(((a + b)))$ and $a + b$ represents the same expression. The structure of the parantheses are represented by a tree. Static semantics defines bindings of names, typing etc. Dynamic semantics defines state transitions. The latter can be formalized by ASMs and is based on the abstract syntax and static semantics. The following example demonstrates these ideas and its formalization.

Figure 1(a) shows the abstract syntax - as a context-free grammar - of a little while language containing declarations of identifiers, having two types **integer** and **boolean**, assignments, operators $-$, $>$, $=$, and **not**, a while-loop, a (one-sided) if-statement, and a write- and a read-statement. There is a constructor and sort for each language concept, cf. Fig. 1(b). ID is the sort of identifiers and INT the sort of integers as specified by the programming language. Both informations are lexical information. As it can be seen from the signatures of the operation symbols, there are sorts such as EXPR and STAT that classify the language concepts expressions and statements, respectively. This can be formalized by sub-sort relations as shown in Fig. 1(c).

Remark 5. In general, the constructor ground-terms of an abstract syntax correspond to an abstract syntax tree. Figure 2 shows an abstract syntax tree for a little while language where the nodes are named with the constructors.

$\langle prog \rangle$ ::= 'prog' $\langle decls \rangle$ 'begin' $\langle stats \rangle$ 'end'
$\langle decls \rangle$ ::= $\langle decl \rangle$
$\langle decls \rangle$ ::= $\langle decls \rangle$';' $\langle decl \rangle$
$\langle decl \rangle$ ::= $\langle vardecl \rangle$
$\langle vardecl \rangle$ ::= id ':' $\langle type \rangle$
$\langle type \rangle$ ::= id
$\langle stats \rangle$ ::= $\langle stat \rangle$
$\langle stats \rangle$ ::= $\langle stats \rangle$';' $\langle stat \rangle$
$\langle stat \rangle$ ::= $\langle assign \rangle | \langle while \rangle | \langle if \rangle | \langle read \rangle | \langle write \rangle$
$\langle assign \rangle$ ::= $\langle des \rangle$ ':=' $\langle expr \rangle$
$\langle while \rangle$ ::= 'while' $\langle expr \rangle$ 'do' $\langle stats \rangle$ 'end'
$\langle if \rangle$ ::= 'if' $\langle expr \rangle$ 'then' $\langle stats \rangle$ 'end'
$\langle read \rangle$ ::= 'read' $\langle des \rangle$
$\langle write \rangle$ ::= 'write' $\langle expr \rangle$
$\langle des \rangle$ ::= $\langle name \rangle$
$\langle name \rangle$ ::= id
$\langle expr \rangle$ ::= $\langle des \rangle | \langle minus \rangle | \langle gt \rangle | \langle eq \rangle | \langle not \rangle | \langle const \rangle$
$\langle minus \rangle$ ::= $\langle expr \rangle$ '-' $\langle expr \rangle$
$\langle gt \rangle$ ::= $\langle expr \rangle$ '>' $\langle expr \rangle$
$\langle eq \rangle$ ::= $\langle expr \rangle$ '=' $\langle expr \rangle$
$\langle not \rangle$ ::= 'not' $\langle expr \rangle$
$\langle const \rangle$ ::= intconst

The operator '-' is left-associativ, all other binary operators are not associative. The priorities of the operators are 'not' $\prec$ '=' $\prec$ '>' $\prec$ '-'

(a) Syntax of the while language

$mkProg$:   DECLS × STATS → PROG
$mkNodecl$:   → DECLS
$addDecl$:   DECLS × DECL → DECLS
$mkVarDecl$: ID × TYPE   → DECL
$mkType$:   ID   → TYPE
$mkNostat$:   → STATS
$addStat$:   STATS × STAT → STATS
$mkAssign$:   DES × EXPR   → ASSIGN
$mkWhile$:   EXPR × STATS → WHILE
$mkIf$:   EXPR × STATS → IF
$mkRead$:   DES   → READ
$mkWrite$:   DES   → WRITE
$mkName$:   ID   → DES
$mkMinus$:   EXPR × EXPR → MINUS
$mkGt$:   EXPR × EXPR → GT
$mkEq$:   EXPR × EXPR → EQ
$mkNot$:   EXPR × EXPR → NOT
$mkConst$:   INT   → CONST

(b) Constructors for Abstract Syntax Trees

ASSIGN $\sqsubseteq$ STAT, WHILE $\sqsubseteq$ STAT, IF $\sqsubseteq$ STAT,
READ $\sqsubseteq$ STAT, WRITE $\sqsubseteq$ STAT,
DES $\sqsubseteq$ EXPR, MINUS $\sqsubseteq$ EXPR, GT $\sqsubseteq$ EXPR
EQ $\sqsubseteq$ EXPR, NOT $\sqsubseteq$ EXPR, CONST $\sqsubseteq$ EXPR,

(c) Sub-Sort Relations in the Abstract Syntax

$prog$:   → PROG
pc :   → OCC
mem : LOC →?VALUE
env :   → ENV
val :   OCC →?VALUE
inp :   → STREAM
out :   → STREAM

In addition there is static function
$bind$: ENV × ID →?LOC
and a sub-sort relation LOC $\sqsubseteq$ VALUE

(e) Dynamic Function Symbols

occ :   PROG × OCC →?NODE
$deftab$:   PROG × OCC →?DEFTAB
$isDefined$:   DEFTAB × ID → BOOLEAN
$identifyDef$: DEFTAB × ID →?DEF
$typeOf$:   VARDEF   → TYPE
pri :   PROG × NODE → TYPE
$post$:   PROG × NODE → TYPE
$isValue$:   PROG × OCC →?BOOL
$isAddr$:   PROG × OCC →?BOOL
$first$:   PROG   → OCC
$next$:   PROG × OCC →?OCC
$first$:   PROG × OCC →?OCC
$cond$:   PROG × OCC →?OCC
lop :   PROG × OCC →?OCC
rop :   PROG × OCC →?OCC
opd :   PROG × OCC →?OCC

DECLS $\sqsubseteq$ NODE, DECL $\sqsubseteq$ NODE, STATS $\sqsubseteq$ NODE,
STAT $\sqsubseteq$ NODE, EXPR $\sqsubseteq$ NODE, VARDEF $\sqsubseteq$ DEF

(d) Static Function Symbols

initial $pc := first(prog)$
        $env := initenv(prog)$
        $\neg D(mem(l))$
        $\neg D(val(o))$

if Ct **is** ASSIGN **then** $mem(val(Lop)) := val(Rop)$
                     Proceed
where $Ct \triangleq occ(prog, pc)$, $Lop \triangleq lop(prog, pc)$,
$Rop \triangleq rop(prog, pc)$, and $Proceed \triangleq pc := next(occ, pc)$

if Ct **is** WHILE $\wedge$ $Cond \doteq true$ **then** $pc := first(prog, pc)$
where $Cond \triangleq val(cond(prog, pc))$

if Ct **is** WHILE $\wedge$ $Cond \doteq false$ **then** $Proceed$

if Ct **is** IF $\wedge$ $Cond \doteq true$ **then** $pc := first(prog, pc)$

if Ct **is** IF $\wedge$ $Cond \doteq false$ **then** $Proceed$

if Ct **is** READ **then** $mem(val(Opd)) := front(inp)$
                    $inp := tail(inp)$
                    Proceed
where $Opd \triangleq opd(prog, pc)$

if Ct **is** WRITE **then** $out := add(out, val(Opd))$
                    Proceed

if Ct **is** NAME $\wedge$ $isValue(prog, pc)$ **then**
      $val(pc) := mem(Loc)$
      Proceed
where $Loc \triangleq bind(env, id(ct))$

if Ct **is** NAME $\wedge$ $isAddr(prog, pc)$ **then**
      $val(pc) := Loc$
      Proceed

if Ct **is** MINUS **then** $val(pc) := val(Lop) - val(Rop)$
                    Proceed

if Ct **is** GT **then** $val(pc) := val(Lop) > val(Rop)$
                    Proceed

if Ct **is** EQ **then** $val(pc) := val(Lop) = val(Rop)$
                    Proceed

if Ct **is** NOT **then** $val(pc) := not(val(Opd))$
                    Proceed

if Ct **is** CONST **then** $val(pc) := const(Ct)$
                    Proceed

(f) Initial State and State Transitions

Fig. 1. A small while-language

The static semantics provides static informations of the program, i.e., informations and conditions that can be determined without executing the program. It is not only the static semantics as defined in the language definition but it also prepares the dynamic semantics, in particular the control-flow and use-def-relations of values. These informations are defined by static functions on the abstract syntax tree and by axioms that need to be satisfied. Figure 1(d) shows these functions for the while language. We omit the axioms and describe informally the meaning of these functions. These functions need to address nodes of abstract syntax trees which are modeled as occurrences. This is a list of natural numbers that specifies which child has to be chosen. Figure 2 demonstrates this concept. For example the list $[1, 0, 1, 1, 1, 0]$ addresses the root of the condition of the second if-statement. Note that the addressing of a node is partial function because not each list of natural numbers identifies a node in an abstract syntax tree. The function symbol *deftab* represents the contextual information at each node, i.e. the definitions that are visible at the node. *isDefined* checks whether there is a binding for a name. *identify* identifies a definition for a name. For example the identifier at occurrence $[1, 0, 1, 0, 0, 0]$ in Fig. 2 identifies the definition stemming from the declaration of i which contains the information that i is a variable with type **integer**. *identify* is partial because undefined identifiers cannot be identified. The function *pri* defines the a-priori type of an expression, i.e., the type determined by the structure of an expression. E.g., in Fig. 2, the expression at node $[1, 0, 1, 1, 1, 0]$ has the a priori type **boolean**. The function *post* defines the type expected by the context of an expression. E.g., in Fig. 2, the expression at node $[1, 0, 1, 1, 1, 0]$ has the a posteriori type **boolean**, because an if statement expects a boolean type for its condition. In general, read- and write statements define the a posteriori type **integer** for its operands. For the while-Language in Fig. 1, the static semantics requires that the a priori type of an expression is equal to its a posteriori type. The functions *isValue* and *isAddr* determine whether a value of a designator or an address of a designator is required for the evaluation. For example, the designator i at the read instruction requires an address, i.e., $isAddr(prog, [1, 0, 0, 0, 1, 0])$ holds. In contrast, the designator i at the write instruction requires the value of i, i.e., $isValue(prog, [1, 1, 0])$. If the occurrence is not a designator, then these functions are undefined.

Remark 6. Each programming language with static binding of names requires a static function analogous to *deftab*. For languages with dynamic binding of names, this function would become a dynamic function as part of the dynamic semantics. In general a definition contains all information necessary for specifying the static conditions and required for the static semantics.

In general, the a priori type of an expression needs not to be equal to its a posteriori type. In this case, the value of the expression needs to be converted to a corresponding value of the a posteriori-type. Strongly typed languages usually introduce a type system where each value of a sub-type is converted automatically to a value of its super-type.

For the preparation of the dynamic semantics of a programming language it is necessary to identify the first instruction of a program (function *first*), the next instruction (function *next*), and access to operands of expressions (*lop, rop, opd*). For example, in Fig. 2, the first instruction is at occurrence $[1, 0, 0, 0, 1, 0]$ which is the computation of the address of variable i. In Fig. 2 the next instruction of

```
prog
   i:integer;
   j:integer
begin
   read(i);
   read(j);
   while not i=j do
      if i>j then i:=i-j end;
      if not i>j then i:=i-j end;
   end;
   write(i)
end
```

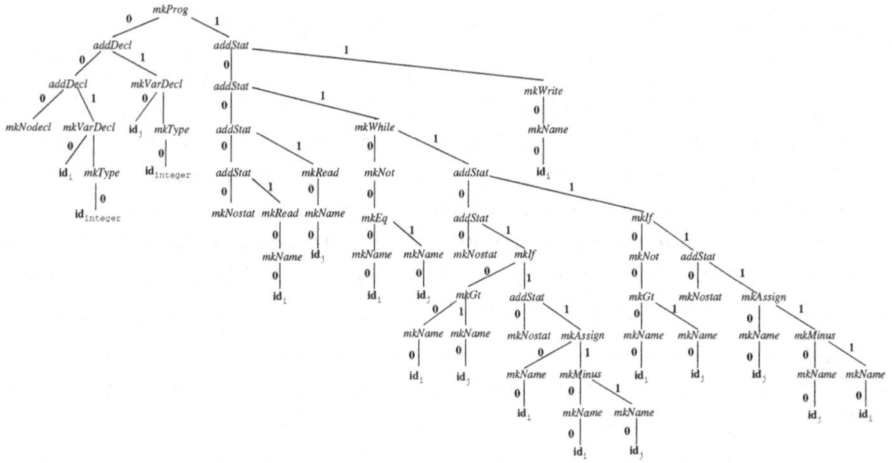

Fig. 2. An example program of the while-language in Fig. 1

$[1, 0, 1, 1, 1, 1, 1]$ (which addresses the assignment in the second if-statement) is $[1, 0, 1, 0, 0, 0]$ (which is computing the address and value of the left-hand side of the equal-expression in the condition of the while-loop). In addition, for the while-language it is necessary to have access to the condition of the while loop and if-statement (*cond*), to the first statement of the body of the while-loop and the then-part of an if-statement (*first*). Note that these functions are all partial since they operate on occurrences.

Figure 1(e) shows the dynamic functions. The constant *prog* is the program to be executed. *pc* is the program counter containing the occurrence of the instruction to be executed next. *mem* is the memory where LOC is the sort of addresses and VALUE the sort of values.

Remark 7. Note, that in many languages, addresses are also values. Therefore, it is not necessary to distinguish between values and addresses in the context of saving intermediate results in evaluating expressions. If a value of a designator has been computed, then the result of the computation of its address is used. For languages where addresses are not values, then VALUE becomes a super-sort for a (new) sort of values VAL and the sort LOC.

Furthermore, there is a constant *env* of sort ENV, the *environment*. In particular, an environment binds addresses to names of variables which is formalized by the static function *bind* that looksup an address for a variable. Finally, each occurrence that is a variable has an address, and each occurrence that is an expression has a value. In general both depend on current state, i.e., the function *val* is also a dynamic function. The functions *inp* and *out* are the input and output streams, respectively, which are lists of integers (for the example language).

Remark 8. For the while languages it is sufficient to define LOC $\triangleq$ ID and BOOL $\sqsubseteq$ VALUE and INT $\sqsubseteq$ VALUE where BOOL is the sort of booleans and INT is the sort of integers of the size defined by the programming language. These sorts have usual constructors and operation symbols. In most programming languages there are more values (e.g. addresses) which can be specified analogously by sub-sort relations. In general, programming languages with static scoping allow different definitions for the same name as long as they are in different scopes (*local declarations*). Hence, the binding to an address depends on the scope currently being executed. Therefore, in contrast to the while language, the environment may change during execution.

Finally Fig. 1(f) shows the rules of the ASM specifying the dynamic semantics of the while language. Expressions are evaluated from left to right. We use macros for abbreviations of frequently used terms and for better readability. These macros are expanded analogous to the macros #define of the C preprocessing language. The guards of the rules check the type of an instruction, respectively, and the corresponding updates specify the state transitions for this instruction. The functions *id* : NODE $\rightarrow$?ID and *const* : NODE $\rightarrow$?VALUE define the identifier for a name and the the value of a constant, respectively.

4   Non-recursive and Recursive Procedures

We first show procedure calls and returns for parameterless procedures. The rules in this section can be combined with the rules in other sections. In this section, we first introduce procedure calls (and returns) for programming languages without recursion. We assume that each procedure has one entry-point, i.e., a unique instruction being executed upon the procedure call. This is usual in all modern programming languages. We also assume that there is an explicit return statement. In programming languages that return after executing the last statement of the procedure body, the abstract syntax implicitly adds a return statement as the last statement of the procedure body. Hence, this assumption is no restriction.

If recursion is not allowed then each procedure has a static local environment, i.e., the addresses of its local variables (if this concept exists) and its local procedures (if this concept exists) are known. In addition the environment of a procedures must store the program counter of the caller (or the program counter of the instruction to be executed after the call). This is being restored when returning from a procedure. Usually, the non-recursion is not dynamically checked.

$mkProcDecl$: NAME × BODY $\rightarrow$ PROCDECL
$mkBody$:   DECLS × STATS $\rightarrow$ BODY
$mkProcCall$: NAME $\rightarrow$ PROCCALL
$mkReturn$: $\rightarrow$ RETURN
$first$:   PROCDEF $\rightarrow$ OCC
id :   PROCCALL $\rightarrow$ ID
$proc$:   PROG × OCC $\rightarrow$?ID
$getEnv$:   ENV × ID $\rightarrow$?ENV
$setpc$:   ENV × ID × OCC $\rightarrow$?ENV
$getpc$:   ENV × ID $\rightarrow$ OCC
PROCDECL $\sqsubseteq$ DECL, PROCCALL $\sqsubseteq$ STAT,
RETURN $\sqsubseteq$ STAT, PROCDEF $\sqsubseteq$ DEF

(a) Static Functions

if Ct is PROCCALL then
  $env := LocEnv$
  $pc := first(ProcDef)$
where
  $Id \triangleq id(prog, pc)$
  $LocEnv \triangleq setpc(getEnv(env, Id), Id, pc)$
  $ProcDef \triangleq identifyDef(deftab(prog, pc), Id)$

if ct is RETURN then
  $pc := next(prog, OldPc)$
  $env := OldEnv$
where
  $Cp \triangleq proc(prog, pc)$
  $OldPc \triangleq getpc(env, Cp)$
  $OldEnv \triangleq getEnv(env, Cp)$

(b) State Transitions for Call and Return

Fig. 3. Non-recursive procedure call and return

Figure 3 shows the formal definition of the ASM-rules for procedure calls and returns. Figure 3(a) shows the operation symbols for the static functions. The functions $mkProcDecl$, $mkBody$, $mkProcCall$, and $mkReturn$ are the constructors for the abstract syntax trees. Procedure calls and returns are statements as shown in the sub-sort relations, respectively.

The sort PROCDEF is the sort of procedure definitions and contains information on the first instruction being executed when executing the procedure body. The function $first$ gets this occurrence from a procedure definition. The function id obtains the identifier of the procedure being called, and the function $proc$ obtains for an occurrence in a procedure body the identifier of this procedure. The other static functions are functions on the environment required in the state transitions for saving and restoring the caller ($setpc$ and $getpc$), and obtaining the local environment of a procedure ($getEnv$). The state transitions in a procedure call changes the environment to the procedure being called. In particular it stores the program counter of the call. Furthermore, the program counter is set to the occurrence of the first instruction in the procedure body, and it stores that the procedure is in execution. These updates are only being executed if the called procedure is not in execution. When returning from a procedure then the old program counter ($OldPc$) and the old environment ($OldEncv$) is being restored and the procedure is no more in execution.

Remark 9. This is almost the semantics of non-recursive procedures in Fortran. Fortran allows to declare local procedures with a restricted nesting depth. The old versions of Fortran allow more than one entry point. Then each entry point is modeled as a separate procedure with a different name but the same local environment as the procedure. Then the formalization is the same as in Fig. 3. This semantics is the semantics for static local variables (SAVE-variables), i.e., if a procedure p is called afterwards, then the local variables of p have the value at the last return from p. For non-static variables, these values are not preserved and the memory at the addresses of these local variables become undefined. It is straightforward to add the corresponding update using **forall** and the pre-defined identifier $\bot$ for undefining a value.

In *COBOL* all identifiers are global, i.e. a procedure does not contain local declarations and parameters. Therefore one environment is sufficient. The PERFORM-statement calls a procedure. The semantics is the same as in Fig. 3.

If a programming language checks dynamically on recursion, then for each procedure, the environment contains a symbolic state for each procedure indicating whether it is currently being in execution. The condition in the rule for a call is extended by the condition that the called procedure is currently not in execution. The formal semantics of *Java* [13] uses a similar formalization for class initialization with three symbolic states: uninitialized, initialization in progress, and initialized. The initialization of classes is only being called for uninitialized classes.

If recursive procedures are allowed, a recursively called procedure might have several calls that are in progress. Hence, the local variables of a procedure must be bound to different addresses for each call in execution. The call-return-mechanism follows the Last In-First Out-principle. Hence, the environment is a stack of procedures where each stack element (*frame*) stores the bindings and the program counter of the caller. The bindings include the bindings of the environment of the procedure being called (*global environment*). If no local procedures are allowed each procedure is bound to the bindings of the global variables. If local procedures (i.e., procedure declarations in a procedure body), the callee extends the environment of the caller by its local variables and local procedures. Thus, the binding of a procedure is a binding to its environment. Similarly, in object-oriented languages, the global environment of an object method is the bindings of its attributes and its methods. Analogously, the global binding of a static method of a class is the binding of the class' static attributes. Note, that this also applies to local methods and local classes.

Figure 4 shows the ASM-rules formalizing procedure calls and returns. We added the rule for declarations which allocates a new location for the declared variable. The rules use all static functions of the non-recursive version except *getEnv*. Instead, the two static functions *push* and *pop* are introduced for pushing and poping frames from the environment. The function *getpc* now obtains the program counter of the top frame of an environment. *getBinding* obtains the bindings from the top frame of an environment. The static function *addBinding* adds a binding of a variable to an address to the bindings of the top frame of an environment. The functions *bindVar* and *bindProc* obtain the address of the object bound to a variable in an environment and the bindings bound to a procedure, respectively. *bind* obtains a binding of a variable within a frame, and $f \in e$ is true iff f is a frame in an environment.

A procedure call pushes a new frame with empty bindings and the program counter of the caller to the environment and proceeds with the first instruction of the procedure body. Since programming languages might also have function calls which behave slightly different, a common supersort CALL is introduced in order to specify the common state transitions for procedure and function calls in the ASM rule for calls. The specific state transitions for procedure calls (i.e., the change of the environment) is specified in the rule for procedure calls. Note that the first instruction might be a variable declaration or procedure declaration. The execution of a variable declaration obtains an unused address and binds it to the variable. An address is used if it is somewhere bound to a variable in a frame in the environment or it is contained as a value in the memory. The execution of a procedure declaration binds the procedure to the current bindings. If the object associated with a variable declaration is created upon execution of its declaration, then the declarations are executed in their order. If the objects of all local declarations are accessible within the whole procedure body, then the first procedure declaration is executed after the last variable declaration. This issue can be modelled by the static function $next$. On procedure return, the old program counter and the environment of the caller are restored and the execution proceeds as usual with the instruction after the call.

$$
\begin{array}{lll}
push: & \text{ENV} \times \text{FRAME} & \rightarrow \text{ENV} \\
pop: & \text{ENV} & \rightarrow \text{ENV} \\
mkFrame: & \text{BINDING} \times \text{OCC} & \rightarrow \text{FRAME} \\
getBindings: & \text{ENV} & \rightarrow \text{BINDING} \\
mkBinding: & & \rightarrow \text{BINDING} \\
addBinding: & \text{ENV} \times \text{ID} \times \text{LOC} & \rightarrow \text{ENV} \\
addBinding: & \text{ENV} \times \text{ID} \times \text{BINDING} & \rightarrow \text{ENV} \\
bindVar: & \text{ENV} \times \text{ID} & \rightarrow ?\text{LOC} \\
bindProc: & \text{ENV} \times \text{ID} & \rightarrow ?\text{BINDING} \\
bind: & \text{FRAME} \times \text{ID} & \rightarrow ?\text{LOC} \\
\cdot \in \cdot: & \text{FRAME} \times \text{ENV} & \rightarrow \text{BOOL} \\
\end{array}
$$

PROCCALL $\sqsubseteq$ CALL, CALL $\sqsubseteq$ STAT

if Ct **is** VARDECL **then**
   $Proceed$
   choose l : LOC $\bullet \neg Used(l)$ **in**
     $env := addBinding(env, Id, l)$
  where
    $Used(l) \triangleq \exists f : \text{FRAME} \bullet \exists id : \text{ID} \bullet f \in env \land bind(f, v) \doteq l \lor \exists l' : \text{LOC} \bullet mem(l') \doteq l$

if Ct **is** PROCDECL **then**
   $env := addBinding(env, Id, Bindings)$
   $Proceed$
where $Bindings \triangleq getBindings(env)$

if Ct **is** CALL **then**
   $pc := first(ProcDef)$

if Ct **is** PROCCALL **then**
   $env := push(env, Frame)$
where
   $Id \triangleq id(prog, pc)$
   $Frame \triangleq mkFrame(bindProc(env, Id), pc)$
   $ProcDef \triangleq identifyDef(deftab(prog, pc), Id)$

if Ct **is** RETURN **then**
   $env := pop(env)$
   $pc := next(prog, getpc(env))$

Fig. 4. Recursive procedure call and return

Remark 10. The functions and rules in Fig. 4 are a framework for modeling the concepts of many programming languages which allow recursion. Languages such as *C* don't allow local procedures. On a global level, all global variables are associated with objects which exist during the whole life time for the program. Hence procedure declarations are executed after the last variable declarations before starting main. Local variable declarations are executed in their order since the objects associated with a local variable exists from its declaration. This also holds for local variables of methods in *Java, C++, C#, Eiffel,* and *Sather* (no matter whether they are static or not).

   Static variables (also local static variables) in *C* exist during the whole execution of the program. Hence, the declaration of static variables are executed when executing procedure declarations. These variables are added to the bindings of the procedure. The declaration of a static variable is not executed when executing a procedure, i.e., the rule for the execution of variable declaration does not apply to local declaration of static variables. The SAVE-variables of *Fortran* for recursive procedures and the **own**-variables in *Algol60* have the same semantics.

Languages such as *Algol60, Algol68, Pascal, Modula,* and *Ada* allow local procedure declarations. For these languages all objects associated to local variables are alive when a local procedure is being called. Hence, first all local variable declarations and then all local procedure declarations are being executed.

Similarly in object-oriented languages such as *Java, C++, C#, Eiffel,* and *Sather* the bindings of the object attributes and object methods of an object are created upon object creation and are visible in all methods, i.e., first the object attribute declarations are executed and then the method declarations are being executed (eventually also the initializers and the constructors). The binding of *this* can be viewed as an implicit parameter of a method being bound when executing parameter passing. For static attributes and static methods the bindings are computed before the first object is being created. In *Java* and *C#* static attribute declarations and static method declaration are executed when the class initialization starts. In *Sather, C++,* and *Eiffel* these declarations are being executed when the program starts, i.e., before calling main. All objects associated to static variable declarations of a class are visible in all static methods of this class. This semantics also applies to local classes. In case of inheritance, the execution of the declarations of an object starts first by executing the bindings of the attribute declarations of the super classes, then inheritance is being resolved (including overriden methods), and finally the method declarations are being executed. Hence, these variants differ just in the order of the instructions being executed (formalized by the static function *next*) but it does not affect the state transition for the single instruction.

5   Parameter Passing

Usually, procedures have parameters. The parameters are local variables of the procedure. The most common parameter passing mechanisms in programming languages are *call-by-value, call-by-result, call-by-reference,* and *call-by-value-and-result* although programming languages usually do not offer all of them. The semantics of the *call-by-value* parameter passing mechanism evaluates the argument to a value and assigns it to the corresponding parameter at the time of the procedure call. The *call-by-result* parameter passing mechanism assigns the value of the parameter to the corresponding argument at the time of the procedure return. *call-by-value-and-result* combines *call-by-value* and *call-by-result.* Finally, *call-by-reference* passes the address of the argument to the address of the parameter, i.e., the argument and the parameter refer to the same object.

It remains to discuss the meaning of an argument corresponding to a parameter. Often, it is *positional,* i.e., the i-th argument is passed to the i-th parameter.

However, many languages also allow non-positional correspondences of arguments. In this case, the correspondence is explicitly specified in the procedure call by associating the argument with the corresponding parameter.

After calling a procedure, the parameters are being passed, i.e., the first instruction is the passing of the first parameter. We assume that for *call-by-value* and *call-by-value-and-results* parameters, the corresponding arguments are all evaluated when calling the procedure. We assume that *call-by-value-and-result* parameters and their corresponding arguments are duplicated such that it is *call-by-value* and a *call-by-result*-parameter[4]. We further assume that for all *call-by-reference*-parameters and *call-by-result*-parameters the addresses of the corresponding arguments are computed when calling a procedure, i.e., the functions *val* is defined for these occurrences, respectively, cf. Fig. 1.

In Fig. 5(a), the functions *mkProcDecl*, *mkPars*, *mkNoPars*, and *mkAddPar* are constructions for procedure declarations in the abstract syntax. The first sub-sort relation in Fig. 5(c) specifies that procedure declarations are declarations. Similarly, the functions *mkValPar*, *mkRefPar*, and *mkResPar* specify the constructors of the parameters passed by *call-by-value*, *call-by-reference*, and *call-by-result*, respectively. The sub-sort relations in Fig. 5(c) specify the sort PAR which subsumes all specific sorts for parameters including the passing mechanism for a parameter. Finally, the functions *mkProcCall*, *mkNoArgs*, and *mkAddArg* are the constructors for the abstract syntax of procedure calls. The last sub-sort relation in Fig. 5(c) specifies that procedure calls are statements.

The correspondence between parameters and arguments of procedure calls is modeled by the static function *corr* where the second argument is the occurrence of the procedure call, the third argument the parameter, and the result the argument corresponding to this parameter. *call-by-value*, and *call-by-result* parameters are local variables, i.e., the state transitions for local variables also be executed.

Remark 11. The static function *corr* can be extended straightforwardly to non-positional parameter passing. If parameters are passed non-positionally, then the procedure call contains the parameter of the callee the argument to be passed to this parameter, i.e., *corr* remains a static function. Some languages such as e.g. *C* or *C++* have the concept of varying number of parameters (called ellipsis in *C* or *C++*). However, the types of the arguments of each call can be determined and therefore, the possible numbers of arguments are known statically. In this case, a procedure with varying number of parameters is considered as overloading the procedure with the required numbers of parameters, respectively. An example is the procedure `printf` in *C* or *C++*.

The transition rules for passing arguments to *value*-parameters on procedure call are straight forward. Reference parameters behave differently than variable declarations since there is no need for a new address. Instead the address of the corresponding argument is bound to the parameter as shown in the transition rule for the reference parameter.

---

[4] A parser can create abstract syntax trees with this duplication.

$mkProcDecl$: NAME × PARS × BODY : → PROCDECL

$mkNoPars$:   → PARS

$mkAddPar$:   PARS × PAR → PAR

$mkValPar$:   NAME   → VALPAR

$mkRefPar$:   NAME   → REFPAR

$mkProcCall$: NAME × ARGS   → PROCCALL

$mkNoArgs$:   → ARGS

$mkAddArg$:   ARGS × EXPR   → ARGS

$corr$:   PROG × OCC × OCC   →?OCC

$addBinding$: ENV × OCC × LOC   → ENV

$bindArg$:   ENV × OCC   →?LOC

(a) Static Functions

$entry$: OCC →?BOOL

(b) Dynamic Function

PROCDECL ⊑ DECL,

VALPAR ⊑ PAR,

REFPAR ⊑ PAR,

RESPAR ⊑ PAR,

(c) Subsorts

if Ct **is** CALL **then**
  $entry(pc) := true$

if Ct **is** VALPAR **then**
  Proceed
  choose l : LOC • ¬$Used(l)$ **in**
    $env := addBinding(env, Id, l)$
    $mem(l) := val(Arg)$
  where $Arg \triangleq corr(prog, OldPc, pc))$

if Ct **is** REFPAR **then**
  $env := addBinding(env, Id, val(Arg))$
  Proceed

if Ct **is** RESPAR ∧ $entry(OldPc)$ **then**
  Proceed
  choose l : LOC • ¬$Used(l)$ **in**
    $env := NewEnv$
where
  $NewEnv \triangleq addBinding(addBinding(env, Id, l), Arg, val(Arg))$

if Ct **is** RESPAR ∧ $entry(OldPc) \doteq false$ **then**
  $mem(Loc) := val(pc)$
  Proceed
where $Loc \triangleq bindArg(env, oldPc, Arg)$

(d) State Transition Rules

Fig. 5. Parameter passing

A special role is played by *result parameters* since they are also used upon return. In both cases, result parameter passing is also executed upon procedure call. Thus, it is required to know whether it passing a result parameter is executed on procedure entry or on procedure return. This is modeled by the dynamic function *entry*. In the state transition of the last parameter, the update $entry(proc(prog, pc)) := false$ is executed. Furthermore, there might be recursive calls using result parameters. The problem is that the execution of a recursive procedure p may destroy the address information of the previous call of p. For example if a procedure p(out x) calls recursively p(x), then the variable x stemming from the parameter of the recursive call and the argument x have different locations, but the address information is being destroyed upon the recursive call since the address of x becomes the address of the parameter of the recursively called procedure. The solution is to save the addresses of the arguments of the corresponding *call-by-result* parameters upon the call of a procedure in the frame of the callee. This is modeled by the additional static function *addBinding* shown in Fig. 5. It adds a binding for the occurrence of an argument to its address to the bindings of the top of the stack. The function *bindArg* looks up this binding.

6   Functions

Function calls usually have the same semantics as procedure calls including parameter passing. However, function calls are expressions and therefore have a value, i.e., the value that is returned after execution of the function body. However, there is also another consequence of function calls being expressions:

function calls might be sub-expressions of an expression e, and then addresses or values of other sub-expressions of e might be computed but not yet used as the following example in *Java* demonstrates:

Example 1 (Function calls in Java*).* Consider the expression x+y*f(z) where f is a function and x, y, and z are variables. *Java* has a strict left-to-right expression evaluation order. Hence, when f is called, the values of x and y are computed. However, these values are being used after the return from f has been executed, since the value of y*f(z) and the value of the complete expression can only be computed after the value of f(z) is available. Similarly, when executing the assignment x=f(y), the address of x has been computed (due to the left-to-right evaluation order), but it is used after returning from f, since storing the return value to the address of x cannot be executed before the value of the function call is known.

In many programming languages, the evaluation order is unspecified. [6] introduces a framework for expression evaluation. However, in any language, there might be addresses or values being computed that are used after returning from a function call within the same expression. If recursion is allowed, then these values need to be stored. For example, if the expressions in Example 1 are in the body of function f, then the addresses and values are bound to the occurrences of x and y, respectively, which are overwritten when executing the recursive call of f.

Figure 6 shows the state transition rules of function calls. The static functions are extended by a static function *addBindings* which adds bindings of values of occurrences to (the top frame) of an environment and bindings of adresses of occurrences to the environment, and the functions *bindVal* selects these values from a binding, respectively. The sorts OCCS and VALS are lists of occurrences and values, respectively. The static function *ith* select the i-th element of a list, respectively, and the static function *length* the length of theses lists. The dynamic function *evaluated* contains the occurrences of all sub-expressions which are evaluated but whose values or addresses are not yet used upon a function call. The macro *Save* saves all values and addresses and addresses in an environment. Similarly, the macro *Restore* restores these values and addresses.

In Fig. 6, the updates for a function call as well as for a function return are added to the updates of a procedure call and return, respectively. Therefore, we have the sub-sort relations shown in Fig. 6, since then the guards of the rules for procedure calls and returns are also satisfied for function calls and returns, respectively. The updates for function calls saves computed but unused values. The updates for function returns restore this values and store the return value (modeled by the static function *opd*) at the occurrence of the function call.

$mkFunDecl$:	NAME × PARS × TYPE × BODY	→ FUNDECL
$mkFunCall$:	NAME × ARGS	→ FUNCALL
$mkFunReturn$:	EXPR	→ FRET
$addBindings$:	ENV × OCCS × VALS	→ ENV
$bindVal$:	BINDING × OCC	→?VAL
$getBinding$:	FRAME	→ BINDING
ith :	OCCS × NAT	→?OCC
ith :	VALS × NAT	→?VAL
$length$:	OCCS	→ NAT
$length$:	VALS	→ NAT

(a) Static Functions

$evaluated$: → OCCS

(b) Dynamic Functions

FUNCALL ⊑ CALL,
FUNCALL ⊑ EXPR
FRET ⊑ RETURN,
FUNDECL ⊑ PROCDECL

(c) Sub-sorts

> **if** Ct **is** FUNCALL **then**
>   **choose** $vals$: VALS • $Values(vals)$ **in**
> $env := Save(vals, push(env, Frame))$
> **where**
> $Values(vs) \triangleq length(vs) \doteq length(evaluated) \wedge$
> $\forall i : \text{NAT} : \bullet i < length(evaluated) \Rightarrow ith(vs) \doteq val(ith(evaluated(i)))$
> $Save(vs, e) \triangleq addBindings(evaluated, v, e)$
>
> **if** Ct **is** FRET **then**
> $val(OldPc) := val(Opd)$
> $Restore(getBinding(top(env)))$
> **where**
> $Opd \triangleq opd(prog, pc)$
> $Restore(b) \triangleq$ **forall** o : OCC • $D(bindVal(o))$ **do** $val(o) := bindVal(b, o)$

Fig. 6. Function call and return

Remark 12. Some languages don't distinguish conceptually between procedures and functions. For example, in *Algol68*, *C*, *C++*, *Java*, and *C#*, procedures are functions with return type void. However, the return statements distinguish between returning a value and returning no value with the same semantics as in Figs. 4 and 6. These languages have the concept of an expression being statements, i.e., in the abstract syntax it also holds EXPR ⊑ STAT. A call is always an expression (and can be used as a statement). Hence, there is no distinction between a function call and a procedure call. If a procedure (or function) is called as statement, then there are no evaluated expressions or designators whose result is being used after the call.

7   Procedures as Values

In some programming languages, it is possible to pass procedures and functions to parameters or to assign procedures to variables. In both cases, a procedure is a *value*. The value is the *closure* of a procedure or function [11], i.e., it consists of the parameters and the body of the procedure or function, respectively, and its bindings in the current environment. For the parameters and the body, it is sufficient to provide the first instruction to be executed (which is the first parameter, if the parameter list is not empty). In Fig. 7 this is modeled by the static function *closure* and the subsort relation CLOSURE ⊑ VALUE. Hence, procedure values can be stored in the memory, i.e., it is possible to assign procedures to variables or pass them as parameters. The static functions *first* and *getBindings* access the first instruction and the bindings of a closure, respectively. The static function *proc* yields the occurrence of the designator specifying the procedure to be called.

$proc$:   PROG × OCC   →?OCC   **if** Ct **is** CALL **then**
$closure$:   OCC × BINDING → CLOSURE $pc := first(Closure)$
$first$:   CLOSURE   → OCC   where $Closure \triangleq val(proc(prog, pc))$
$getBindings$: CLOSURE → BINDING
CLOSURE ⊑ VALUE $Frame \triangleq mkFrame(getBindings(Closure), pc)$

if Ct **is** PROCDECL **then**
 $Proceed$
 choose l : LOC • ¬$IsUsed(l)$ **in**
  $env := addBinding(env, Id, l)$
  $mem(l) := closure(first(ProcDef), getBindings(env))$

Fig. 7. Procedures as values

Passing procedures is passing the closure by the passing mechanism *call-by-value* to the parameter, i.e., a procedure parameter is simply a *call-by-value* parameter of a procedure type. However, there are two possibilities to access a variable (or parameter) of a procedure type: computing the value (the closure) and calling the closure. For the former, the closure of a procedure declaration is stored in the memory and the procedure name is bound to a new location, i.e., the ASM-rules for procedure and function declarations need to be changed but the rules for the evaluation of designators remain unchanged. For the latter, the ASM-rules for procedure calls and function calls in Figs. 4 and 6 need to be changed in order to access the closure of the procedure or function to be called. These rules are identical to those in Figs. 4 and 6 except that the macro *FRAME* has been changed into the definition in Fig. 7.

Remark 13. Languages that don't have the concept of local procedures such as *C* and *C++* use function pointers instead of procedure parameters. This is sufficient because the bindings visible at the function declarations are the global names and these are statically known. At first glance, it seems that this also holds for *Java* and *C#*. However, these languages know the concept of anonymous functions (in *Java* these are called lambda-expressions). The value of an anonymous function is a closure and the bindings are those of the current environment. Thus, anonymous functions in *Java* and *C#* require the formal model of this section.

The concept of procedures and functions as values are already introduced in LISP [11] - however with dynamic bindings. *Algol68* introduces the concept of procedures/functions as values and the concept of anonymous functions with static binding. It has the semantics as introduced in this section. *Pascal, Modula,* and *Ada* don't have the possibility to assign closures to variables but only to pass them as parameters, called *procedure parameters* in these languages. It also has the semantics as introduced here.

8 Asynchronous Procedures

If an asynchronous procedure is being called, then the caller and the callee execute concurrently. Thus, each call starts a new thread. Within this thread it is possible to call synchronous procedures, i.e., a stack frame is pushed to the thread environment as usual. On the other hand, the execution may call an asynchronous procedure. This call forks a new thread from the callers frame. Therefore, the environment has the behaviour of a *cactus stack* [8] (or called *tree of stacks* by [5]), cf. Fig. 8. Each thread can non-deterministically continue its execution. Before we consider explicit synchronizations, we consider the more simple case that each procedure can only return if all called asynchronous procedures have returned. Otherwise, calling a procedure and returning from a procedure behave as usual. In particular, the parameter passing mechanism has the same behaviour as discussed in Sect. 5.

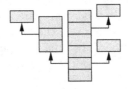

Fig. 8. A cactus stack with 5 threads

mkAsyncProc :	NAME × PARS × BODY	→ APROCDECL
mkAsyncCall :	NAME × ARGS	→ APROCCALL
push :	ENV × FRAME × THREAD	→ ENV
top :	ENV × THREAD	→ FRAME
pop :	ENV × THREAD	→ ENV
fork :	ENV × FRAME × THREAD	→ ENV
terminate :	ENV × THREAD	→ ENV
forked :	ENV × THREAD × THREAD	→ BOOL
mkFrame :	BINDING × OCC × THREAD	→ FRAME
addBinding :	THREAD × ENV × ID × LOC	→ ENV
addBinding :	THREAD × ENV × ID × BINDING	→ ENV
addBinding :	THREAD × ENV × OCC × LOC	→ ENV
bind :	THREAD × ENV × ID	→?LOC
bindArg :	THREAD × ENV × OCC	→?LOC
getpc :	THREAD × ENV	→?OCC
main :		→ THREAD
threads :	ENV	→ THREADS
· ∈ · :	THREAD × THREADS	→ BOOL
selectThread :	ENV × THREAD	→?ENV
APROCDECL ⊑ DECL		

(a) Static Functions and subsorts

val :	THREAD × OCC	→?VALUE
pc :	THREAD	→?OCC
entry :	THREAD × OCC	→?BOOL
evaluated :	THREAD	→ OCCS
thread :		→ THREAD

(b) Dynamic Functions

initial $pc(main) := first(prog)$
$\qquad \neg t \doteq main \Rightarrow \neg D(pc(t))$
$\qquad \neg D(mem(l))$
$\qquad \neg D(val(t, o))$
$\qquad thread := main$

(c) Initial State

Fig. 9. Static and dynamic functions, initial state

In a first step, we extend the previous semantics to the execution on threads. For this, a dynamic constant *thread* that contains the thread currently being executed is introduced, cf. Fig. 9(b). Furthermore, all static functions related to environments have an additional parameter of sort THREAD. E.g. *push* must specify the thread where a frame is pushed, cf. Fig. 9(a). There is now one program counter per thread. This is modeled by adding a parameter sort THREAD to *pc*. Since a procedure can be active on different threads, the dynamic functions *evaluated*, *val*, and *entry* also become an additional parameter of sort THREAD, cf. Fig. 9(b). All ASM-rules except the rules for returning from functions and procedures are relative to the dynamic constant *thread*, i.e., it is added as an argument if the corresponding static and dynamic functions require this argument, cf. Fig. 10. The semantics of a procedure or function returns change because it is different for asynchronous procedures.

For asynchronous procedure calls and returns, some more static functions are required. The static functions *fork* and *terminate* fork and terminate a thread, respectively. Furthermore, there is a static function *threads* yielding the set of threads in an environment and *selectThread* for selecting the environment specific to a thread, cf. Fig. 9(a). The constant *main* is the main thread (for execution of a procedure *main* as e.g. in *C*, *C++*, *Java*, or *C#* or a main program as e.g. in *Pascal*, *Modula*, or *Ada*), cf. Fig. 9(a). Figure 11 shows the transition rules for asynchronous procedure calls and returns. The first rule chooses nondeterministically a new active thread that is not waiting. This models the interleaving semantics. A thread is active iff it is contained in the environment and is waiting if it tries to execute a return but it has still some non-terminated forked threads. For an asynchronous procedure call a new thread is created, the program counter of the new thread is updated to the first instruction of the callee, and the program counter of the caller proceeds to the next instruction. The return from a synchronous procedure or synchronous function is the same as before w.r.t. the thread currently being executed. The return from an asynchronous procedure call terminates the corresponding thread. Note that the ASM rules don't distinguish between the case whether in a programming language may offer two possibilities for the distinction between synchronous and asynchronous procedure calls: First, a procedure must be declared as synchronous or asynchronous. Second, the language distinguishes between the concept of a synchronous call and an asynchronous call (as a statement). Since both possibilities can be modeled by the same abstract syntax, the ASM rules capture both variants.

The semantics in Fig. 11 specifies that a procedure (or function) return is possible only if all forked threads are terminated. Thus, before a return of a procedure *p* is being executed, all called asynchronous procedures must execute their return. Thus, this is a synchronization at the execution of a return instruction of *p*. If several asynchronous procedures have been called by *p* the order of returning from these callees is not specified but results from the non-determinism of the interleaving semantics. Furthermore, there is no possibility to access the return value of an asynchronous function call. Explicit synchronization statements allow more control over asynchronous functions. We consider here a barrier synchronization, i.e. the caller of an asynchronous procedure (or function) waits at this synchronization statement until the corresponding thread terminates (i.e., the callee executes its return statement). If the callee is a function, then the

if Ct is READ then
   $mem(val(thread, Opd)) := front(inp)$
   $inp := tail(inp)$
   $Proceed$
where $Pc \triangleq pc(thread)$
$Ct \triangleq occ(prog, Pc)$
$Proceed \triangleq Pc := next(prog, Pc)$
$Opd \triangleq opd(prog, Pc)$
if Ct is WRITE then
   $out := add(out, val(thread, Opd))$
   $Proceed$
if Ct is NAME $\wedge$ $isValue(prog, Pc)$ then
   $Proceed$
   $val(thread, Pc) := mem(Loc)$
where $Loc \triangleq bind(thread, env, Id)$
$Id \triangleq id(prog, Pc)$
if Ct is NAME $\wedge$ $isAddr(prog, Pc)$then
   $val(thread, Pc) := Loc$
   $Proceed$
if Ct is VALPAR then
   $Proceed$
   choose $l : \mathrm{LOC} \bullet \neg Used(l)$ in
      $env := addBinding(thread, env, Id, l)$
      $mem(l) := val(threadArg)$
where $Arg \triangleq corr(prog, OldPc, Pc)$
$OldPc \triangleq next(prog, getpc(thread, env))$
if Ct is REFPAR then
   $env := addBinding(thread, env, Id, val(Arg))$
   $Proceed$
if Ct is RESPAR $\wedge$ $entry(OldPc) \doteq false$ then
   $mem(Loc') := val(thread, pc)$
   $Proceed$
where $Loc' \triangleq bindArg(thread, env, OldPc, Arg)$
if Ct is FUNCALL then
   choose $vals : \mathrm{VALS} \bullet Values(vals)$ in
      $env := Save(vals, NewEnv(env, thread))$
where
$Values(vs) \triangleq$
   $length(vs) \doteq length(evaluated(thread)) \wedge$
   $\forall i : \mathrm{NAT} \bullet i < length(evaluated(thread))$
      $\Rightarrow ith(vs) \doteq val(thread, ith(evaluated(thread), i)))$
$NewEnve(e, t) \triangleq push(e, Frame(t), t))$
$Save(vs, e) \triangleq addBindings(evaluated(thread), vs, e)$

if Ct is ASSIGN then
   $mem(val(thread, Lop)) := val(thread, Rop)$
   $Proceed$
where $Lop \triangleq lop(prog, pc(t))$
$Rop \triangleq rop(prog, pc(t))$
if Ct is WHILE $\wedge$ $Cond(t) \doteq true$ then
   $Pc := first(prog, Pc)$
where $Cond \triangleq val(thread, cond(prog, Pc))$
if Ct is WHILE $\wedge$ $Cond(t) \doteq false$ then $Proceed$
if Ct is IF $\wedge$ $Cond(t) \doteq true$ then
   $Pc := first(prog, Pc)$
if Ct is IF $\wedge$ $Cond(t) \doteq false$ then $Proceed$
if Ct is MINUS then
   $val(thread, Pc) := val(thread, Lop) - val(thread, Rop)$
   $Proceed$
if $Ct(t)$ is CONST then
   $val(thread, Pc) := const(Ct)$
   $Proceed$
if Ct is CALL then
   $Pc := first(Closure)$
   $entry(thread, Pc) := true$
where $Closure \triangleq val(thread, proc(prog, Pc))$
if Ct is PROCCALL then
   $env := push(env, Frame(thread), thread)$
where
$Frame(t) \triangleq mkFrame(getBindings(Closure), Pc, t)$
if Ct is PROCDECL then
   $Proceed$
   choose $l : \mathrm{LOC} \bullet \neg isUsed(l)$ in
      $env := addBinding(thread, env, Id, l)$
      $mem(l) := DefClos(first(ProcDef), thread, env)$
where $DefClos(ot, e) \triangleq closure(o, getBindings(t, e))$
if Ct is RESPAR $\wedge$ $entry(thread, OldPc)$ then
   $Proceed$
   choose $l : \mathrm{LOC} \bullet \neg Used(l)$ in
      $env := NewEnv'(thread, env, l, Arg)$
where $Env_1(t, e, l) \triangleq addBinding(t, e, Id, l)$
$NewEnv'(t, e, l, a) \triangleq$
   $addBinding(t, Env_1(t, e, l), a, val(t, a))$

Fig. 10. Asynchronous execution of threads

choose $t : \mathrm{THREAD} \bullet CanProceed(t)$ in
   $thread := t$
where $IsActive(t) \triangleq t \in threads(env)$
$IsWaiting(t) \triangleq \exists t' : \mathrm{THREAD} \bullet forked(env, t, t')$
$CanProceed(t) \triangleq IsActive(t) \wedge \neg IsWaiting(t)$
if Ct is APROCCALL then
   $Proceed$
   choose $t : \mathrm{THREAD} \bullet \neg isActive(t)$ in
      $env := fork(env, Frame(t), thread)$
      $pc(t) := first(Closure)$
      $entry(t, pc(t)) := true$

if Ct is RETURN $\wedge$ $\neg IsWaiting(thread)$then
   if $OldPc$ is PROCCALL then
      $env := pop(env, thread)$
      $Pc := next(prog, OldPc)$
   if $OldPc$ is APROCCALL then
      $env := terminate(env, thread)$
if Ct is FRET then
   $val(thread, OldPc) := val(thread, Opd)$
   $Restore(getBinding(top(env, thread)))$
where
$Opd \triangleq opd(prog, Pc)$
$Restore(b) \triangleq$
   forall $o : \mathrm{OCC} \bullet D(bindVal(b, o))$ do
      $val(thread, o) := bindVal(b, o)$

Fig. 11. Procedure returns and asynchronous procedure calls

synchronization is an expression and its value is the return value of the function. Hence, it must also be possible to declare threads.

$mkThread :$ NAME $\rightarrow$ THREADDECL
$mkSyncProc :$ NAME $\rightarrow$ SYNC
$mkSyncFun :$ NAME $\rightarrow$ SYNCEXPR
THREADDECL $\sqsubseteq$ VARDECL, SYNC $\sqsubseteq$ STAT
SYNCEXPR $\sqsubseteq$ EXPR, SYNCEXPR $\sqsubseteq$ SYNC
THREAD $\sqsubseteq$ VALUE
$IsWaiting(t) \triangleq$
  $Cmd(t)$ is SYNC $\wedge \neg Sync(t)$ is RETURN$\vee$
  $\exists t' :$ THREAD $\bullet$ $forked(env, t, t')$
where $Cmd(t) \triangleq occ(prog, pc(t)))$
$Synchronize(t) \triangleq val(t, id(prog, pc(t)))$
$Sync(t) \triangleq Cmd(Synchronize(t))$

if Ct is APROCCALL then
then
  $Proceed$
  choose $t :$ THREAD $\bullet$ $\neg IsActive(t)$ in
    $env := fork(env, Frame(t), thread)$
    $pc(t) := first(Closure)$
    $entry(t, pc(t)) := true$
    $val(thread, Pc) := t$
if Ct is SYNC $\wedge$ $Sync(thread)$ is RETURN then
  $env := terminate(env, Synchronize(t))$
  $Proceed$
if Ct is SYNCEXPR $\wedge$ $Sync(thread)$ is FRET then
  $val(thread, Pc) = RetVal(Synchronize(thread))$
where $RetVal(t) \triangleq val(t, opd(prog, Sync(t)))$

Fig. 12. Explicit barrier synchronization

Figure 12 shows the semantics of a synchronization statement and expression. The static functions only need to be extended by the constructors for the declaration of threads, and the synchronization statements and synchronization expressions. The main difference between synchronization statements and synchronization expressions is that the latter has a value, the return value of the asynchronous function. Furthermore, threads are values, i.e., they can be stored in the memory. Due to the sub-sort relations, the thread declarations can be used analogous to variables, i.e., no new transition rules are required for thread declarations, accessing thread variables, assigning threads to thread variables. Even passing threads to value or reference parameters is possible. However, the choice of selecting active, non-waiting threads for execution is slightly different to Fig. 11 since in addition, threads might wait at a synchronize-statement/expression for the return statement of the callee. This is formalized by the revision of the macro *IsWaiting*. The transition rule for asynchronous procedure call in Fig. 12 extends the corresponding rule in Fig. 11 by storing t as value of the current call in the current thread. The transition rule for synchronization checks whether the asynchronous procedure to be synchronized reached the return statement. This is the only possibility to return from asynchronous procedures. If a synchronization synchronizes with a function, then value of the synchronization becomes the return value of the asynchronous function to be synchronized.

Remark 14. It is also possible to specify other synchronization mechanisms. Locking variables can be modeled by a dynamic function $lock :$ LOC $\rightarrow$ BOOL and writing to $mem(l)$ is possible only if l is not locked. This affects the transition rules for assignments, and passing parameters by value or by result. Also note, that threads or procedures might be locked since these are considered as values too. Another synchronization mechanism are critical sections. If a critical section is being executed, then no critical section of any other thread can be executed concurrently. It is possible to specify the semantics of critical sections, by specifying whether a thread enters and leaves a critical section. If a critical section is being entered then the execution of the thread waits until no other thread is in a critical section. Hence, a dynamic function $inCriticalSection$: THREAD $\rightarrow$ BOOL is required. This specification is independent of whether there are explicit statements for entering and leaving critical sections or the critical sections are statically declared.

Threads in *Java* and *C#* are bound to objects and created upon object creation. Thus, the environments are bound to these objects.

9   Related Work

There is a lot of work on defining programming language semantics using Abstract State Machines, see [3] for an overview. Usually, the language semantics is modular in the sense that starting from a kernel language, the language is extended concept per concept without changing the ASM rules of the previous definitions. A nice example of this approach is the language definition of *Java* in [13].

Montages [1,2] is a visual model of the abstract syntax, and control- and data-flow. We used static functions for this purpose, and their formal (algebraic) specification can be derived from *Montages* specifications. Similarly, the static semantics can be specified using attribute grammars [10] and these specifications can be systematically transformed into static functions and axioms.

Most of the formalization concepts for ASM transition rules in this paper can be found in the works mentioned in [3]. In addition, we have shown the difference between non-recursive and recursive procedures, local procedures, the transition rules for all usual parameter passing mechanisms, and in particular procedures as values. It therefore covers many programming languages. The modeling of concurrency in [13] (and similarly the concurrency of C# [14]) is similar to our approach. [6] discusses for expression evaluation semantics that covers many programming languages. It is in the same spirit as this work for procedures and functions.

The idea of using asynchronous procedures and base the execution model on cactus stacks stems from our work on protocol conformance checking of Web Service Compositions [9] or deadlock analysis of Web Service Compositions [16]. [15] contains an operational semantics of synchronous and asynchronous procedure calls based on inference rules.

[12] has a similar motivation as this work, i.e., it provides a modular framework for modeling programming language semantics. In contrast to this work it is based on rewrite systems. The framework provides possibilities for specifying configurations which are closely related to states of ASMs. In contrast to [12] we also consider local procedures within a local scope and asynchronous procedures.

10   Conclusions

In this paper, we have shown the formalization of the operational semantics using Abstract State Machines for frequently used concepts related to procedures and functions. Usually not all formalizations are needed for a formal semantics in a programming language. However, our results provide a toolkit for the semantics of procedures and functions: just look at the concepts and take the corresponding rules as specified in this paper. Details are discussed in the remarks at the end of each section. The model in Sect. 7 is sufficient for all variations of synchronous procedures and functions: Procedures are variables that store closures as procedure values. If a programming language has no procedure variables or procedure parameters, then this behaves like a constant declaration since the execution of a procedure declaration of a procedure p stores the closure in the location associated to p, but within the language it is not possible to declare variables of a procedure type. Thus, the closure of p remains unchanged during the lifetime of p. Issues like correspondence between arguments of a procedure call and the parameters of the callee can be modeled by a static function - even if non-positional parameter passing is possible. If only parameterless procedures are allowed all rules related to parameter passing can be removed. Similarly, only the rules for the parameter passing mechanisms offered by the programming language are needed. A similar remark applies to asynchronous procedures.

The formalizations are rather independent of the formalization of the other statements and expression evaluation. Thus, they can be combined with similar toolkits and frameworks for other variants of programming language concepts. One concept that requires a special focus is exception handling as this concept may abort a current execution (including procedure and function calls). Another concept that may affect the environment is the classical unconditional jump since this may remove stack frames upto the stack frame where the jump target is visible.

Acknowledgement. The first author thanks Egon very much for the hint to use evolving algebras in the *Verifix* project. We planned at an early stage (if the first author remembers correctly, this was around 1993/94) to use an operational semantic for the languages involved in a compiler. Egon's hint to evolving algebras - later renamed to ASMs - was a key decision towards the success of *Verifix*. Furthermore, both authors would like to thank Egon for his fruitful discussions on cactus stacks and their execution model during his visit in October and November 2019 in Halle (Saale). These discussions inspired in particular the results in Sect. 8. Egon, thank you very much.

References

1. Anlauff, M.: XASM- an extensible, component-based abstract state machines language. In: Gurevich, Y., Kutter, P.W., Odersky, M., Thiele, L. (eds.) ASM 2000. LNCS, vol. 1912, pp. 69–90. Springer, Heidelberg (2000). https://doi.org/10.1007/3-540-44518-8_6

2. Anlauff, M., Kutter, P.W., Pierantonio, A.: Enhanced control flow graphs in montages. In: Bjøner, D., Broy, M., Zamulin, A.V. (eds.) PSI 1999. LNCS, vol. 1755, pp. 40–53. Springer, Heidelberg (2000). https://doi.org/10.1007/3-540-46562-6_4

3. Börger, E.: The abstract state machines method for modular design and analysis of programming languages. J. Log. Comput. **27**(2), 417–439 (2017)

4. Börger, E., Stärk, R.: Abstract State Machines. Springer, Heidelberg (2003). https://doi.org/10.1007/978-3-642-18216-7
5. Dahl, O.-J., Nygaard, K.: SIMULA: an ALGOL-based simulation language. Commun. ACM **9**, 671–678 (1966)
6. Zimmermann, W., Dold, A.: A framework for modeling the semantics of expression evaluation with abstract state machines. In: Börger, E., Gargantini, A., Riccobene, E. (eds.) ASM 2003. LNCS, vol. 2589, pp. 391–406. Springer, Heidelberg (2003). https://doi.org/10.1007/3-540-36498-6_23
7. Goos, G., Znnmerrnaun, W.: Verification of compilers. In: Olderog, E.-R., Steffen, B. (eds.) Correct System Design. LNCS, vol. 1710, pp. 201–230. Springer, Heidelberg (1999). https://doi.org/10.1007/3-540-48092-7_10
8. Hauck, E.A., Dent, B.A.: Burroughs' B6500/B7500 stack mechanism. In: AFIPS 1968 (Spring): Proceedings of the Spring Joint Computer Conference, 30 April–2 May 1968, pp. 245–251. ACM (1968)
9. Heike, C., Zimmermann, W., Both, A.: On expanding protocol conformance checking to exception handling. SOCA **8**(4), 299–322 (2013). https://doi.org/10.1007/s11761-013-0146-2
10. Knuth, D.E.: Semantics of context-free languages. Math. Syst. Theory **2**(2), 127–145 (1968). https://doi.org/10.1007/BF01692511
11. McCarthy, J.: Recursive functions of symbolic expressions and their computation by machine, part I. Commun. ACM **3**(4), 184–195 (1960)
12. Rosu, G., Serbănută, T.F.: An overview of the K semantic framework. J. Log. Algebraic Program. **79**(6), 397–434 (2010)
13. Stärk, R., Schmid, J., Börger, E.: Java and the Java Virtual Machine. Springer, Heidelberg (2001). https://doi.org/10.1007/978-3-642-59495-3
14. Stärk, R.F., Börger, E.: An ASM specification of C# threads and the.NET memory model. In: Zimmermann, W., Thalheim, B. (eds.) ASM 2004. LNCS, vol. 3052, pp. 38–60. Springer, Heidelberg (2004). https://doi.org/10.1007/978-3-540-24773-9_4
15. Weißbach, M.: Deadlock-analyse service-orientierter softwaresysteme. Ph.D. thesis, Institut für Informatik, Martin-Luther-Universität Halle-Wittenberg (2019)
16. Weißbach, M., Zimmermann, W.: On limitations of abstraction-based deadlock-analysis of service-oriented systems. In: Fazio, M., Zimmermann, W. (eds.) ESOCC 2018. CCIS, vol. 1115, pp. 79–90. Springer, Cham (2020). https://doi.org/10.1007/978-3-030-63161-1_6
17. Winter, K.: Model checking for abstract state machines. J. Univ. Comput. Sci. **3**(5), 689–701 (1997)
18. Winter, K., Duke, R.: Model checking object-Z using ASM. In: Butler, M., Petre, L., Sere, K. (eds.) IFM 2002. LNCS, vol. 2335, pp. 165–184. Springer, Heidelberg (2002). https://doi.org/10.1007/3-540-47884-1_10
19. Wirsing, M.: Algebraic specification. In: van Leeuwen, J. (ed.) Handbook of Theoretical Computer Science, vol. B, pp. 615–788. MIT-Press/Elsevier (1990)

Author Index

Aït-Ameur, Yamine   1
Ambos-Spies, Klaus   14
Arcaini, Paolo   215

Banach, Richard   29
Batory, Don   63
Beierle, Christoph   82
Benavides, David   63
Bitterlich, Martin   239
Bodenmüller, Stefan   239
Bombarda, Andrea   215
Bonfanti, Silvia   215
Borgert, Stephan   283
Bowen, Jonathan P.   96

Covino, Emanuele   187

Elstermann, Matthes   283

Fantechi, Alessandro   121
Ferrarotti, Flavio   135
Fleischmann, Albert   283

Gargantini, Angelo   215
Gnesi, Stefania   121
González, Sénen   135

Haldimann, Jonas   82
Heradio, Ruben   63

Kern-Isberner, Gabriele   82

Laleau, Régine   1
Leuschel, Michael   147

Makowsky, J. A.   173
Méry, Dominique   1

Oh, Jeho   63

Pani, Giovanni   187
Prinz, Andreas   199

Reif, Wolfgang   239
Riccobene, Elvinia   215

Scandurra, Patrizia   215
Schellhorn, Gerhard   239
Schewe, Klaus-Dieter   266
Semini, Laura   121
Singh, Neeraj Kumar   1
Stary, Christian   283

Thalheim, Bernhard   301

Weißbach, Mandy   326
Wolski, André   283

Zhu, Huibiao   29
Zimmermann, Wolf   326

Printed in the United States
by Baker & Taylor Publisher Services